AF323135

RENEWABLE ENERGY - UTILISATION AND SYSTEM INTEGRATION

Edited by **Wenping Cao** and **Yihua Hu**

Renewable Energy - Utilisation and System Integration
http://dx.doi.org/10.5772/59389
Edited by Wenping Cao and Yihua Hu

Contributors

Jérôme Cros, Maxim Bergeron, Stéphanie Rakotovololona, Philippe Viarouge, Jessy Mathault, Bouali Rouached, Mathieu Kirouac, Horizon Gitano, Sergio Vazquez Y Montiel, Fermin Granados Agustin, Lizbeth Castañeda Escobar, Dorin Bica, Mircea Dulau, Lucian Ioan Dulau, Marius Muji, Jean Patric Da Costa, Emerson Giovani Carati, Everton Luiz De Aguiar, Carlos Marcelo De Oliveira Stein, Rafael Cardoso, Jose Pelegri-Sebastia, Marcelo Gustavo Molina, Jean-François Toubeau, François Vallée, Jacques Lobry, Vasiliki Klonari, Wenping Cao, Yihua Hu, Maria De P. Pablo-Romero, Rocio Roman, Antonio Sánchez-Braza, Rocio Yniguez, Mostafa Abarzadeh, Hossein Madadi Kojabadi, Liuchen Chang, Karin Hammarlund Kramer, Marina Frolova, Anna Branhult

CBS, Edition **2017**

Published by InTech

Janeza Trdine 9, 51000 Rijeka, Croatia

Notice

Publishing Process Manager Iva Simcic

Technical Editor InTech DTP team

Cover Designer InTech Design Team

Additional hard copies can be obtained from orders@intechopen.com

Renewable Energy - Utilisation and System Integration, Edited by Wenping Cao and Yihua Hu
p. cm.
Print ISBN 978-953-51-2407-8
Online ISBN 978-953-51-2408-5

Contents

Contents

Preface

Utilisation of renewable energy is a key move towards establishing low-carbon economy worldwide. Among existing renewable energy systems, wind, solar, and biomass have been playing a pivotal role in generating substantial electricity and the technical challenges are still significant. As the renewable energy typically needs to connect to the power grids of some kinds, their generating systems will give rise to impact on the power grids, especially those weak ones.

This book presents 12 chapters, in an attempt to fill some gaps in understanding wind power conversion, photovoltaic control and fault detection, biomass producer and frequency energy harvesting, as well as the renewable energy integration issues.

The book is organised into four sections: Wind Energy, Solar Energy, Biomass and Other Sources, and System Integration. The reader will be exposed to these broad topics in technical depth, including modelling, control, electrical generators and power converters, and social and political aspects. Some case studies are also given to provide unique perspective of countries that develop these renewable energies. The editors are grateful to all contributing authors and especially to the Publishing Process Manager, Ms. Iva Simcic, for her tremendous patience and management skills.

Dr. Wenping Cao
Aston University, Birmingham, UK

Dr. Yihua Hu
Liverpool University, Liverpool, UK

Wind Energy

Wind Power Development and Landscape – Social Participation, Opportunities and Challenges

Karin Hammarlund, Marina Frolova and Anna Brånhult

Additional information is available at the end of the chapter

http://dx.doi.org/10.5772/63424

Abstract

This chapter describes and exemplifies how different methods for landscape analysis have been applied for wind farm development and planning in Sweden and Spain. In our case studies from Sweden and Spain we discuss the possibilities, strengths and shortcomings of use of different methods of landscape assessment for wind power development in relation with the definition of landscape by the European Landscape Convention (ELC). The Swedish case study is an example of how the proposed guidelines for wind development in the master plan of Uddevalla has been contested and how dialogue-based landscape analysis has redirected wind energy planning. This has been possible through disentangling and illuminating social values in the description of landscape characters as a basis for new planning guidelines. The Spanish case of the Valle de Lecrín and Alpujarra demonstrates limitation of a study based only on expert analysis of visual impact of wind farms. By complementing these expert methods with dialogue, local stakeholder's perspective was revealed and the role of wind turbines in the construction of a local landscape was understood. We show that the role of the landscape planner should be understood as an interpreter and mediator, rather than that of an expert and prescriber.

Keywords: Energy landscapes, ELC, landscape assessment methods, public participation, wind power planning

1. Introduction

The European Landscape Convention (ELC) [1] definition of landscape as "an area, as perceived by people, whose character is the result of the action and interaction of natural and/ or human factors" highlights the importance of developing landscape policies dedicated to the management and creation of landscapes. This includes planning procedures that allows the general public and other stakeholders to participate in policy creation and implementation. Identification and assessment of landscapes thus constitute a central element of the ELC.

According to some recent approaches to landscape [2, 3], landscape can be considered a material, social, institutional and political process that is constantly changing and dynamic. As such, a *deliberative* landscape analysis performed as a mutual learning process between concerned actors may help us to understand interlinks between physical, social, economic and cultural processes. This is most evident when we consider the transition from a conventional energy system to a renewable energy system, demanding more space in areas holding high natural and cultural values. This has proven to cause infected conflicts between stakeholders, making this landscape transition very difficult. A deliberative landscape analysis highlight conflicts but can also show how the assessment of landscapes can promote the intersubjectivity necessary for any dialogue concerning incorporation of new energy infrastructure in to a sustainable ecosystem.

This chapter aims to explore how we can assess landscapes and prepare grounds for decision making, planning and evaluation of impact from change through different methods of landscape assessment and how public participation can become its integrated part. In our case studies from Sweden and Spain, we discuss the possibilities, strengths and shortcomings of these assessment methods in relation to the definition of landscape by the ELC. Our case studies are drawn from both practice and research on wind power planning and landscape analysis in Sweden and Spain.

The landscape analysis exemplified by the Swedish case study is an integration of the above-mentioned methods; however, these have been revised in relation to the needs of the specific project and deriving data from and in dialogue with people affected by wind developments. The Spanish case study is based on critical review of landscape analysis made in the municipality of Lecrín Valley in 2004–2006 contrasted with new data obtained through field observations and in-depth interviews with different stakeholders.

2. Landscape and environmental impact assessment

Landscape is a salient issue in the development of wind power. It is commonly accepted that the most frequent public concern when weighing up the costs and benefits of wind power concern landscape values [4]. Normative planning processes, often with a markedly hierarchical structure and reliant on existing landscape norms/values or classifications, have been shown to direct wind power deployment towards non-protected, allegedly 'less sensitive' areas and to increase social or environmental injustice [5]. Cowell [6] and Nadaï and Labussiere [7] have highlighted that in wind power planning processes the ways in which landscape qualities entered in a selective way the planning rationales, favoring qualities that were formally mapable or even measurable 'at a distance'.

Conversely, cases of planning approaches based on participation or responsive to specific local situations have been shown to contribute to the emergence of new landscape representations and norms described by the authors as energy landscapes—or landscapes of which renewable energy infrastructures are perceived and treated as part, even if these developments face opposition [8, 5]. The practices and values associated with landscape by different stakeholders play a role that requires further analysis in different contexts.

Depending on what tradition and method of landscape analysis is chosen, the landscape might be viewed differently: either as a visual surface that is possible to evaluate in an objective manner through character assessment or as an ongoing interaction between physical and social forces. The physical and social aspects of landscape analysis should be equally important to present a qualified description of landscape impact and valid mitigation measures. Descriptions (representations) of landscape characters, qualities and functions need to be communicated in a stringent process open to public participation resulting in a deliberated and well-informed evaluation of impacts.

In many countries protected areas and preservation policies and plans are used as point of departure when describing impacts of change [9]. Landscape is often treated as scenery and impacts on this scenery are described as mainly visual. The rest of impact analysis is devoted to environmental aspects, i.e. flora, fauna and cultural history. Environmental impact assessments (EIAs) disconnect the real-time landscape and most change is consequently deemed as a threat to protected areas or species. To conserve something in constant change such as a landscape is of course a contradiction in terms and the disconnection of nature from culture leads to fragmented ecosystems that are not viable, yet that is how we assess impacts in accordance with the law. A dynamic and holistic view promoted by the term landscape as defined by the ELC, gives an update on presents day conditions for animals, plants and people illuminating the interconnections necessary in order to reach a multifunctional and sustainable development of environment and human society in interaction constituting the landscape [2,10].

2.1. Rethinking landscape analysis

Landscape analysis performed in relation to wind development has unfortunately been greatly influenced by a misinterpretation of the work of Kevin Lynch [11]. This has permeated landscape analysis and manifested an expert paradigm where the visual values and physical structures of the landscape have been evaluated by the trained expert in a spatially static way fitting nicely in to the established methods for EIAs. It is however possible to "walk in the foot paths" of Kevin Lynch and discover that his attempts to elaborate information from and in conversations with people living in and experiencing a landscape can be refined in order to develop a landscape analysis appropriate for communicative and collaborative planning. Lynch's work was concerned with the orientation, perception and activity of people in an urban landscape. By rethinking Lynch and emphasizing the potential of his notation of structural features in relation to important landscape functions and landscape characters defined through LCA and/or HLC, the Swedish case study shows how a landscape analysis becomes a powerful tool for participation in planning. In this context it is important to note that both LCA and HLC are often applied as expert analysis with no dialogue. The Swedish case study integrates the different methods and complements them with dialogue o illuminate the social construction of landscapes. This way the landscape analysis as a process will act as a link between abstract ideas about landscape characters in different scales and the human perceptual experience of the qualities and functions of these landscapes.

The Spanish tradition of landscape analysis for wind power development is very recent and was also marked by urban perspective. Before 2006 landscape impacting evaluation of energy development was not explicitly required by the Spanish laws on Environmental Impact Assessment, although it normally was part of the process. Only in 2006 the National Law 9/2006 on Strategic Environmental Assessment (based on Directive 2001/42/CE) established that Reports on Environmental Sustainability should evaluate landscape impacts of any project and plan.

The most common types of landscape analysis for wind power development in Spain have been based until quite recently on visual impact assessment through landscape elements inventory, photo-visualizations or/and GIS viewshed studies. Only during last decade landscape characters defined through LCA has been introduced in some methodological proposals for landscape analyses, although it did not become a commonly used method [12, 13].

In Spain landscape analysis is not generally hold in dialogue with people affected by wind development, although landscape analysis through dialogue with local residents as presented by the Spanish examples of Valle de Lecrín and Alpujarra (Andalusia) demonstrate that the expert needs to take into consideration different stakeholders' relations to wind turbines and how these as artifacts inscribe into their use of the territory and landscape.

2.2. National and regional landscape policies

In both countries landscape conservation policies seem to be connected to the meaning of the concept of landscape in respective languages.

The concept of landscape (*landskap* in Swedish) has in germanic languages historically a double meaning as either a smaller territory defined through common culture and history, or merely as the visual surface of things [14]. Sweden is divided into 25 'landskap' (in the sense of territory), based on common culture and history as well as a shared geography. These 'landskap' play an important role in people's self-identity, but they do no longer serve any political or administrative purposes. In Spanish the concept of landscape is generally defined as visual surface of things, but the word for landscape in Spanish as in the other Roman languages is *paisaje*, in which 'pais' means 'land' in the sense of the bounded area of a region or country. 'Land' does not mean soil, but refers rather to a historically constituted place [2]. The meaning of landscape has roots going deep into the identities of the historical regions from which the Mediterranean Landscape Charter and the Landscape Convention emerged (*Idem.*). Spain is divided into 17 autonomous regions (Comunidades Autónomas); many of them are historical regions based on common cultural, historical and/or language identity. This is the case of Andalusia.

In Sweden there are many tools for conservation; however, they are focused on conservation of areas, species of plants and animals (the Environmental Code) or the conservation of ancient remains, churches, other buildings, and place names (the Heritage Conservation Act). The term 'landscape' is only rarely mentioned in regulations and in a legal sense referred to as either environment or the image of the landscape in an aesthetic sense (*landskapsbild*). In the environmental code landscape has lost its old meaning of landscape as territory and identity.

The everyday landscape is to a certain extent protected by general rules of consideration in the environmental code; however, many decisions and policies that have an effect on landscapes are not regulated by Swedish law [15]. There is a general lack in EIAs regarding adequate public participation, comprehensive overview of landscape impact, and stakeholder coordination [16]. There is no mention of the significance of the landscape to people and conflicts of interest often arise because of unclear or overlapping laws [9]. The Swedish Energy Agency has pointed out areas of national interest for wind energy developments. The assessment of these areas suitability was made primarily with regard to factors such as average wind speed and safety distance to housing areas with no concern to landscape character or public interests. These areas are now being revised and the criteria for areas of national interest for wind are heavily debated.

Unlike Sweden, in Spain, energy planning is the responsibility of the Central Government, although the regions play a very important role in the decision-making process. Local governments (municipalities) on the other hand play only a secondary role in the authorization procedure, which at times has resulted in a lack of awareness of project development and the absence of strong opposition to renewable power projects [17–19].

As for landscape regulation in Spain, the situation is similar to the Swedish one. Although there are many national and regional tools which directly or indirectly concern landscape conservation and the term "landscape" is used in many of the recent laws approved after signing the ELC, energy planning systems, which are often based on engineering and economic considerations, are difficult to match with landscape management on a local level.

In spite of the new laws, landscape impacts have only secondary role in decision-making process concerning wind farms. As some recent research have demonstrated [20], local authorities in Spain do not have capacity to introduce landscape impacts of wind farms as an substantial factor into decision-making process, since they have limited power in this process and because in practice there are only three requirements for giving wind farm license by these authorities: wind turbines should be situated out of any protected natural area, they should be close to electric evacuation line and should correspond to a territory officially defined as 'wind resource area'. The process of planning and authorization in Spain does not give enough power to landowners and the general public in the EIA process. Information about projects and procedures lack transparency and clarity. Whereas, the EIA in Sweden by some is understood as giving too much concessions to those opposing to wind power projects [19].

Although public participation in the decision-making process on wind power development has become a common procedure in Spain, due to the adoption of the Aarhus Convention and the ELC, public engagement is generally seen as a one-way communication with an end result determined in advance. The tendency towards a top-down, technocratic, hierarchical way of thinking about how the planning system has to be shaped, inherited from the period of centralized policies before 1978, persists in Spain [21]. Public participation is understood as public approval (or validation) and consultation does not involve stakeholders in decision-making processes. Also, public surveys does not affect the decision making process. Therefore, social participation processes are viewed as a way of conducting political control of perceptions and/or indoctrination of stakeholders who are not convinced in the suitability of wind power projects for their territory dominates in Spain [22].

The siting of wind power schemes in Spain has not been decided at a local level, which creates an important contradiction between the intentions of Spanish landscape policy and the actual mechanics of decision-making processes. In spite of essential changes due to application of the ELC, which encouraged some Spanish regions to incorporate landscape as an important issue in land use regulation, landscape policies are still out of step with the development of renewable energy policies.

3. Case studies

3.1. Swedish case study: Uddevalla municipality

3.1.1. Background

Uddevalla municipality is located on the Swedish west cost in the metropolitan area of Gothenburg (Figure 1). The municipality has a varied landscape including forested highlands, fjords, grabens and mosaic landscapes. The varied topography and the, for Swedish conditions, high population density are challenges that the guidelines for wind power development in the municipality has to handle.

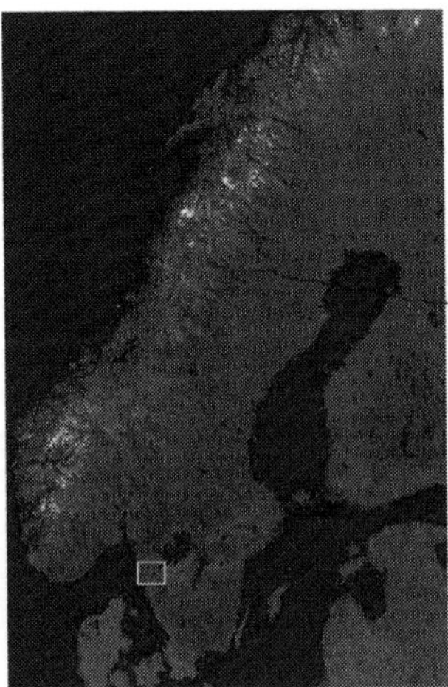

Figure 1. Uddevalla municipality, Sweden.

The proposed guidelines for wind power development were at the start of the case study a result of planning guidelines excluding areas from wind power development based on legislative restrictions. Meaning that areas with planning restrictions due to high cultural or natural values were regarded as unsuitable to wind development without further investigations as to if the present day status of these values were viable and relevant. Also, the possibility that not protected areas might prove more functional habitats or hold important recreational and social values were never investigated. These guidelines led to that the coastal areas and the highlands were eliminated, leaving the mosaic landscapes in the central and eastern parts of the municipality open for possible wind development. Since the guidelines identified areas that both authorities and local citizens objected to this became the start of a discussion whether the values protected by legislative restrictions were relevant for wind power development in Uddevalla municipality. The situation became very infected and affected the trust of the public in officials and politicians. Conflicts between wind developers and the effected public also increased as wind developments were proposed in areas with no restrictions for wind development. There were also no directives or recommendations regarding how to adapt wind developments to the landscape conditions.

Because of these conflicts there was a general need to evaluate the social aspects of landscape in connection to demarcation of areas as suitable for wind developments. Some officials suggested to politicians that a landscape analysis based on public dialogue and participation might redeem the confidence of the public and reveal an alternate strategy for wind development. The landscape analysis undertaken was in line with the ELC, focusing on illuminating the values resting with the everyday landscape as well as disclosing if the present day conditions of restricted areas were relevant in relation to wind power development. The objective of the landscape analysis was hence to create a present day description of landscape functions and values in order to better socially anchor the wind power planning.

3.1.2. A description of the landscape as perceived by people

Expert methods for landscape analysis such as Landscape Character Assessment (LCA) [23], Historical Landscape Characterization (HLC) and Structural analysis in accordance with Kevin were used to collect and present information on geophysical conditions, historical time depth, ecological functions, and to provide a basis for dialogue. These methods were combined and revised to fit the purpose of the landscape analysis.

After an identification of geophysical generic landscape types, walk through evaluations with the public were made in order to confirm the relevance of the description of landscape types and elaborate these in to landscape character areas describing the social functions as well as naming these areas in accordance with how they are referred to in an everyday context. An additional objective of the walk through evaluations was to discuss the ongoing force and pace of change as well as the landscapes sensitivity to this change.

The character areas that were extracted from the public input on landscape types as collected during walk through evaluations and were then elaborated in workshops were the public, officials and politicians collaborated (Figure 2). This procedure created a collective learning process regarding the possibilities as well as difficulties with implementing wind energy in a specific landscape type and character with its specific social context.

Figure 2. A walk through landscape evaluation in Uddevalla.

3.1.3. Revised guidelines for wind power development

The landscape analysis did not state expert evaluations regarding suitable and non-suitable areas for wind power development, but described possibilities and difficulties resting with the landscape type and character in relation to wind power developments. These possibilities and difficulties created a basis for revision of guidelines for wind power development as perceived by people based on a positive planning approach stating that wind power developments correctly located in relation to important landscape functions, both ecological and social, can even restore a fragmented and exploited landscape.

The character areas with least difficulties and most possibilities were studied further through desktop analysis and field surveys. At this stage legislative restrictions were reintroduced a part of the assessment and added as one layer of information regarding identification of suitable and non-suitable areas for wind development.

The revised planning guidelines are hence the result of a socially anchored description of landscape values and functions as well as an expert assessment/evaluation of suitability in relation to legislative regulations and as well an elaboration of methods for landscape analysis acknowledged by the ELC. The expert is transformed into a process leader with the objective to reconnect disconnected landscape values and support different perspectives in evaluations of possibilities and restrictions connected to a wind development. The landscape as an arena for dialogue and deliberation on changes repowers an understanding of important aspects of a shared social life in relation to specific processes of change, a rite of passage that equip us to redress the conflicts we must overcome in places where we live.

The Swedish case study is an example of how the proposed guidelines for wind development in the master plan of Uddevalla has been contested and how landscape analysis as basis for wind energy planning and development has disentangled the landscape values and illuminated social values in the description of landscape characters as a basis for new socially anchored wind power planning.

3.2. Spanish case study: Valle de Lecrín and Alpujarra

3.2.1. Background

The Spanish case study is located in the neighbouring *Comarcas* (or small regions) of Valle de Lecrín and Alpujarra of Granada in the Autonomous Region of Andalusia in the South-East Spain. The study area belongs to the mountain system of Sierra Nevada and the adjoining valleys. The *Comarcas* of Valle de Lecrin and Alpujarra consist of 38 municipalities with a population of 52 thousand inhabitants. These Comarcas have a traditional agrarian character and its agriculture is based mostly on olive and citric production in the lower parts and on vine, olive and almond production in the upper parts (Figure 3).

Figure 3. Valley of Lecrín. General view. M.Frolova.

Since the farming sector is currently in deep crisis, this area has recently been developing rural tourism owing to its exceptional landscapes and natural, ethnological and cultural resources, and the neighboring Natural and National Parks of Sierra Nevada. The important landscape values of Sierra Nevada have led to the emergence of conservationist policies and to the granting of protected heritage status to many of the landscapes in the study area, via various different declarations protecting natural and cultural heritage [24].

3.2.2. First landscape assessment for conflictive wind farms development

Wind-farms installed in some of the municipalities of our study area since 2004, have had a considerable impact on the landscape of the mid and low-mountains, raising new issues regarding landscape practices and values. While the installation of wind power is difficult in the Alpujarra due to the Sierra Nevada National Park protection measures, and only two wind farms were constructed, the neighbouring Valle de Lecrín has developed five wind farms with installed capacity of about 34 MW, 20 wind turbines and about 1,69 MW installed per turbine [13]. The beginning of wind power implementation in this area dates back to the early 2000s raising issues regarding landscape practices and different tensions and conflicts. An important anti-wind power initiative was launched by the British company Mirador Andalusian Developoment S.L. that wanted to build a rural hotel complex in the Lecrín Valley, with some local stakeholders involved [24]. A detailed study on landscape and environmental impacts had been carried out in order to demonstrate an important value of the Lecrín Valley land-scapes. The results of this study, called "Landscape and environmental viability criteria for wind farms localization", were published in 2006 when some of the wind farms were already constructed [25]. This study referred to the ELC as an important basis for landscape conser-vation in Spain. The study was based on analysis of visual impact, effects on visual landmarks and on view shed analysis using Geographical Information Systems (GIS). At the same time landscapes with high quality were defined through inventory of different cultural and natural elements, and landscape fragility of the Lecrín Valley was mapped. Finally, the map of viability of wind power project was produced which demonstrated that the most part of the Lecrín Valley corresponded to exclusion areas or apt areas with environmental, landscape, etc. constraints (Figure 4) (*Idem.*). All this study was made by experts, without taking into consid-eration the local residents' landscape perception.

As a consequence of this study the number of wind power projects for which permission was granted was reduced by local administrations.

3.2.3. A study of wind power project perception by stakeholders

In order to complete landscape assessment developed in 2006 twenty in-depth interviews were made between 2011, 2012 and 2013 with different local stakeholders involved in the develop-ment of wind power projects, whether opposing or supporting it.

The acceptance of wind power projects by the local population was determined by several factors related not so much to their esthetic, but mostly with inscription of wind projects into local economy and its compatibility with local activities. First of all wind turbines provide substantial income for some rural land owners and town councils. Secondly, they are compatible with traditional rural activities such as agriculture and livestock-grazing, and with most other local business activities. Moreover local ranchers perceive positively the roads and passes constructed for windfarms maintenance as facilitating access to some isolated territories. Our interviews with local residents revealed a strong connection between the public acceptance of landscape changes brought by renewables development and the economic gains this development brings. Thus, the local residents are more prepared to accept landscape changes if they participate in renewable power development in terms of the economic benefits to be reaped from it. Finally some stakeholders perceived wind turbines as an element integrated into the local landscape. Interestingly windmills have already appeared on some representations of local landscape (information panels) for tourists. In the interviews we conducted with tourists, we also discovered that some tourists had come to take pictures of wind farms, attracted by their "powerful image". [24]

Opposition to wind farms was mainly conditioned by two factors. On one hand, many local residents felt themselves excluded from the decision-making process. On the other hand, some opposition to wind farms aroused out of disconnection between territorial development models chosen by neighboring municipalities. For example, Town Council and residents of

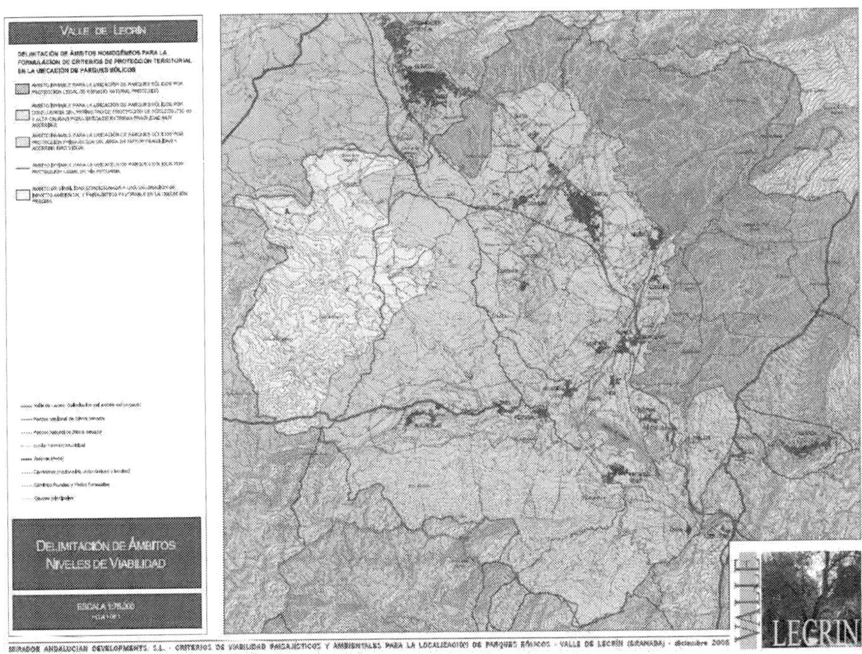

Figure 4. Delimitation of homogeneous zones for formulation of criteria of landscape protection for wind farms siting.
Source: Medina Barbero, R., Sánchez del Árbol, M.A. (dir.), 2006.

Nigüelas, one of the Valle de Lecrín villages, opposed to construct wind farms on their territory in order to preserve their landscapes (*Idem.*). Now, ten years after installation of wind farm in a neighboring village many interviewees complain that landscapes of Nigüelas had been spoilt and another village had received all the economic benefits brought by the new wind farm. Therefore, if local population could have chosen now they would have accepted wind farms in their territory. In this case landscape impact acceptation is also connected with the economic benefits of renewable projects for local residents.

The Spanish case study demonstrate that landscape assessment for wind power developments based only on an expert perspective emphasizing visual impact has important limitations, since landscape reflects emotional, economical, etc. relations of local populations with their place. Our analysis of results of the interviews has shown that wind turbines may not be considered a problem in the local landscape, they might even add to the construction of local identity. In order to take into account local place aspects a dialogue with the local community is required. The relationships of the local population with wind power do not emerge through its purely visual perception, but require a multidimensional view of the wind power landscape that brings together local practices, experiences and perceptions as well as physical properties.

4. Conclusion

Landscape has progressively been endowed with a new meaning in the assessment of renewable energy projects. The important shift can be spotted moving away from an expert paradigm focusing impact at a great distance for the local living context, to a more participative and deliberative planning processes. This important progress is aided by landscape assessment methods that offer new opportunities to take people's perceptions and use of landscape into consideration. Landscape Character Assessment and Historical Landscape Characterization are methods for landscape analysis that preferably include dialogue in accordance with the ELC to account for new dimensions of landscapes, counteracting the attempts to objectivize social dimensions through classification and/or quantification.

The Swedish case study is an example of how the proposed guidelines for wind development in the master plan of Uddevalla have been contested and how dialogue-based landscape analysis has redirected wind energy planning. This has been possible through disentangling and illuminating social values in the description of landscape characters as a basis for new planning guidelines. Walk-through evaluations have evaluated and complemented the descriptions of geophysical conditions (landscape types) with social functions elaborating these in to character areas with well-recognized names. The walk-through evaluations were complemented by workshops and these dialogue methods also complemented the structural analysis in accordance with Lynch original work, however, now in a rural and peri-urban context.

The Spanish case study demonstrates limitations of a study, based only on expert analysis of visual perception and accessibility of landscape, of visual landmarks and on viewshed analysis, without taking into consideration the local residents' landscape and wind power perception. By complementing these expert methods with dialogue, local stakeholder's perspective was revealed and the role of wind turbines in the construction of a local landscape was understood.

Both case studies show that there is still a substantial lack in the ambition to take on board people's perceptions in planning and impact assessment practices in accordance with the ELC. The prevailing methods of landscape assessment rely on the trained experts definition and evaluation of landscape values. In conclusion both the Swedish and Spanish case studies show that the role of the landscape planner should be understood as an interpreter and mediator, rather than that of an expert and prescriber.

Acknowledgements

This chapter is part of the research project CSO2011-23670, which has been funded by the Spanish Ministry of the Economy and Competitiveness. We are grateful to Kira Gee, Jenny Nord, Miguel Ángel Sánchez del Árbol and Belén Pérez for their valuable comments. We are also grateful to Ramboll Sweden for their research cooperation concerning the Swedish case study as well as the municipality of Uddevalla.

Author details

Karin Hammarlund[1], Marina Frolova[2*] and Anna Brånhult[3]

*Address all correspondence to: mfrolova@ugr.es

1 Department of Landscape Arcitecture, University SLU Alnarp, Sweden

2 Department of Regional and Physical Geography, University of Granada, Granada, Spain

3 Ramböll Sverige AB, Malmö, Sweden

References

[1] Council of Europe [Internet]. European Landscape Convention, Treaty Series 176; 2000 [cited 2014 March 1]. Available from: http://conventions.coe.int/Treaty.

[2] Olwig K. Landscape, Nature, and the Body Politic. Madison, Wisconsin: The University of Wisconsin Press; 2002.

[3] Nadaï A, van Der Horst D. Introduction: Landscapes of Energies. Landsc Res. 2010; 35(2): 235–257.

[4]   Wolsink M, Breukers S. Contrasting the core beliefs regarding the effective implementation of wind power. An international study of stakeholder perspectives. J Environ Plan Manag. 2010; 53(5): 535–558.

[5]   Frolova M, Prados M.-J, Nadai A, editors. Renewable Energies and European Landscapes: Lessons from Southern European Cases. New York London: Springer; 2015.

[6]   Cowell R. Wind power, landscape and strategic, spatial planning-the construction of 'acceptable locations' in Wales. Land Use Policy. 2009; 26(3): 222–232.

[7]   Nadaï A, Labussière O. Wind power planning in France (Aveyron): from State regulation to local experimentation. Land Use Policy. 2009; 26(3): 744–754.

[8]   Nadaï A, Labussière O. Playing with the line, channelling multiplicity – wind power planning in the Narbonnaise (Aude, France). 2013; Environ Plan D 31(1): 116–139.

[9]   Swedish National Heritage Board. Proposals for Implementation of the European Landscape Convention in Sweden. 2008; [cited 2014 March 1]. Available from: http://www.raa.se/publicerat/rapp2008_17.pdf.

[10]   Oles T, Hammarlund K. The European Landscape Convention, Wind Power, and the Limits of the Local: Notes from Italy and Sweden. Landsc Res. 2011; 36(4): 471–485.

[11]   Lynch K. The image of the city. Cambridge: MA, MIT Press; 1960.

[12]   Frolova M, Bejarano JF, Torres A, Lucena M, Pérez B. Valoración social. In: Ghislanzoni M, editor. Guía de integración paisajística de parques eólicos en Andalucía. Sevilla: Consejería de Medio Ambiente y Ordenación del Territorio; 2014. p. 65–78.

[13]   Ghislanzoni M, editor. Guía de integración paisajística de parques eólicos en Andalucía. Sevilla: Consejería de Medio Ambiente y Ordenación del Territorio; 2014.

[14]   Olwig K. Recovering the Substantive Nature of Landscape. Ann Assoc Am Geogr. 1996; 86(4): 630–653.

[15]   Lerman P. Landskap i svensk rätt. Underlag för Landskapskonventionens genomförande. Lagtolken; 2007.

[16]   Antonsson H. The treatment of landscapes in a Swedish EIA process. Environ Impact Assess Rev. 2011; 31(3): 195–205.

[17]   Frolova M, Pérez Pérez B. New landscape concerns in the development of renewable energy projects in South-West Spain. In: Roca Z, Claval P, Agnew J, editors. Landscapes, Identities and Development. Farnham: Ashgate Publishing; 2011. p. 389–401.

[18]   Iglesias G, Del Río P, Dopico JA. Policy analysis of authorization procedures for wind energy deployment in Spain. Energy Policy. 2011; 39: 4067–4076.

[19]   Redondo F. The situation of wind power and human perception and attitude towards it. A comparison between Sweden and Spain. Master Thesis. Halmstad University; 2014.

[20] Zografos C, Saladié S. La ecología política de conflictos sobre energía eólica. Un estudio de caso en Cataluña. Documents d'Anàlisi Geogràfica. 2012; 58(1): 177–192.

[21] Frolova M. Landscapes, Water Policy and the Evolution of Discourses on Hydropower in Spain. Landsc Res. 2010; 35(2): 235–257.

[22] Florido del Corral D. Los molinos de viento de Trafalgar. Paradojas, ambivalencia y conflictos respecto a las relaciones humano-ambientales en el sur de Andalucía. In: Pascual J, Florido D, editors. ¿Protegiendo los recursos? Áreas protegidas, poblaciones locales y sostenibilidad. Sevilla: Fundación el Monte; 2005. p. 209–242.

[23] The Countryside Agency and Scottish Natural Heritage. Landscape Character Assessment. Guidance for England and Scotland (and Topic Papers). Prepared by Carys Swanwick, Department of Landscape, University of Sheffield and Land Use Consultants, 2002; [cited 2014 March 1]. Available from: http://www.naturalengland.org.uk/ourwork/landscape/englands/character/default.aspx.

[24] Frolova M, Jiménez-Olivencia Y, Sánchez-del Árbol M-A, Requena-Galipienso A, Pérez-Pérez B. Hydropower and landscape in Spain: emergence of the energetic space in Sierra Nevada (Southern Spain). In: Frolova M, Prados M.-J, Nadai A, editors. Renewable Energies and European Landscapes: Lessons from Southern European Cases. New York London: Springer; 2015. p. 117–134.

[25] Medina Barbero R, Sánchez del Árbol MA, editors. Landscape and environmental viability criteria for wind farms localization - Valle de Lecrín (Granada). Granada: Grupo Najarra Lencom-Mirador Andalucian Developments, SL; 2006.

Chapter 2

Study of Novel Power Electronic Converters for Small Scale Wind Energy Conversion Systems

Mostafa Abarzadeh, Hossein Madadi Kojabadi and Liuchen Chang

Additional information is available at the end of the chapter

Abstract

This chapter proposes a study of novel power electronic converters for small scale wind energy conversion systems. In this chapter major topologies of power electronic converters that used in wind energy converter systems have been analysed. Various topologies of DC/AC single stage converters such as high boost Z-source inverters (ZSI) have been investigated. New proposed schemes for inverters such as multilevel and Z-source inverters have been studied in this proposed chapter. Multilevel converters are categorized into three major groups according to their topologies which are diode clamped multilevel converters (DCM), cascade multilevel converters (CMC) with multiple isolated dc voltage sources and flying capacitor based multilevel converters (FCMC). Z-source inverters are divided to ZSI, qZSI and trans-ZSI types. Trans-ZSI is mostly used for high step-up single stage conversions.

Keywords: Small scale wind energy conversion system, Multi-level converters, Z-source inverters, single stage converter

1. Introduction

Nowadays, renewable energy based systems have expeditious growth and enhanced technological development due to increasing worldwide electricity demand, environmental concerns and financial aspects. Due to environmental issues of using fossil fuel in power generation and increasing the demand of energy in world, renewable energy resources are much more considered to provide power energy demand. Renewable energy resources are environmentally friendly, socially beneficial, and economically competitive for many applications. Power electronic converters optimize renewable energy conversion system performance to operate on maximum power point. Furthermore, mentioned converters prepare suitable and standard

output power to supply loads. Thus power electronic converters enable efficient interconnection between renewable energy systems and loads [1, 2].

Wind turbines, photovoltaic systems and fuel cells are the main resources for renewable resource generation systems [1]. In comparison with other renewable energy resources, wind power is more appropriate for couple of applications with relatively lower expenditure [3, 4]. As long as most major companies are concentrating on implementation of large-scale wind turbines, small-scale wind turbines as small distributed power generators have significant development in order to utilize in micro grid systems, small faraway sites and small remote consumers [5]. Recently, with regards to increased global concern about negative effects of fossil-fuel based energy conversion systems, small-scale wind generation systems are much more noticed as alternative source for renewable energy conversion applications [6]. Small scale wind turbines (<100kW) can deliver power directly to homes, farms, schools, businesses, and industrial firms, compensating the need to purchase some portion of the host's electricity from the grid; such wind turbines can also provide power to off-grid sites [4-6]. A small-scale wind energy conversion system comprises a generator, a power electronic converter, and a control system. Among different types of small-size wind turbine, permanent magnet (PM) generator is widely used because of its high reliability, high efficiency and simple structure [1-3]. The power electronic converter in WECS has been used to provide suitable and applicable output power in order to supply stand-alone loads or to connect to power grid. Control systems are implemented to regulate the small-scale wind turbines shaft speed in order to operate at desired speed which leads to extract maximum power from wind [1-5].

For faraway communities and remote areas, it is applicable and considerable to use the stand-alone small-scale wind energy conversion system including the battery bank to provide reliable and stable power supply system for these areas [2, 7-10]. For the stand-alone wind energy conversion system, battery storage system acts as load to WECS system. Since the voltage remains almost constant, but the current flows through it can be varied, the battery can be also considered as a load with a various resistance. The battery can absorb any level of power as long as the charging current does not exceed its limitation. [2, 11, 12].

There is increasing demand for a grid connected small wind generating system (without battery storage systems) in order to provide renewable energy sources in distributed power systems. Small-scale WECS are exploited as undeniable parts of micro grids to provide distributed power resources. In this case the generated power form the wind generator is supplied to the utility grid. The AC grid can also be hybrid grid by using another power generation resources such as photovoltaic systems, fuel cells and diesel generators. Applicable grid connected inverter topology which exploits maximum power even at low wind speeds for small-scale WECS are considered to provide alternative power resources for AC grids and micro grids. Standard grid connected wind turbines use a charging controller to charge the batteries to provide reliable and stable power sources. The output grid connected inverter also is used to interface small-scale WECS with the power grid [5].

Authors proposed novel Z-source inverter that can produce very high output voltage gains. With the same duty ratio the proposed topology produces higher boost factor than diode assisted QZSI, alternate cascaded ZSI, and SL-ZSI. In other word the new topology requires lower duty ratio for the same boosting voltage. Lower duty ratio will lead in higher modulation index and lower THD at output voltage. In renewable energy conversion systems both voltage buck and boost operating capabilities are needed, so before VSI an additional dc–dc converter is required, that this two stage configuration, increases the cost and reduces the efficiency of the system. By proposed ZSI, buck-boost capabilities is achieved with a single-stage power conversion. This unique feature also increases the immunity of the inverters to the EMI noises, which may cause uncontrolled shoot-zero (or open circuit) to destroy the conventional VSIs and CSIs, respectively. Meanwhile, in ZSI, both switches in a leg can be turned-on simultaneously to eliminate dead-time and to improve the quality of output waveform.

This chapter presents novel power electronic converters for small scale wind energy conversion systems for small-scale wind generators. Small-scale wind turbine consists of permanent magnet synchronous generator (PMSG), power electronic conversion system and load. Power electronic conversion system consists of AC/DC rectifier, DC/DC converter and DC/AC inverter. The chapter is organized as follows. In part 2, the small-scale wind turbine system is briefly described. In part 3, the conventional power electronic conversion system for wind turbine is proposed. Part 4 describes the multi-level converters and their applications in wind energy conversion systems (WECS). Finally, Z-source inverters are proposed in part 5.

2. Small-scale wind turbine system

The conventional small wind turbine comprises rotor, electric generator, control system and power conversion system. Rotor with a variable number of blades is utilized to convert the wind speed to mechanical rotational speed. Electric generator is also used to convert rotor mechanical rotation to electric power, control and power conversion systems are implemented to supply electricity into a battery bank or power grid [1, 13].

The electric generator is the major part of a small WECS which converts the mechanical rotation of wind rotor shaft to electrical power. There are two major types of utilized electrical generators for WECS which are self-excited induction generators (SEIG) and permanent magnet synchronous generators (PMSG). In order to provide applicable rotational speed to generators, gearbox must be utilized between wind turbine rotor and generator shaft. The gearbox converts the low rotational speed of the turbine rotor to the high rotational speed of generator shaft. Considering the fact that utilizing gearbox in WECS system leads to increase system overall weight, gearbox elimination can be achieved by utilizing the axial flux permanent magnet synchronous generators (AFPMSG) multiple-pole generator systems [1, 2, 13].

In order to energize the magnetic circuits in self-excited induction generators, parallel capacitors bank is equipped at the generator terminals to provide its necessary reactive power. Due to use the fixed capacitors bank, the reactive power and output terminal voltage might not be directly controlled. Consequently, terminal voltage fluctuation problem is emerged in self-excited induction generators [1, 2]. Therefore, the multiple-pole axial flux permanent magnet synchronous generators (AFPMSG) driven by a wind-turbine shaft without gearbox are much more considered. Small scale wind conversion system may be connected to loads or power systems with full rated power electronic converters. The wind turbines with a full scale power converter between the generator and load will lead to improved performance. The gearbox driven self-excited induction generator based small-scale WECS is depicted in Fig. 1(a). The direct-couple axial flux permanent magnet synchronous generator based WECS is shown in Fig.1(b). The full range power conversion system is equipped between generator and load for providing standard output to load.

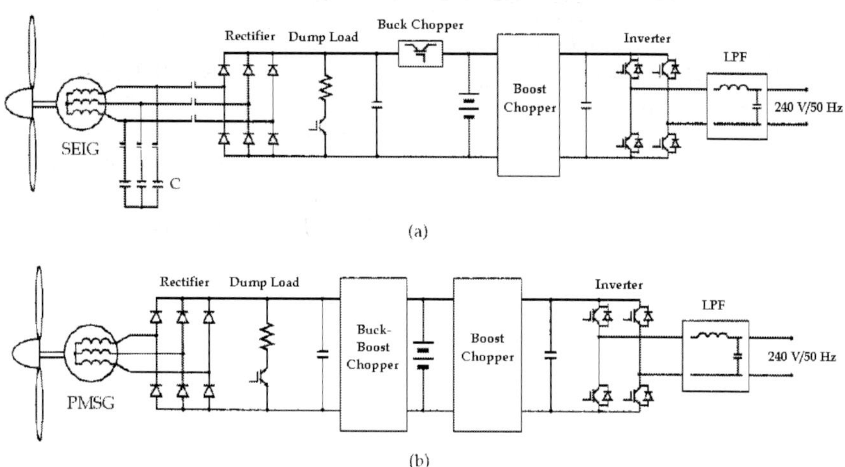

Figure 1. Small scale wind generation system. (a) self-excited induction generator with gearbox (b) direct coupled axial flux permanent magnet synchronous generator [1, 2]

The power protection during higher wind speeds in small scale wind turbines will be done by furling control [1, 2, 14]. Auto furling system is a passive mechanism utilized to protect the small-scale wind turbine in higher wind speeds [13]. There are several methods to mechanically control a wind turbine, like tilt up turbines, changing blade angle, and spoilers. Upwind rotor structure with passive yaw control by using tail vane is utilized in major part of small wind turbines. Auto furling system is operated by interaction between gravity and wind force. It operates based on interaction between wind turbine tail degree from wind turbine rotor and effective wind stream aerodynamic force. Eventually, during wind speeds above the nominal wind speed the tail vane starts to hinge which leads to decrease the wind turbine rotor speed. The over-speed protection is resulted during this condition. During higher wind speeds, generated output power of wind turbine could be bigger than the nominal power of wind generator which may damage the wind rotor or energy conversion unit. Hence, during high wind speeds, furling system deflects the turbine from the wind stream to protect whole system from probable damages. During furling conditions, there is a sharp decrease in the wind turbine output power curve [1, 13-16]. The typical furling system is depicted in Fig. 2.

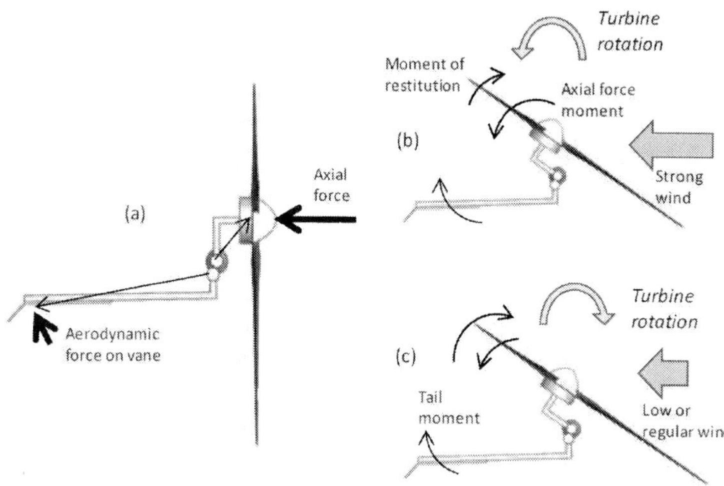

Figure 2. Overview of the operating principles of a furling system. (a) Aerodynamic forces. (b) Furling movement in strong winds. (c) Restitution of normal (aligned) operation upon reduction of the wind speed [13].

3. Conventional power electronic conversion system for wind turbine

In order to achieve maximum efficiency of wind energy conversion system the AC/DC/AC topology almost always used in small scale systems. This topology results maximum efficiency and full range working range which is applicable for WECS systems. To convert the variable AC voltage which is generated by AFPMSG, the diode rectifier should be used. Due to the fact that active power flows uni-directionally from the AFPMSG to power converter and load, only a simple diode rectifier can be applied to the generator side converter [5, 6]. In order to regulate and track maximum power point of wind turbine in various wind speeds, DC/DC converter is used [17]. In order to supply standard output AC loads, the inverter should be implemented at the WECS output.

3.1. AC/DC bridge rectifier

A three-phase bridge rectifier which gives six-pulse ripples on the output voltage is commonly used in small-scale WECS. The pair of diodes which are having the maximum amplitude of line-to-line voltage will conduct [18]. The typical three-phase bridge rectifier is depicted in Fig.3.

If we assume that V_m is the maximum value of the phase voltage, then the average output voltage can be considered as:

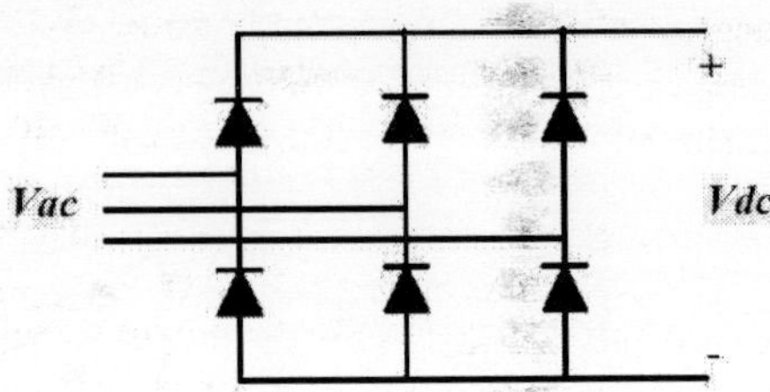

Figure 3. Three-phase bridge rectifier.

$$V_{dc} = \frac{2}{2\pi/6} \int_0^{\pi/6} \sqrt{3}\, V_m \cos\omega t\, d(\omega t) = \frac{3\sqrt{3}}{\pi} V_m = 1.654 V_m \tag{1}$$

3.2. DC/DC converters

Dc/dc converters are switch-mode converters which regulate variable voltage to a desired reference dc output voltage. The mostly used dc-dc converters which are defined as three terminal regulators are explained in this section [19]. They are characterized by the use of a choke rather than a transformer between the input and output lines, resulting in higher efficiencies and wider voltage ranges. There are four basic topologies of non-isolated dc to dc converters which can produce higher, lower or reversed polarity to the input.

3.2.1. Buck regulators

The buck regulator which is shown in Fig. 4 provides the output voltage lower than input dc voltage with the same polarity of input dc voltage [18]. The fundamental circuit of buck dc-dc converter is shown in Fig. 4.

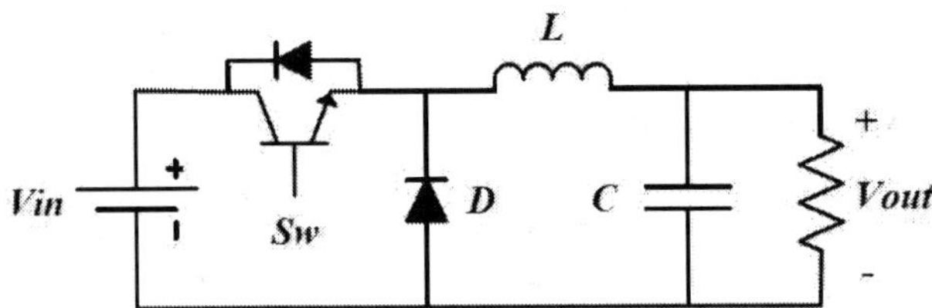

Figure 4. Basic circuit of buck dc-dc regulator.

The output voltage of buck converter is calculated by

$$V_{out} = k \times V_{in} \tag{2}$$

where V_{out} is output voltage, V_{in} is input voltage and k is duty cycle of switch. The output voltage can be controlled by varying the duty cycle. The buck converter is also known as a step-down converter, since its output must be less than the input voltage [18-21].

3.2.2. Boost regulators

The boost regulator which is depicted in Fig. 5 provides output voltage higher than input dc voltage with the same polarity of input dc voltage. Due to the existence of right-half-plane zero in boost converter's transfer function, it is hard to control it during high step up gain between output and input voltage [18-21]. The basic circuit of boost dc-dc converter is shown in Fig.5.

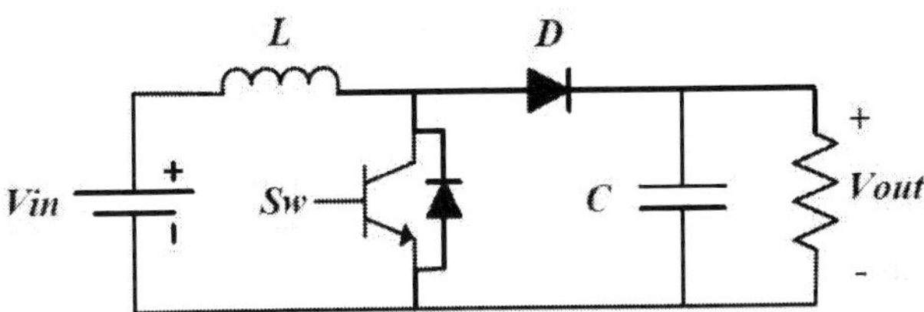

Figure 5. Basic circuit of boost dc-dc regulator.

The output voltage of boost converter is calculated by

$$V_{out} = \frac{1}{1-k} V_{in} \tag{3}$$

3.2.3. Buck-boost regulators

In order to achieve both step-up and step-down capability of dc-dc converter, buck-boost converter which is combination of buck and boost topologies is provided. The output voltage has opposite polarity to the input, but its value may be higher, equal, or lower than that of the input. This regulator also has right-half-plane zero in its transfer function which limits the step up gain of this converter [18-21]. The basic circuit of buck-boost dc-dc converter is shown in Fig.6.

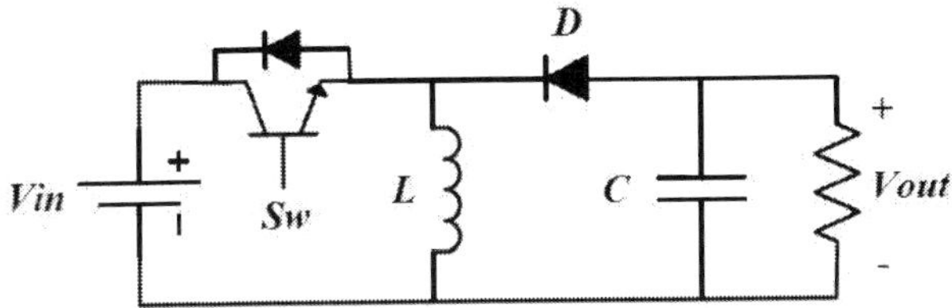

Figure 6. Basic circuit of buck-boost dc-dc regulator.

The output voltage of buck-boost converter is calculated by

$$V_{out} = -\frac{k}{1-k} V_{in}$$

(4)

3.2.4. Cuk regulators

Since buck-boost regulator has reverse polarity to the input dc voltage, so the new class of buck-boost regulator named cuk converter has been proposed. In this converter, the output voltage has the same polarity of input, but its value may be higher, equal, or lower than that of the input. This converter also has the right-half-plane zero in its transfer function [18-21]. The basic circuit of cuck dc-dc converter is shown in Fig.7.

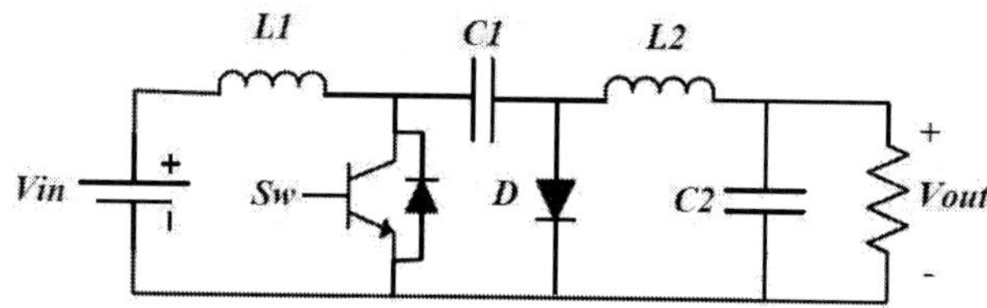

Figure 7. Basic circuit of cuck dc-dc regulator.

The output voltage of cuk converter is calculated by

$$V_{out} = \frac{k}{1-k} V_{in}$$

(5)

3.3. DC/AC inverters

DC-to-ac converters are defined as inverters which are used to convert input dc voltage to symmetrical output ac voltage in order to supply ac load or connect to grid. The inverter output voltage must satisfy specific standards which are defined as desired voltage amplitude, frequency and maximum allowable total harmonic distortion (THD) value. The ideal ac voltage waveforms should be sinusoidal. However, due to practical inverters work based on switching devices, output ac voltage of inverters is non-sinusoidal and contain certain harmonics [18, 22].

There are various switching methods which are applied to control the inverters. The pulse width modulation (PWM) control method is mostly used to provide suitable switching signals to control inverter switches. There are two major types of inverter which are classified as voltage source inverters (VSIs) and current source inverters (CSIs). The VSI inverters which are widely used in small scale WECS applications independently control the ac output voltage waveform. While, the CSI inverters independently control the ac output current waveform [18].

The single-phase VSI is shown in Fig. 8. It comprises four switches which are S_1, S_2, S_3 and S_4. This type of inverter has three output voltage states which are V_d, 0, $-V_d$. V_d is resulted at inverter output when S_1 and S_2 are on. When S_3 and S_4 are on, the inverter output voltage is $-V_d$. If all switches are off, the output voltage is 0. The root mean square (RMS) output voltage can be found from

$$V_O = \sqrt{\left(\frac{2}{T_0}\int_0^{T_0/2} V_d^2 dt\right)} = V_d \tag{6}$$

where V_O is rms value of output voltage. Output voltage can be represented in Fourier series. The rms value of fundamental component as

$$v_O = \sum_{n=1,3,\dots}^{\infty} \frac{4V_d}{n\pi}\sin n\omega t \xrightarrow{n=1} V_1 = \frac{4V_d}{\sqrt{2}\pi} = 0.90V_d \tag{7}$$

where v_O is output voltage and V_1 is rms value of fundamental component of v_O.

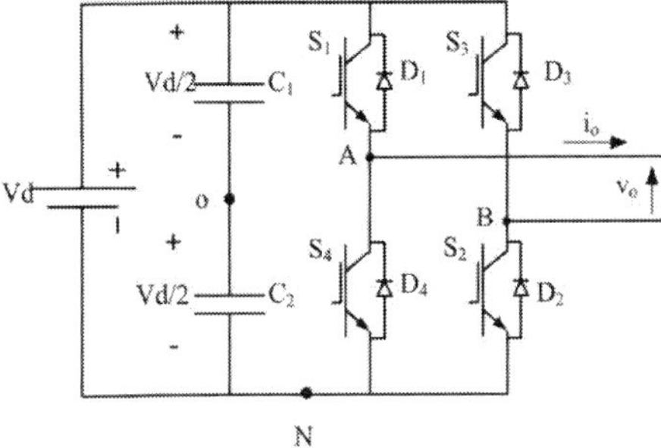

Figure 8. Single-phase full bridge inverter

4. Multi-level converters

The two-level VSI provides three output voltage levels which are 0 or $\pm V_d$. In order to improve the output voltage or current quality, it is required to apply higher switching frequency and enhanced PWM techniques which are restricted by switching device operating specifications and control systems implementation requirements [18].

The most remarkable superiority of multilevel converters in comparison with two-level inverters are synthesizing an output staircase voltage waveform which improves harmonic content significantly, also reduces output $\frac{dv}{dt}$, electromagnetic interference and filter inductance, etc [18, 24-27]. By increasing the number of voltage levels, the output waveform will be a staircase form and this will lead in significant reduction of output voltage THD.

Multilevel converters divide to three major types as below:

- Diode-clamped multilevel (DCM) converters

- Cascaded multilevel (CM) converters [26].

- Flying capacitor multilevel (FCM) converters [24, 27].

4.1. Diode-clamped multilevel (DCM) converter

The n-level DCM converter which comprises $(n-1)$ capacitors, $2(n-1)$ switches and $(n-1)(n-2)$ clamping diodes provides n-levels output voltage. Considering dc bus voltage as E, the voltage across each capacitor is $\frac{E}{2}$, and each switch stress is restricted to one capacitor voltage level $\frac{E}{2}$ through clamping diodes.

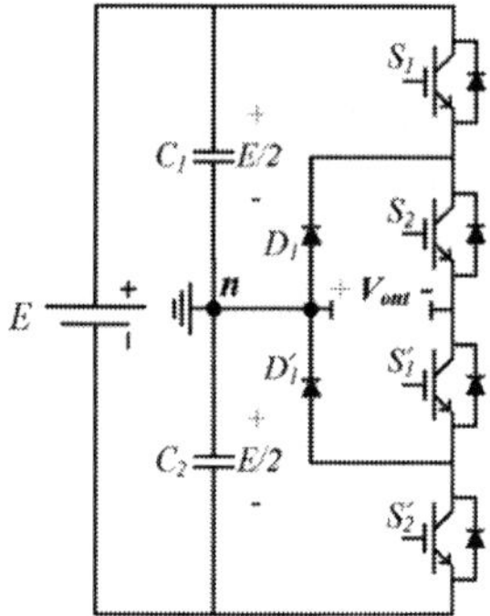

Figure 9. 3-level diode-clamped converter.

In order to provide staircase output voltage in DCM converter, if the S_1 and S_2 power switches are on, the staircase output voltage is $V_{out} = \frac{E}{2}$. When S'_1 and S'_2 power switches are on, the output voltage is $V_{out} = -\frac{E}{2}$. For output voltage $V_{out} = 0$, S'_1 and S_2 power switches must be on [18, 23].

The considerable advantages of DCM converter are as follows:

- Output voltage THD is significantly decreases by increasing the voltage levels which leads to reduction or elimination of output filter size.

- With regards to the fact that switches operates at the fundamental frequency, the Inverter efficiency is high.

- The implemented control procedure is simple.

The notable disadvantages of DCM inverter are as follows

- By increasing the number of voltage levels, the number of clamping diodes is increased drastically.

- It is complicated to control the real power flow of DCM converter in multi converter systems [18, 23].

4.2. Cascade multilevel (CM) converter

A CM inverter comprises series connections of H-bridge inverters which are supplied by isolated dc sources. The aforementioned isolated dc sources may be provided by WECS systems, batteries, or other voltage sources. The configuration of CM converter with isolated dc voltage sources is depicted in Fig.10.

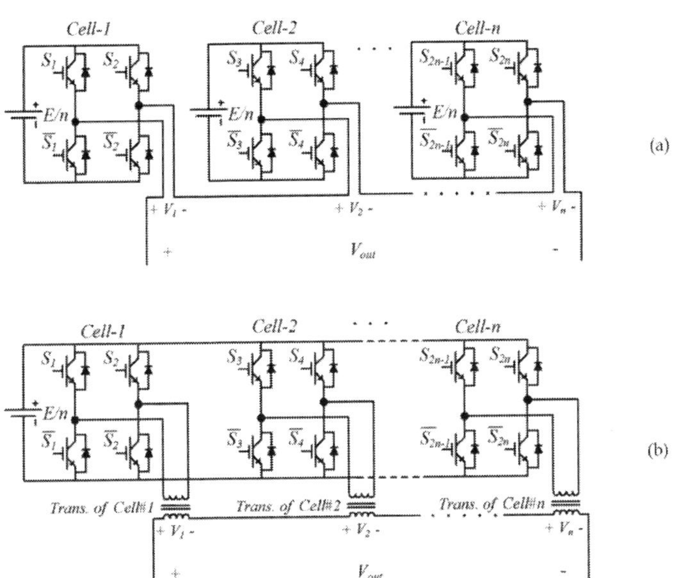

Figure 10. The 2n+1 levels cascade multilevel inverter: (a) with separated dc voltage sources. (b) with one dc voltage source and isolator transformers.

Each h-bridge inverter provides three output voltage levels which are 0 or $\pm E$. All five switch states for h-bridge inverter are shown in table 1. The output voltage of CM converter is obtained by the sum of inverter outputs. Hence, the CM inverter output voltage can be expressed as

$$v_{out} = \sum_{i=1}^{n} v_i \tag{8}$$

where n is the number of cells and v_i is the output voltage of cell i.

For n number of cells, the output voltage of CM converter has $2n + 1$ levels. For instance, in order to provide five level output voltage, two h-bridge cascaded converters with isolated dc voltage sources are needed. Switch states of mentioned five-level CM converter is depicted in table 2 [18, 23, 26].

The advantages of the CM inverter are as follows:

- CM needs the minimum number of components to provide the same number of staircase voltage levels in comparison with DCM and FCM converters.

- In order to reduce switching losses and device stresses, soft-switching techniques can be implemented on CM converter.

The disadvantage of the CM inverter is requirement of isolated dc voltage sources for real power conversions [18, 23, 26].

Output Voltage Level	State of (S1 , S2 , S3 , S4)	Number of States
$+\frac{2}{2}E$	$(1,0,1,0)$	1
$+\frac{1}{2}E$	$(1,0,1,1),(1,0,0,0),(1,1,1,0),(0,0,1,0)$	4
0	$(1,1,1,1),(1,1,0,0),(0,0,0,0),(0,0,1,1)$	4
$-\frac{1}{2}E$	$(0,1,1,1),(0,1,0,0),(1,1,0,1),(0,0,0,1)$	4
$-\frac{2}{2}E$	$(0,1,0,1)$	1

Table 1. Switches states for a five-level CM inverter.

4.3. Flying capacitor multicell (FCM) converter

The FCM converters comprise ladder connection of FC cells. Each FC cell in FCM converter consists of a flying capacitor and a pair of switches with a complimentary state. In order to provide various levels at the inverter output, the commutation between adjacent cells with their FCs charged to the desired values is needed [24]. Hence, the voltage balancing of flying capacitors is indispensable issue in FCM converters. Appropriate operation of FCM converter depends on suitable voltage balancing of FCs [23, 24]. Applying the phase-shifted carrier pulse-width modulation (PSC-PWM) technique in FCM converter control system guarantees the FCs self-balancing process [24]. Typical configuration of FCM converter and the phase-shifted carrier pulse-width modulation (PSC-PWM) technique for five-level FCM inverter are shown in Fig.11 and Fig.12, respectively.

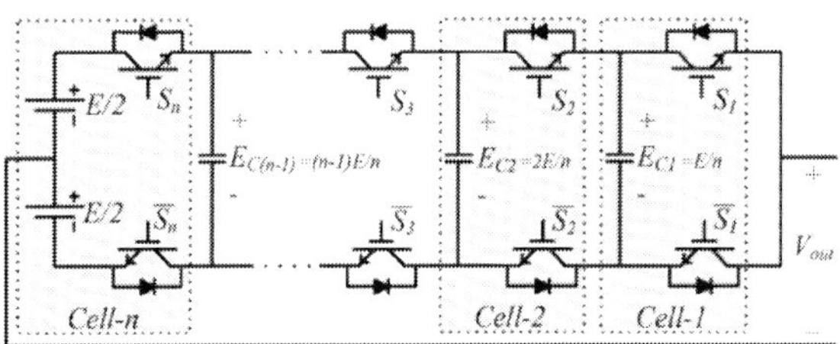

Figure 11. The n cells ($n + 1$ levels) FCM inverter.

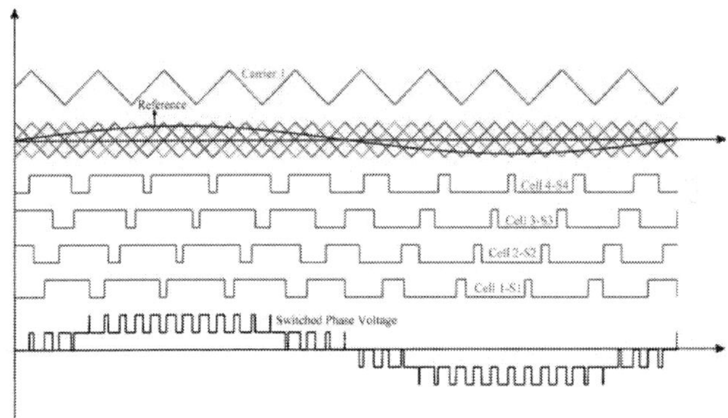

Figure 12. The phase-shifted carrier pulse-width modulation (PSC-PWM) technique for five-level FCM inverter.

The advantages of the FCM inverters are:

- FCM inverter does not require to isolated dc voltage sources. Hence, it has transformer-less operation capability.

- Clamping diodes are not needed to use in FCM inverters.

- There are plenty of redundant switching states which result inherent self-balancing property of the voltage across FCs.

- Switching stress is equally distributed between power switches.

The major disadvantages of the FCM inverter are:

- Numerous FCs is needed to provide high staircase output voltage levels.

- Implementing switching and control pattern can be more sophisticated and complicated [18, 23, 24].

The switches states of five-level FCM inverter is expressed in table 2.

Output Voltage Level	State of (S4 , S3 , S2 , S1)	Number of States
$+\frac{2}{4}E$	$(1,1,1,1)$	1
$+\frac{1}{4}E$	$(1,1,1,0),(1,1,0,1),(1,0,1,1),(0,1,1,1)$	4
0	$(1,1,0,0),(1,0,0,1),(0,0,1,1),(0,1,1,0)$	4
$-\frac{1}{4}E$	$(0,0,0,1),(0,0,1,0),(0,1,0,0),(1,0,0,0)$	4
$-\frac{2}{4}E$	$(0,0,0,0)$	1

Table 2. Switches states for a five-level FCM inverter

4.4. Multilevel converters applications in wind energy conversion systems

Multilevel inverters provide more than two voltage levels. A desired output voltage waveform can be synthesized from the multiple voltage levels with significantly improved harmonic content, decreased output dv/dt, reduced electromagnetic interference, smaller filter induc-tance, and increased carrier ratio (R_c). Furthermore, low switching and conduction power losses, increased efficiency, fault tolerant, and high modularity are other valuable features of multilevel converters. There are a couple of topologies that provide multilevel voltages and are suitable for industrial applications.

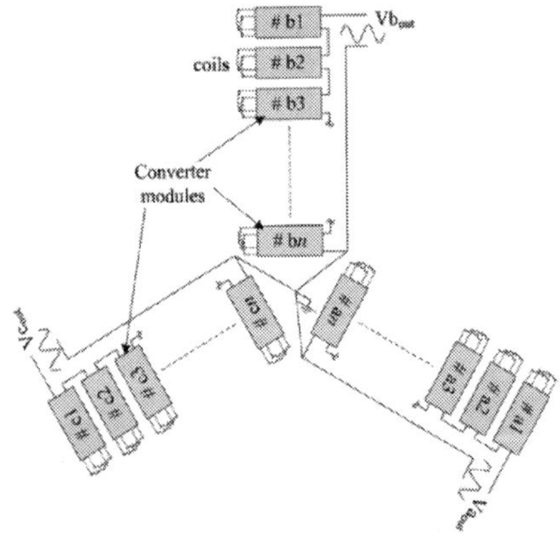

Figure 13. Electrical configuration of multiple winding stator[25].

With regards to above-mentioned multilevel converter topologies, it is clear that using multilevel converters will lead in increased complexity to control and implementation of power electronic conversion system. Among multilevel converters, DCM is widely used for wind energy conversion system because it needs just one power supply to provide desired output voltage. On the other hand, increasing in voltage levels of DCM causes expeditiously increase in number of clamping diodes and complexity of control system. Thus, using CHB and FCM converters have been mentioned in couple of papers. In order to overcome the required isolated power supplies in these converters, input DC sources elimination methods have been proposed in some papers. Also, using novel structure of AFPMSG generators with variety of stator windings has been mentioned. For instance, in [25] new multi level conversion system is proposed. In order to provide multiple DC source to multilevel converter, multiple winding stator has been used in [25]. Electrical configuration of mentioned system and circuit diagram of proposed structure of converter are shown in Fig.13 and Fig.14, respectively.

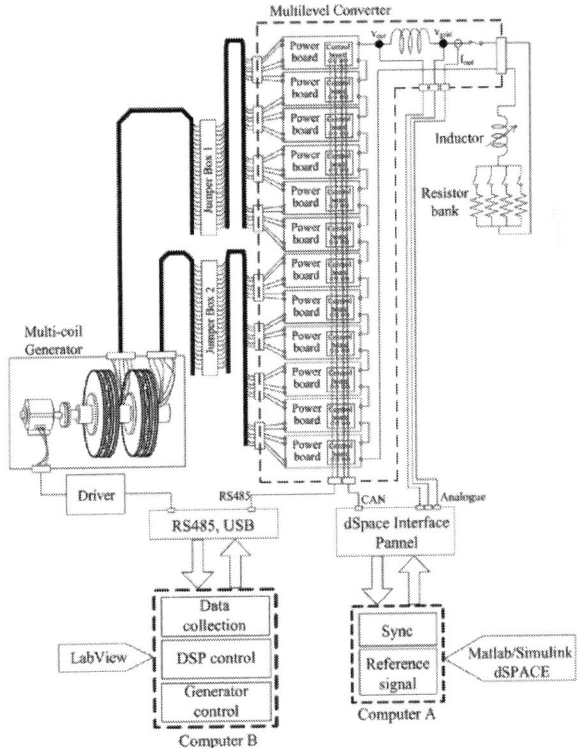

Figure 14. Circuit diagram of multilevel converter based on multiple winding stator [25].

5. Z-source inverters

V-source inverter is widely used in wind energy conversion systems. V-source inverters have the following demerits as theoretical and experimental point of view:

- The ac output voltage is limited below and cannot exceed the dc-rail voltage or the dc-rail voltage has to be greater than the ac output voltage. Therefore, the V-source inverter is a step-down inverter. For couple of applications where the available dc voltage is limited, increasing voltage is demanded to have wide-range operation capability. Hence, additional dc-dc boost converter is needed to obtain a desired ac output which increases system cost and decreases system efficiency.

- The upper and lower switches of each phase leg cannot be turned on at the same time either by purpose or by electromagnetic interference (EMI) noise. Otherwise, phase leg would be short-circuited. The shoot-through problem is emerged by EMI noises. Providing dead time between upper and lower switches which is used to prevent phase leg short-circuiting, will result in high THD at output voltage waveform.

- An output LC filter is needed for providing a sinusoidal voltage, which causes additional power loss and control complexity [28].

In order to overcome the above problems of the traditional V-source converters, [28] presents an Z-source converter for implementing dc-to-ac power conversion. Typical Z-source converter structure is depicted in Fig. 15. In this type of converter, unique impedance network is used to couple the converter main circuit to the power source, load, or another converter, for providing superior performances in comparison with traditional V-source converters where a capacitor is used. Hence, the Z-source converter dominates the aforementioned disadvantages and restrictions of the traditional V-source converter and provides a novel power conversion configuration. In Fig. 15, a two-port network that consists of a split-inductor L_1 and L_2 and capacitors C_1 and C_2 connected in X shape is employed to provide an impedance source (Z-source) coupling the inverter to the dc source, load, or another converter. The dc source can be a battery, diode rectifier, thyristor converter, fuel cell, a capacitor, or a combination of those. Fig. 15 shows two three-phase Z-source inverter configurations. The inductance L_1 and L_2 can be provided through a split inductor or two separate inductors.

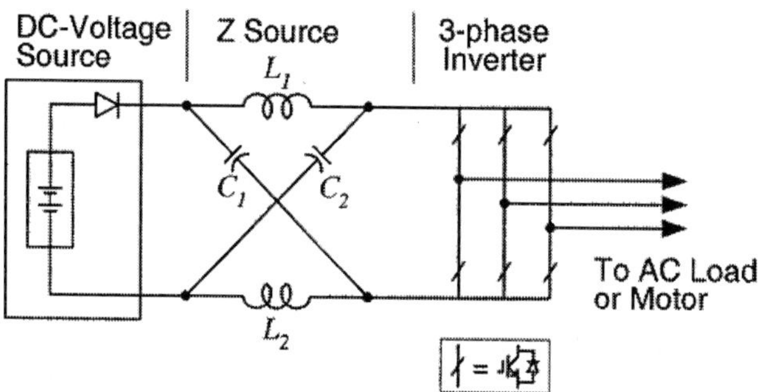

Figure 15. Z-source inverter configuration.

Fig.1 shows the traditional two-stage power conversion for wind energy conversion applications. Because wind turbine usually produce a voltage that changes widely depending on wind speed. For wind power generation, a dc–dc boost converter is needed because the V-source inverter cannot produce an ac voltage that is greater than the dc voltage. Z-source inverter can directly produce an ac voltage greater and less than the wind turbine voltage. So Z-source inverter is a combination of dc-dc boost converter and traditional VSI.

5.1. Quasi Z-source inverters

Multiple structures have been derived from main structure of Z-source inverter that is mentioned above. In this structure, the current drawn from the source is discontinuous. This causes couple of limitations in some applications, and a decoupling capacitor bank at the front end is sometimes used to prevent current discontinuity and protect the energy source. In order to prevail the frailties in the classical ZSI, quasi-Z-source inverters (qZSIs) have been proposed [29-31]. The qZSI provides a lower voltage stress on capacitors. The ratio between the dc-link voltage across the inverter bridge, v_{PN}, and the input dc voltage, V_{dc} (the boost factor of the classical ZSI and qZSI), can be expressed as

$$B = \frac{v_{PN}}{V_{dc}} = \frac{1}{1 - 2\frac{T_0}{T}} = \frac{1}{1 - 2D} \tag{9}$$

where T_0 is the interval of the shoot-through state during the switching period T and $D = T_0/T$ is the duty cycle ratio of shoot-through state.

The boost factor from (9) indicates that when D is between 0 and 0.5, B varies in $(1, \infty)$. The infinite value of B is not achievable in practical because of the parasitic effects found in the physical components. Therefore, the ZSI and qZSI are not appropriate for applications where both a high boost voltage and a buck voltage gain are required. To overcome the boost limitations in (9), some papers have recently focused on increasing the boost factor of the ZSI by adding the inductors, capacitors, and diodes to the Z-impedance network in order to produce a high dc-link voltage for the main power circuit from a very low input dc voltage. They are, namely, continuous-current diode/capacitor-assisted extended-boost quasi-ZSIs [32], switched-inductor ZSIs [33], switched-inductor quasi-ZSIs [34] and single-stage boost inverters with couple inductor [35].

5.1.1. Extended-Boost ZSI Topologies

In [32], new converter topologies are mainly categorized as diode-assisted boost or capacitor-assisted boost topologies. All these topologies can be modulated using the modulation methods proposed for the original ZSI. The other main advantage of these topologies is their expandability which was not possible with the original ZSI. In these topologies, in order to increase the boosting range, another stage can be cascaded at the front end without increasing the number of active switches. The only additions for each added new stage would be one inductor, one capacitor, and two diodes for the diode-assisted case and one inductor, two capacitors, and one diode for the capacitor-assisted case. For each added new stage, the boost factor is increased by a factor of $1/(1 - D_s)$ in the case of the diode-assisted topology, where D_S is shoot-through duty ratio. In the case of capacitor-assisted topology, the boost factor is changed to $1/(1 - 3D_S)$ in comparison with $1/(1 - 2D_S)$ in the traditional topology. Furthermore, hybrid type of topologies can be achieved by combination of capacitor- and diode-assisted techniques, if the number of stages is larger than two [32]. Diode-assisted extended-boost qZSI topologies are shown in Fig.16.

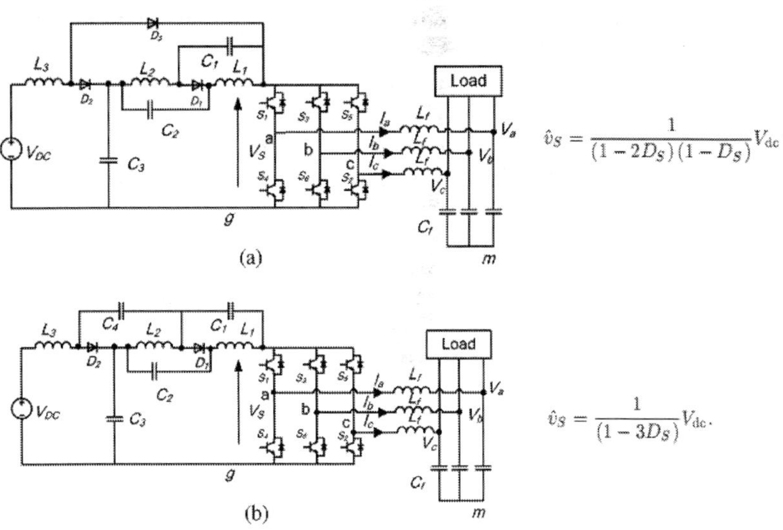

Figure 16. Extended-boost ZSI inverter (a) Diode-assisted extended-boost qZSI (b) Capacitor-assisted extended-boost qZSI [32].

In Fig.16, $\hat{v}_S$ is the output voltage of impedance network of extended-boost ZSI inverter.

5.1.2. Switched Inductor Z-Source Inverter

Recently significant improvement of dc-dc conversion techniques such as switched capacitor (SC), switched inductor (SL), hybrid SC/SL, voltage multiplier cells and voltage lift techniques have been introduced in [33, 36-38], which are used to achieve the high step-up capacity in transformer-less and cascade configurations. The main purpose is to reach a high efficiency, high power density, and simple structures. In [33], the combination with the Z-source inverter and advanced dc–dc improvement techniques has been proposed. In [33], the concept of the SL techniques has been combined into the classical Z-source impedance network, and consequently, a new SL Z-source impedance network has been proposed [33]. The new proposed topology is then defined the SL Z-source inverter is depicted in Fig.17.

The main characteristics are summarized as:

- The basic X-shape configuration is preserved.

- Only six diodes and two inductors are added in comparison with the traditional ZSI.

- The boost factor has been increased from $1/(1-2D)$ to $(1+D)/(1-3D)$.

- The new structure is extensible for the further development using the coupled inductor techniques and other potential improving techniques [33].

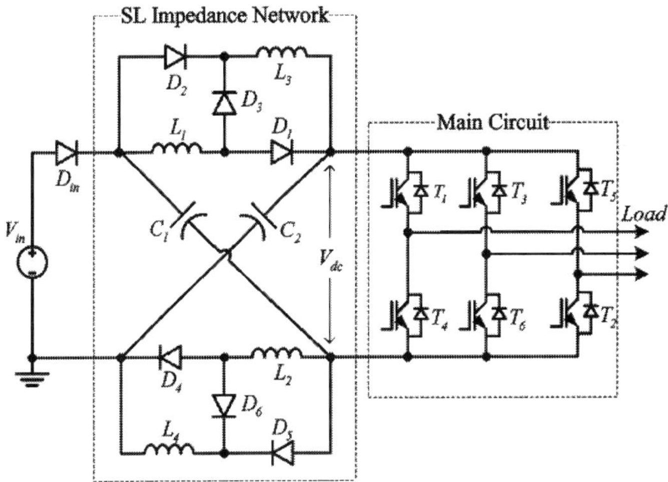

Figure 17. Switched inductor Z-source inverter [33].

5.1.3. Switched-Inductor Quasi-Z-Source Inverter

In order to overcome the shortcomings of the classical ZSI, quasi-ZSIs (qZSIs) have been proposed. qZSIs have many advantages, such as reducing passive component ratings and improving input profiles. In [34], the switched-inductor structure has been applied to the continuous input current quasi-Z-source topology to create a novel topology which is called an SL-qZSI. Compared with the SL-ZSI, the proposed SL-qZSI enhances input current, reduces the passive component count, and enhances reliability. Moreover, the shoot-through current, voltage stress on capacitors, and current stress on inductors and diodes in the proposed SL-qZSI are lower than in the SL-ZSI for the same input and output voltages. The proposed SL-qZSI avoids the inrush current at start-up, which may damage the devices [34]. Fig.18 shows the SL-qZSI topology.

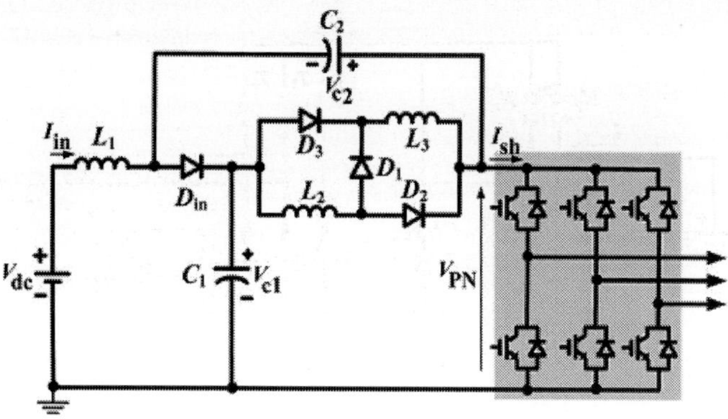

Figure 18. Switched-inductor quasi-Z-source inverter [34].

Regarding that no current flows to the main circuit during start-up, SL-qZSI topology provides inrush current elimination, in contrast with SL-ZSI topology. However, the inductors and capacitors in the proposed switched-inductor qZSI still resonate. Compared with a conventional continuous-current qZSI, the proposed inverter adds only three diodes and one inductor, and the boost factor increases from $1/(1 - 2D)$ to $(1 + D)/(1 - 2D - D^2)$.

5.1.4. Single-Stage Boost Inverter With Coupled Inductor

In [35] single-stage boost inverter with a unique impedance network that consisting coupled inductor has been proposed. The bus voltage can be stepped up by applying shoot-through zero state to accumulate and transfer energy within the impedance network. Similar to ZSI, the single-stage boost inverter completely avoids damaging devices during shoot-through zero states. So, it has enhanced reliability. Second, the inductors and capacitors do not have to be of high consistency which leads to easier circuit parameters design. Third, both shoot-through zero states and coupled inductor's turn ratio can be regulated to control the boost gain. So, the output voltage can be regulated in a wide range and can be stepped up to a higher value. Finally, the bus voltage equals the sum of the capacitor voltages, and it is higher than each capacitor voltage. This ensures status capacitor voltage ratings can be fully utilized. The single-stage boost inverter with coupled inductor is appropriate for applications where the input voltage has wide range variation from relatively low to higher level while output must have a stable ac voltage [35]. Single-stage boost Z-source inverter with coupled inductor is shown in Fig.19.

5.1.5. Enhanced-Boost Z-Source Inverters with Switched Z-Impedance

Fig. 20 shows the proposed Z-source inverter that produces high output voltage gain. In traditional Z-source inverters for high boosting voltage, low modulation index is required, so under these conditions output voltage will have low quality with high total harmonic distor-

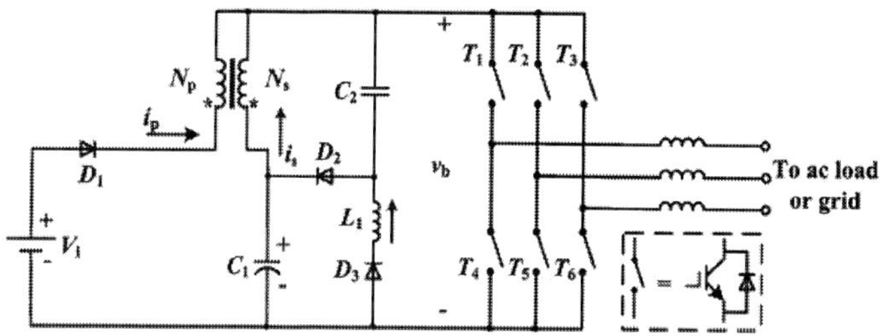

Figure 19. Single-stage boost Z-source inverter with coupled inductor [35].

tion. Compared with the conventional high-boost ZSI topologies, the proposed inverter by authors uses shorter shoot-through duration and larger modulation index to improve the output waveform quality [39].

The boost factor of proposed ZSI can be calculated as:

$$B = \frac{1}{2D^2 - 4D + 1} \tag{10}$$

The output ac voltage can be calculated by:

$$\hat{V}_{ac} = M.B.\frac{V_{in}}{2} = G.\frac{V_{in}}{2} \tag{11}$$

where

$$G = M.B = \left(0 \sim \infty\right) \tag{12}$$

is buck–boost factor. The buck–boost factor (G) can be controlled by changing D and M. Duty cycle of shoot-through, D, is limited by modulation index by the following relation:

$$D \leq 1 - M \tag{13}$$

The boost factor from (10) indicates that when D is between 0 and 0.29, B varies in (1, ∞). However, the infinite value of B is not accessible due to parasitic effects of physical components. Even if a large value of D is used to produce the high boost gain, the modulation index must be small as indicated in (13). However, low modulation index, M will result in poorer spectral performances and low THD. Small M is generally not preferred since it leads to low output waveform quality and high power losses due to lower order harmonics at output waveforms.

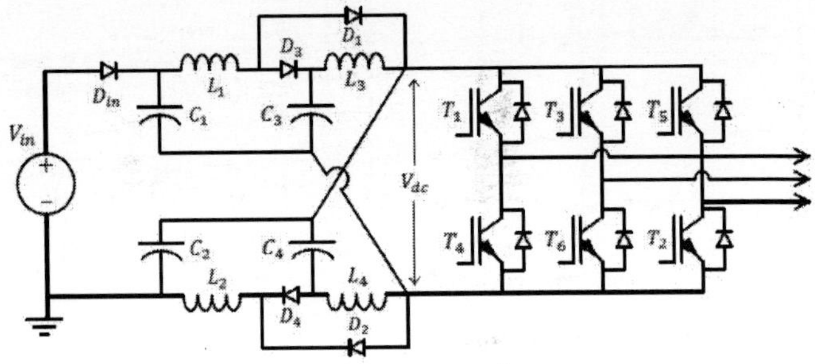

Figure 20. Proposed Z-source inverter with switched Z-impedance [39].

5.1.5.1. Simulation Results

The proposed topology shown in Fig. 20 has been simulated by PSIM software platform. The main circuit parameters for proposed topology are as follows:

$L_1 = L_2 = L_3 = L_4 = 700\mu H$, $C_1 = C_2 = C_3 = C_4 = 500\ \mu F$

The switching frequency $F_s = 10 kHz$

Three phase load R= 60 Ω, L= 5mH,

DC input voltage $V_{in} = 100$

Fig. 21 shows that with the same duty ratio the proposed topology produces higher boost factor than diode assisted QZSI, alternate cascaded ZSI, conventional ZSI, and SL-ZSI. In other word the new topology requires lower duty ratio for the same boosting voltage. Lower duty ratio will lead in higher modulation index of M and lower THD at the output voltage. So this topology is very suitable for renewable energy conversion systems, without requirement of dc-dc converters. Fig. 22 shows the output voltage gain against modulation index for five topologies. In order to produce the same output voltage gain, the proposed topology uses higher modulation index than other two traditional methods.

Fig. 23 shows the simulation results for the proposed ZSI with shoot-through duty cycle of D = 0.18 and modulation index of M = 0.82. From (10), the boost factor, B = 2.9. From simulation results one can observe that the capacitors' voltage V_{c1} and V_{c3} are, respectively, boosted to 195 and 237.5 V in the steady state, and the peak dc-link voltage V_{dc} is boosted to 290 V. So the boosted factor can be calculated to be 290/100 =2.9. Meanwhile, the peak ac output phase voltage $\hat{V}_{ac}$ has been boosted to 119 V. These results clearly verify the high boosting capability of the proposed ZSI.

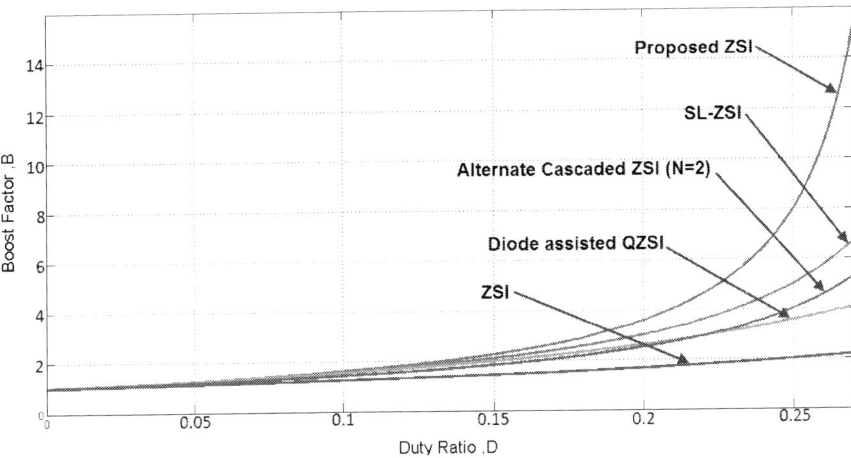

Figure 21. Boost Factor against the duty cycle of shoot-through.

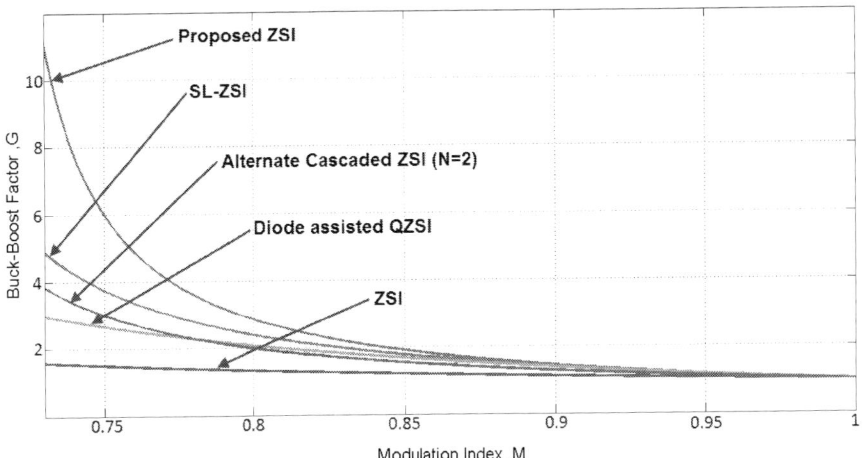

Figure 22. Buck-Boost factor against the modulation index.

5.1.5.2. Experimental Results

Fig. 24 shows the experimental results for proposed ZSI when shoot-through duty cycle D = 0.18 and modulation index M = 0.82. DC link voltage is boosted from 100 V to 268 V, which is slightly less than the calculated value (2.9*100 = 290 V) from (10) due to the voltage reduction on the passive components, diodes, and switching devices. Capacitor voltages V_{c1} and V_{c3} are boosted to 186V, and 219 V, respectively. Meanwhile, the rms value of line to line output

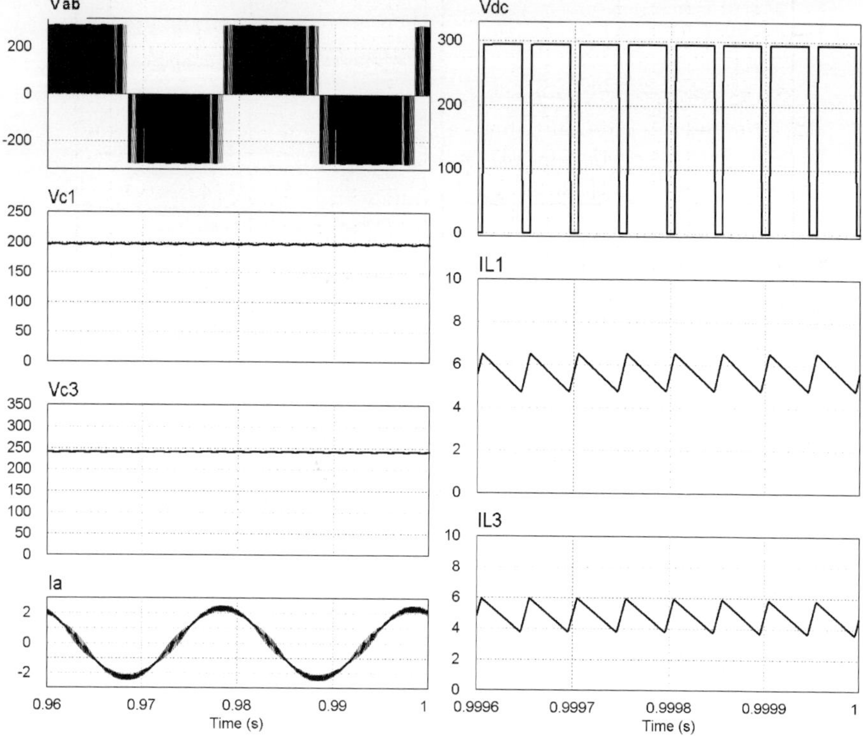

Figure 23. Simulation results for the proposed ZSI, M=0.82 and D=0.18.

voltage is 193V. Fig. 24 further shows the dc-link voltage, inductor current, and load current which clearly are similar to simulations except for some spikes and switching noises superimposed. These spikes and noises are not seen in simulation and are thus likely picked up from the hardware semiconductor switching. Because of proper snubber design, the MOFETs have enough protection against the spikes and noises. From Fig. 24 it is very clear that the peak dc-link voltage maintains constant value during the non-shoot-through states, and it is quite consistent with the theoretical value. The simulation and experimental results of Figs. 23 and 24 match well with each other and those of the theoretical analysis for the proposed inverter.

5.2. Trans Z-source Inverters

The voltage-fed Z-source inverter can have theoretically infinite voltage boost gain. However, the higher the voltage boost gain, the smaller the modulation index has to be used. In wind energy conversion applications, a low-voltage dc source has to be boosted to a desirable ac output voltage. A small modulation index results in a high voltage stress imposed on the inverter bridge. Transformer-based Z-source inverter (trans-ZSIs) has the possibility of

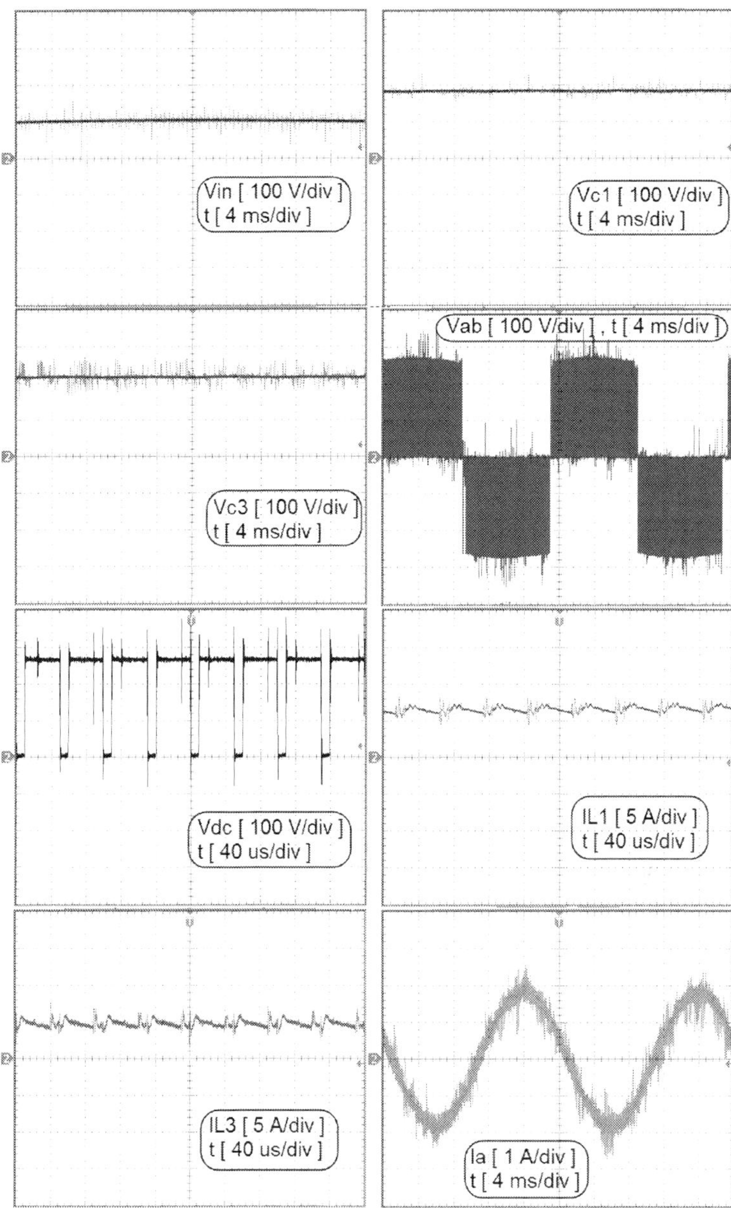

Figure 24. Experimental results for the proposed ZSI, M=0.82 and D=0.18.

increasing voltage gain with the minimum component count [31, 40]. Applying a transformer to the Z-source network has been introduced in the trans-Z-source. Voltage-fed trans-quasi-ZSI proposed in [40], where two inductors in the impedance Z-network are replaced by a transformer with a turn ratio of n:1 to achieve a high voltage gain. It comprises one transformer, one capacitor, and one diode. Fig. 25 depicts transformer-based Z-source inverter circuit diagram. The boost factor of this inverter is increased to

$$B = \frac{1}{1-(1+n)\,{T_0}\big/{T}} = \frac{1}{1-(1+n)D} \tag{14}$$

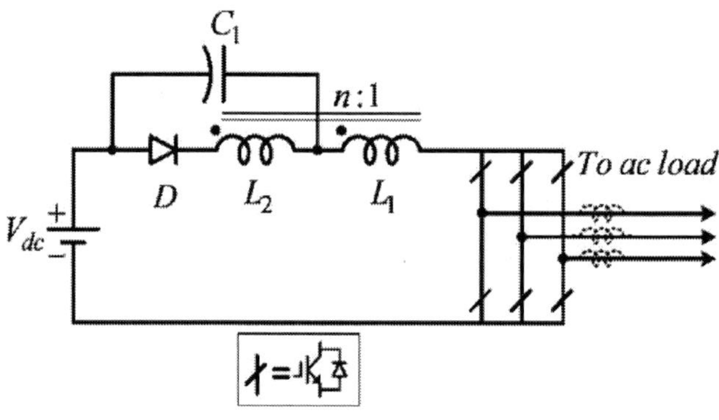

Figure 25. Transformer-based Z-source inverter [39].

5.3. Z-source converters applications in wind energy conversion systems

With regards to wind energy conversion systems that usually requires to increase the dc-link voltage to applicable standard voltage to supply ac load, high step-up voltage conversion must be used. In order to boost the dc link voltage in traditional wind energy conversion systems, power electronic transformer (PET) has widespread usage to convert the low dc-link voltage to applicable dc-voltage that is desirable in dc-to-ac inverters. Circuit diagram of traditional energy conversion system based on PET is depicted in Fig. 26.

In recent wind energy conversion systems, power electronic transformer (PET) is replaced with ZSI, qZSI and trans-ZSI high step-up converters to convert low dc-link voltage to desirable dc link voltage. This will lead in power electronic devices reduction and power conversion stages reduction in wind energy conversion system. Application of ZSI in renewable energy conver-

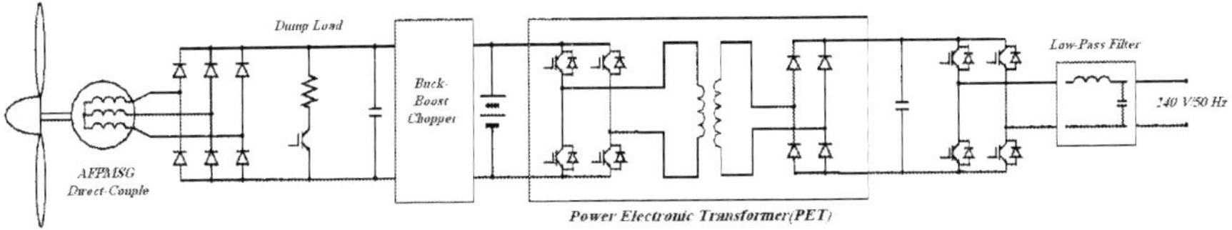

Figure 26. Wind energy conversion system based on power electronic transformer (PET).

sion systems will increase the power conversion reliability and decrease system costs. Fig. 27 shows typical configuration of wind energy conversion system based on ZSI inverters.

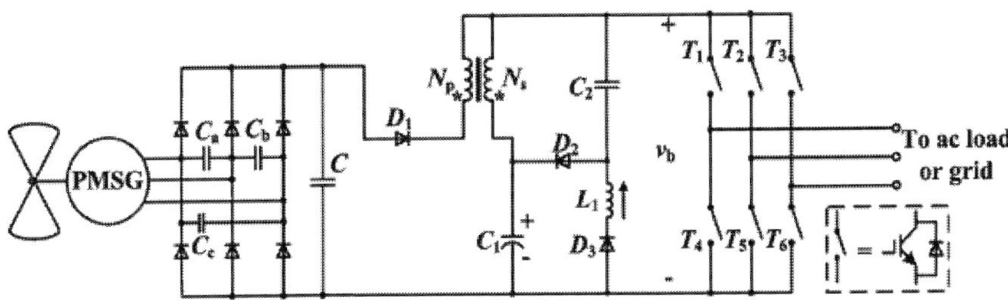

Figure 27. Typical configuration of wind energy conversion system based on single-stage boost inverter with coupled inductor [35].

6. Conclusion

In this chapter, various power electronic circuit topologies for small-scale wind energy conversion systems are investigated and considered. In order to convert wind turbine generator voltage to desirable and suitable output voltage, power electronic converters should be applied between wind turbine generator and load. In order to utilize maximum power point tracking methods on wind turbine to exploit maximum power from wind energy conversion system power electronic conversion system should have high efficiency and variable and high boost capability. To achieve this aim, new generator configuration and various Z-source converters which are proposed in literature are considered in this chapter. Furthermore, novel configuration of ZSI has been proposed and introduced by authors. Compared with the conventional ZSI, diode assisted QZSI, alternate cascaded ZSI(N = 2), and SL-ZSI, the proposed inverter has the highest boosting ability. The proposed inverter uses shorter shoot-through duration and larger modulation index to improve the output waveform quality.

Author details

Mostafa Abarzadeh[1*], Hossein Madadi Kojabadi[1] and Liuchen Chang[2]

*Address all correspondence to: mst.abarzadeh@gmail.com

1 Department of Electrical Engineering, Sahand University of Technology, Tabriz, Iran

2 Department of Electrical and Computer Engineering, University of New Brunswick, NB, Canada

References

[1]   Abarzadeh M., Madadi kojabadi H., Chang L. Small Scale Wind Energy Conversion Systems. In: Al-Bahadly I., editor. Wind Turbines: InTech; 2011. P639-652.

[2]   Abarzadeh M., Madadi kojabadi H., Chang L. Power Electronics in Small Scale Wind Turbine Systems. In: Rupp Carriveau, editor. Advances in Wind Power: InTech; 2012. P305-335.

[3]   Yuan Lo K., Ming Chen Y., Ruei Chang Y: MPPT Battery Charger for Stand-Alone Wind Power System. IEEE Transactions on Power Electronics, 2011; 26(6) 1631 - 1638.

[4]   Smith J., Thresher R., Zavadil R., DeMeo E., Piwko R., Ernst B., and Ackermann T: A mighty wind. IEEE Power and Energy Magazine, 2009; 7(2) 41-51.

[5]   Nayar C., Dehbonei H., Chang L. A Low Cost Power Electronic Interface for Small Scale Wind Generators in Single Phase Distributed Power Generation System. In: Australasian Universities Power Engineering Conference (AUPEC); 25-28 September 2005; Hobart. IEEE; 2005.

[6]   Pathmanathan M, Tang C, Soong W.L, Ertugrul N. Comparison of power converters for small-scale wind turbine operation. In: Australasian Universities Power Engineering Conference (AUPEC); 14-17 December 2008; Sydney. IEEE; 2008. p. 1-6

[7]   Borowy B. S., Salameh Z. M.: Dynamic response of a stand-alone wind energy conversion system with battery energy storage to a wind gust. IEEE Transactions on Energy Conversion. 1997; 12(1) 73–78.

[8]   Billinton R., Bagen, Cui Y.: Reliability evaluation of small standalone wind energy conversion systems using a time series simulation model. Generation, Transmission and Distribution, IEE Proceedings, 2003; 150(1) 96–100.

[9]   Bagen, Billinton R., Evaluation of different operating strategies in small stand-alone power systems. IEEE Transactions on Energy Conversion, 2005; 20(3) 654–660.

[40] W. Qian, F. Z. Peng, and H. Cha, Trans-Z-source inverters, IEEE Transactions on Power Electronics, vol. 26, no. 12, pp. 3453–3463, Dec. 2011.

Solar Energy

Chapter 3

Modelling and Control of Grid-connected Solar Photovoltaic Systems

Marcelo Gustavo Molina

Additional information is available at the end of the chapter

http://dx.doi.org/10.5772/62578

Abstract

At present, photovoltaic (PV) systems are taking a leading role as a solar-based renewable energy source (RES) because of their unique advantages. This trend is being increased especially in grid-connected applications because of the many benefits of using RESs in distributed generation (DG) systems. This new scenario imposes the requirement for an effective evaluation tool of grid-connected PV systems so as to predict accurately their dynamic performance under different operating conditions in order to make a comprehensive decision on the feasibility of incorporating this technology into the electric utility grid. This implies not only to identify the characteristics curves of PV modules or arrays, but also the dynamic behaviour of the electronic power conditioning system (PCS) for connecting to the utility grid. To this aim, this chapter discusses the full detailed modelling and the control design of a three-phase grid-connected photovoltaic generator (PVG). The PV array model allows predicting with high precision the I-V and P-V curves of the PV panels/arrays. Moreover, the control scheme is presented with capabilities of simultaneously and independently regulating both active and reactive power exchange with the electric grid. The modelling and control of the three-phase grid-connected PVG are implemented in the MATLAB/Simulink environment and validated by experimental tests.

Keywords: Photovoltaic System, Distributed Generation, Modeling, Simulation, Control

1. Introduction

The worldwide growth of energy demand and the finite reserves of fossil fuel resources have led to the intensive use of renewable energy sources (RESs). Other major issues that have driven strongly the RES development are the ever-increasing impact of energy technologies on the environment and the fact that RESs have become today a mature technology. The

necessity for having available sustainable energy systems for substituting gradually conventional ones requires changing the paradigm of energy supply by utilizing clean and renewable resources of energy. Among renewables, solar energy characterizes as a clean, pollution-free and inexhaustible energy source, which is also abundantly available anywhere in the world. These factors have contributed to make solar energy the fastest growing renewable technology in the world [1]. At present, photovoltaic (PV) generation is playing a crucial role as a solar-based RES application because of unique benefits such as absence of fuel cost, high reliability, simplicity of allocation, low maintenance and lack of noise and wear because of the absence of moving parts. In addition to these factors are the decreasing cost of PV panels, the growing efficiency of solar PV cells, manufacturing-technology improvements and economies of scale [2-3].

The integration of photovoltaic systems into the grid is becoming today the most important application of PV systems, gaining interest over traditional stand-alone autonomous systems. This trend is being increased due to the many benefits of using RES in distributed (also known as dispersed, embedded or decentralized) generation (DG) power systems [4-5]. These advantages include the favourable fiscal and regulatory incentives established in many countries that influence straightforwardly on the commercial acceptance of grid-connected PV systems. In this sense, the growing number of distributed PV systems brings new challenges to the operation and management of the power grid, especially when this variable and intermittent energy source constitutes a significant part of the total system generation capacity [6]. This new scenario imposes the need for an effective design and performance assessment tool of grid-connected PV systems, so as to predict accurately their dynamic performance under different operating conditions in order to make a sound decision on whether or not to incorporate this technology into the electric utility grid. This implies not only to identify the current-voltage (I-V) characteristics of PV modules or arrays, but also the dynamic behaviour of the power electronics interface with the utility grid, also known as photovoltaic power conditioning system (PCS) or PV PCS, required to convert the energy produced into useful electricity and to provide requirements for connection to the grid. This PV PCS is the key component that enables to provide a more cost-effective harvest of energy from the sun and to meet specific grid code requirements. These requirements include the provision of high levels of security, quality, reliability, availability and efficiency of the electric power. Moreover, modern DG applications are increasingly incorporating new dynamic compensation issues, simultaneously and independently of the conventional active power exchange with the utility grid, including voltage control, power oscillations damping, power factor correction and harmonics filtering among others. This tendency is estimated to augment even more in future DG applications [7].

This chapter presents a full detailed mathematical model of a three-phase grid-connected photovoltaic generator (PVG), including the PV array and the electronic power conditioning system, based on the MATLAB/Simulink software package [8]. The model of the PV array proposed uses theoretical and empirical equations together with data provided by the manufacturer, and meteorological data (solar radiation and cell temperature among others) in order to predict with high precision the I-V and P-V curves of the PV panels/arrays. Since the

PV PCS addresses integration issues from both the distributed PV generating system side and from the utility side, numerous topologies varying in cost and complexity have been widely employed for integrating PV solar systems into the electric grid. Thus, the document includes a discussion of major PCS topologies. Moreover, the control scheme is presented with capabilities of simultaneously and independently regulating both active and reactive power exchange with the electric grid [9].

The modelling and simulation of the three-phase grid-connected PV generating system in the MATLAB/Simulink environment allows design engineers taking advantage of the capabilities for control design and electric power systems modelling already built-up in specialized toolboxes and blocksets of MATLAB, and in dedicated block libraries of Simulink. These features allows assessing the dynamic performance of detailed models of grid-connected PV generating systems used as DG, including power electronics devices and advanced control techniques for active power generation using maximum power point tracking (MPPT) and for reactive power compensation of the electric grid.

2. Photovoltaic Generator (PVG) model

The building block of the PV generator is the solar cell, which is basically a P-N semiconductor junction that directly converts solar radiation into DC current using the photovoltaic effect. The most common model used to predict energy production in photovoltaic cells is the single diode lumped circuit model, which is derived from physical principles, as depicted in Fig. 1. In this model, the PV cell is usually represented by an equivalent circuit composed of a light-generated current source, a single diode representing the nonlinear impedance of the P-N junction, and series and parallel intrinsic resistances accounting for resistive losses [10-11].

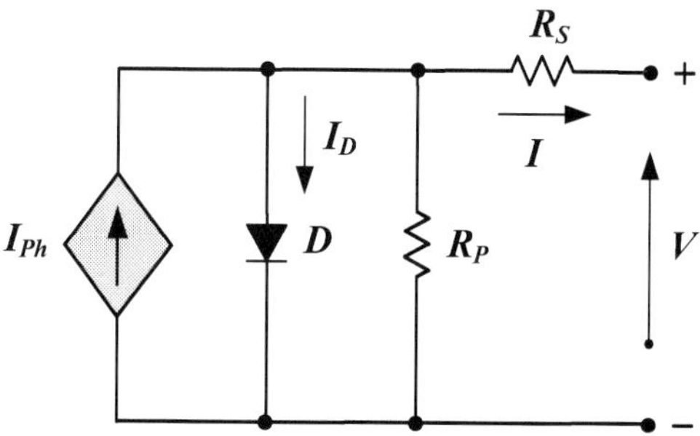

Figure 1. Equivalent circuit of a PV cell

PV cells are grouped together in larger units called modules (also known as panels), and modules are grouped together in larger units known as PV arrays (or often generalized as PV generator), which are combined in series and parallel to provide the desired output voltage and current. The equivalent circuit for the solar cells arranged in N_P-parallel and N_S-series is shown in Fig. 2.

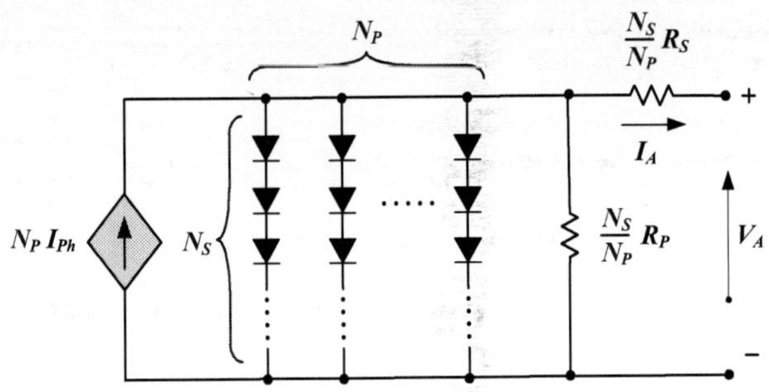

Figure 2. Equivalent circuit of a generalized PV generator

The mathematical model that predicts the power production of the PV generator becomes an algebraically simply model, being the current-voltage relationship defined in Eq. (1).

$$I_A = N_P I_{Ph} - N_P I_{RS} \left\{ \exp\left[\frac{1}{A V_{Th}} \left(\frac{V_A}{N_s} + \frac{I_A R_S}{N_P} \right) \right] - 1 \right\} - \frac{N_P}{R_P} \left(\frac{V_A}{N_s} + \frac{I_A R_S}{N_P} \right),$$
(1)

where:

I_A: PV array output current, in A

V_A: PV array output voltage, in V

I_{Ph}: Solar cell photocurrent, in A

I_{RS}: Solar cell diode reverse saturation current (aka dark current), in A

A: Solar cell diode P-N junction ideality factor, between 1 and 5 (dimensionless)

R_S: Cell intrinsic series resistance, in Ω

R_P: Cell intrinsic shunt or parallel resistance, in Ω

V_{Th}: Cell thermal voltage, in V, determined as $V_{Th} = k T_c/q$

k: Boltzmann's constant, 1.380658e-23 J/K

T_C: Solar cell absolute operating temperature, in K

q: Electron charge, 1.60217733e-19 Cb

This nonlinear equation can be solved using the Newton Raphson iterative method. The parameters I_{Ph}, I_{RS}, R_S, R_P, and A are commonly referred to as "the five parameters" from which the term "five-parameter model" originates. These five parameters must be known in order to determine the current and voltage characteristic, and therefore the power generation of the PV generator for different operating conditions. Thus, in order to obtain a complete model for the electrical performance of the PV generator over all solar radiation and temperature conditions, Eq. (1) is supplemented with equations that define how each of the five parameters changes with solar radiation and/or cell temperature. These equations introduce additional parameters and thus complexity to the model.

The five parameters in Eq. (1) depend on the incident solar radiation, the cell temperature, and on their reference values. Manufacturers of PV modules normally provide these reference values for specified operating conditions known as Standard Test Conditions (STC), which make it possible to conduct uniform comparisons of photovoltaic modules by different manufacturers. These uniform test conditions are defined with a solar radiation of 1000 W/m², a solar cell temperature of 25 °C and an air mass AM (a measure of the amount of atmosphere the sun rays have to pass through) of 1.5.

Actual operating conditions, especially for outdoor conditions, are always different from STC, and mismatch effects can affect the real values of these reference parameters. Consequently, the evaluation of the five parameters in real operating conditions is of major interest in order to provide an accurate mathematical model of the PV generator.

2.1. Dependence of the PV array photocurrent on the operating conditions

The photocurrent I_{Ph} for any operating conditions of the PV array is assumed to be related to the photocurrent at standard test conditions as follows:

$$I_{Ph} = f_{AM_a} f_{IA} \left[I_{SC} + \alpha_{I_{SC}} \left(T_C - T_R \right) \right] \frac{S}{S_R}, \tag{2}$$

where:

f_{AM_a}:   Absolute air mass function describing the solar spectral influence on the photocurrent I_{Ph}, dimensionless

f_{IA}:   Incidence angle function describing the influence on the photocurrent I_{Ph}, dimensionless

I_{SC}: Solar cell short-circuit current at STC, in A

$\alpha_{I_{SC}}$: Solar cell temperature coefficient of the short-circuit current, in A/module/diff. temp (in K or °C)

T_R: Solar cell absolute reference temperature at STC, in K

S: Total solar radiation absorbed at the plane-of-array (POA), in W/m²

S_R: Total solar reference radiation at STC, i.e. 1000 W/m²

The absolute air mass function f_{AM_a} accounting for the solar spectral influence on the "effective" irradiance absorbed on the PV array surface is described through an empirical polynomial function, as expressed in Eq. (3) [10].

$$f_{AM_a} = \sum_{i=0}^{4} a_i \left(AM_a\right)^i = M_P \sum_{i=0}^{4} a_i \left(AM\right)^i ,$$

(3)

where:

$a_0 - a_4$:   Polynomial coefficients for fitting the absolute air mass function of the analysed cell material, dimensionless

AM_a: Absolute air mass, corrected by pressure, dimensionless

AM: Atmospheric optical air mass, dimensionless

M_P: Pressure modifier, dimensionless

The pressure modifier corrects the air mass by the site pressure in order to yield the absolute air mass. This factor is computed as the ratio of the site pressure to the standard pressure at sea level.

The air mass AM is the term used to describe the path length that the solar radiation beam has to pass through the atmosphere before reaching the earth, relative to its overhead path length. This ratio measures the attenuation of solar radiation by scattering and absorption in atmosphere; the more atmosphere the light travels through, the greater the attenuation. As can be noted, the air mass indicates a relative measurement and is calculated from the solar zenith angle which is a function of time.

The incidence angle function f_{IA} describes the optical effects related to the solar incidence angle (I_A) on the radiation effectively transmitted to the PV array surface and converted to electricity through the panel photocurrent. This modifier accounts for the effect of reflection and absorption of solar radiation and is defined as the ratio of the radiation absorbed by the solar cell at some incident angle θ_I to the radiation absorbed at normal incidence. The incidence angle is defined between the solar radiation beam direction (or direct radiation) and the normal to the PV array surface (or POA), as can be seen from Fig. 3. By using the geometric relationships between the plane at any particular orientation relative to the earth and the beam solar radiation, both the incidence angle and the zenith angle can be accurately computed at any time.

An algorithm for computing the solar incidence angle for both fixed and solar-tracking modules has been documented in [11]. In the same way, the optical influence of the PV module surface, typically glass, was empirically described through the incidence angle function [10], as shown in Eq. (3) for different incident angle θ_I (in degrees).

$$f_{IA} = 1 - \sum_{i=1}^{5} b_i \left(\theta_I\right)^i ,$$

(4)

where:

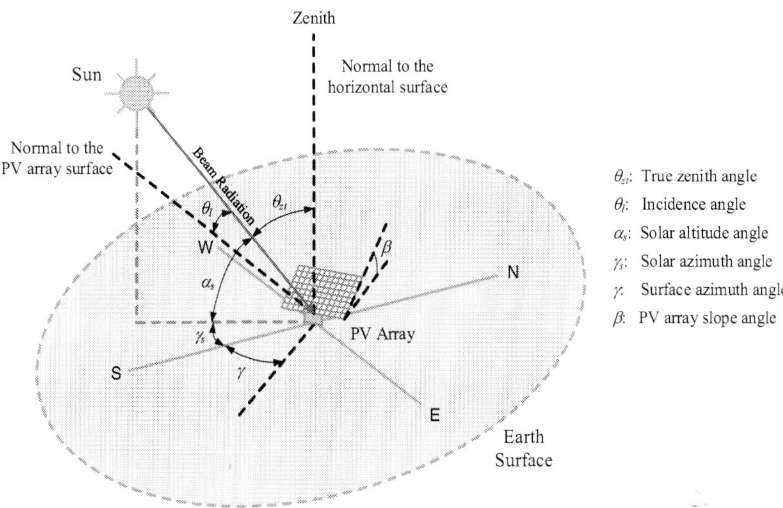

Figure 3. Zenith angle and other major angles for a tilted PV array surface

b_1- b_5: Polynomial coefficients for fitting the incidence angle function of the analysed PV cell material, dimensionless

Even though the correlation of Eq. (4) is cell material-dependant, most modules with glass front surfaces share approximately the same f_{IA} function, so that no extra experimentation is required for a specific module [11]. An alternative theoretical form for estimating the incidence angle function without requiring specific experimental information was proposed in [12].

2.2. Dependence of the PV array reverse saturation current on the operating conditions

The solar cell reverse saturation current I_{RS} varies with temperature according to the following equation [13]:

$$I_{RS} = I_{RR}\left[\frac{T_C}{T_R}\right]^3 \exp\left[\frac{q\,E_G}{A\,k}\left(\frac{1}{T_R} - \frac{1}{T_C}\right)\right] \tag{5}$$

where:

I_{RR}: Solar cell reverse saturation current at STC, in A

E_G: Energy band-gap of the PV cell semiconductor at absolute temperature (TC), in eV

The energy band-gap of the PV array semiconductor, E_G is a temperature-dependence parameter. The band-gap tends to decrease as the temperature increases. This behaviour can be better understood when it is considered that the interatomic spacing increases as the amplitude of the atomic vibrations augments due to the increased thermal energy. The increased interatomic spacing decreases the average potential seen by the electrons in the material, which in turn reduces the size of the energy band-gap.

2.3. Dependence of the PV array series and shunt resistances on the operating conditions

Series and shunt resistances are very significant in evaluating the solar array performance since they have direct effect on the PV module fill factor (FF). The fill factor is defined as the ratio of the power at the maximum power point (MPP) divided by the short-circuit current (I_{sc}) and the open-circuit voltage (V_{oc}). In this way, the FF serves as a quantifier of the shape of the I-V characteristic curve and consequently of the degradation of the PV array efficiency.

The series resistance R_S describes the semiconductor layer internal losses and losses due to contacts. It influences straightforwardly the shape of the PV array I-V characteristic curve around the MPP and thus the fill factor. As the series resistance increases, its deteriorative effects on the short-circuit current will be increased, especially at high intensities of radiation, while not affecting the open-circuit voltage. This unwanted feature causes a reduction of the peak power and thus the degradation of the PV array efficiency. The dependence of the PV array series resistance on the cell temperature can be characterized by Eq. (6).

$$R_S = R_{SR}\left[1 + \alpha_{SR}\left(T_C - T_R\right)\right]$$

(6)

where:

R_{SR} : Solar cell series resistance at STC, in Ω

α_{SR} : PV array temperature coefficient of the series resistance, Ω/module/diff. temp. (in K or °C)

The shunt (or parallel) resistance R_p accounts for leakage currents on the PV cell surface or in PN junctions. It influences the slope of the I-V characteristic curve near the short-circuit current point and therefore the FF, although its practical effect on the PV array performance is less noticeable than the series resistance. As the shunt resistance decreases, its degrading effects on the open-circuit current voltage will be increased, especially at the low voltages region, while not affecting the short-circuit current. The shunt resistance is dependent upon the absorbed solar radiation. As indicated in [14], the shunt resistance is approximately inversely proportional to the short-circuit current, and thus to the absorbed radiation, at very low intensities. As the absorbed radiation increases, the slope of the I-V characteristic curve near the short-circuit current point decreases and then the effective shunt resistance proportionally decreases. In this way, this phenomenon can be empirically characterized by Eq. (7).

$$\frac{R_{PR}}{R_P} = \frac{S}{S_R},$$

(7)

2.4. Dependence of the PV array material ideality factor on the operating conditions

The P-N junction ideality factor A of PV cells is generally assumed to be constant and independent of temperature. However, as reported by [15] the ideality factor varies with temperature for most semiconductor materials by the following general expression, as given in Eq. (8).

$$A = A_R - \left[\frac{\alpha_A T_C^2}{T_C + \beta_D} \right] \tag{8}$$

where:

A_R : Ideality factor of the PV cell semiconductor at absolute zero temperature, 0 K (-273.15°C), dimensionless, assumed 1.9 for silicon cells

α_A : Temperature coefficient of the ideality factor, for silicon 0.789e-3 K^{-1}

β_D : Temperature constant approximately equal to the 0 K Debye's temperature, for silicon 636 K

The analysis of the five parameters I_{Ph}, I_{RS}, R_S, R_P, and A has permitted to complete the detailed five-parameter model representative of the PV solar array for different operating conditions.

3. Photovoltaic Power Conditioning System (PCS) model

Usually, one of the major challenges of grid-connected or utility-scale solar photovoltaic systems is to attain an optimal compatibility of PV arrays with the electricity grid. Since a PV array produces an output DC voltage with variable amplitude, an additional conditioning circuit is required to meet the amplitude and frequency requirements of the stiff utility AC grid and inject synchronized power into the grid. As the output of PV panels are direct current, the PV PCS is typically a DC-AC converter (or inverter) which inverts the DC output current generated by the PV arrays into a synchronized sinusoidal waveform. This PV interface must generate high quality electric power and at the same time be flexible, efficient and reliable.

Another key challenge of grid-connected PV systems is the procedure employed for power extraction from solar radiation and is mostly related to the nature of PV arrays. Each PV module is a nonlinear system with an output power mostly influenced by atmospheric conditions, such as solar radiation and temperature. To transfer the maximum solar array power into the utility grid for all operating conditions, a maximum power point tracking (MPPT) technique is usually implemented. Therefore, each grid-connected PV generating system has to perform two essential functions, i.e. to extract the maximum output power from the PV array, and to inject a sinusoidal current into the grid.

The photovoltaic PCS can be classified with respect to the number of power stages of its structure into three classes, known as single-stage, dual-stage and multi-stage topologies, as depicted in Fig. 4 [16].

The first structure of the PV PCS connects the PV array directly to the DC bus of a power inverter. Consequently, the maximum power point tracking of the PV modules and the inverter control loops (current and voltage control loops) are handled all in one single stage. The second topology employs a DC-DC converter (or chopper) as interface between the PV array and the static inverter. In this case, the additional DC-DC converter connecting the PV panels and the inverter handles the MPPT control. The third arrangement uses one DC-DC converter for connecting each string of PV modules to the inverter. For these multi-stage inverters, a DC-DC converter implements the maximum MPPT control of each string and one power inverter handles the current and voltage control loops.

The two distinct categories of the inverter are known as voltage source inverter (VSI) and current source inverter (CSI). Voltage source inverters are named so because the independently controlled output is a voltage waveform. In this structure, the VSI is fed from a DC-link capacitor, which is connected in parallel with the PV panels. Similarly, current source inverters control the AC current waveform. In this arrangement, the inverter is fed from a large DC-link inductor. In industrial markets, the VSI design has proven to be more efficient and to have higher reliability and faster dynamic response.

Since applications of modern distributed energy resources introduce new constraints of high quality electric power, flexibility and reliability to the PV-based distributed generator, a two-stage PV PCS topology using voltage source inverters has been mostly applied in the literature. This configuration of two cascade stages offers an additional degree of freedom in the operation of the grid-connected PV system when compared with the one-stage configuration. Hence, by including the DC-DC boost converter, various control objectives, as reactive power compensation, voltage control, and power oscillations damping among others, are possible to be pursued simultaneously with the typical PV system operation without changing the PCS topology [17].

The detailed model of a grid-connected PV system is illustrated in Fig. 5, and consists of the solar PV arrangement and its PCS to the electric utility grid [8]. PV panels are electrically combined in series to form a string (and sometimes stacked in parallel) in order to provide the desired output power required for the DG application. The PV array is implemented using the aggregated model previously described, by directly computing the total internal resistances, non-linear integrated characteristic and total generated solar cell photocurrent according to the series and parallel contribution of each parameter. A three-phase DC-AC voltage source inverter is employed for connecting to the grid. This three-phase static device is shunt-connected to the distribution network by means of a coupling transformer and the corresponding line sinusoidal filter. The output voltage control of this VSI can be efficiently performed using pulse width modulation (PWM) techniques [18].

3.1. Voltage source inverter

Since the DC-DC converter acts as a buffer between the PV array and the power static inverter by turning the highly nonlinear radiation and temperature-dependent I-V characteristic curve of the PV system into a quasi-ideal atmospheric factors-controlled voltage source characteris-

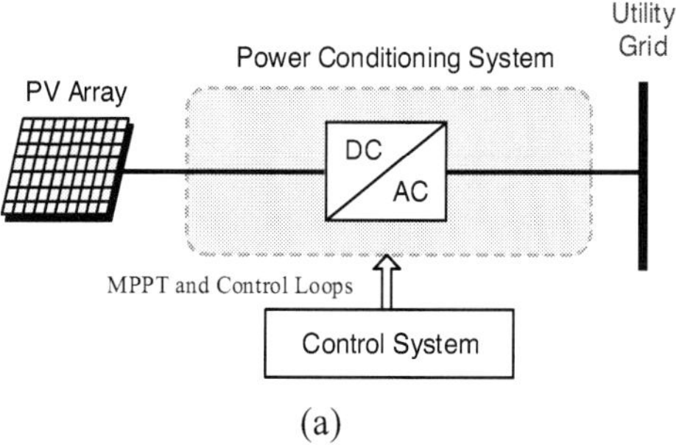

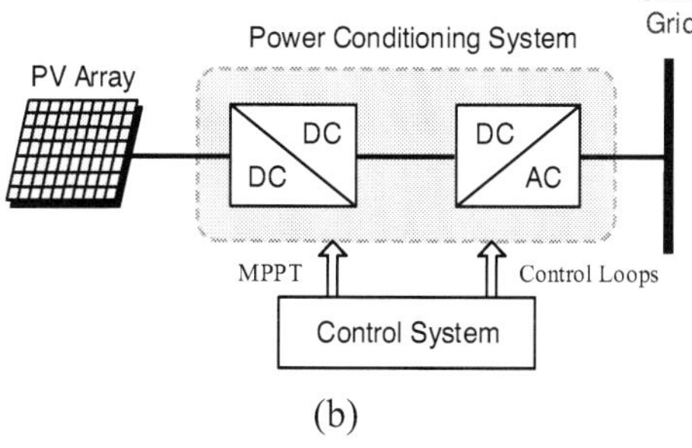

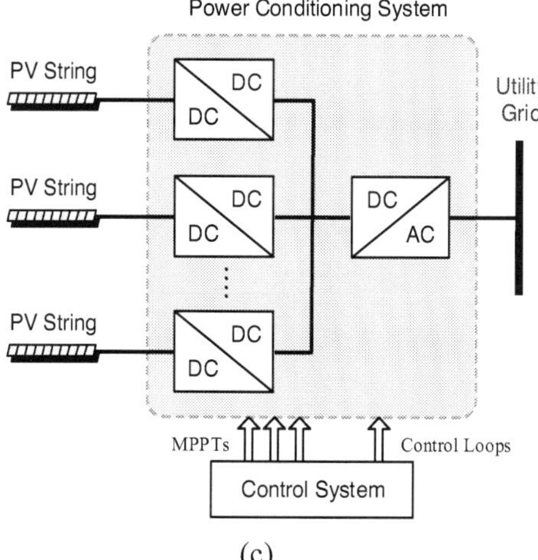

Figure 4. PV PCS configurations: (a) single-stage inverter, (b) dual-stage inverter, and (c) multi-stage inverter.

tic, the natural selection for the inverter topology is the voltage source-type. This solution is more cost-effective than alternatives like hybrid current source inverters (HCSI).

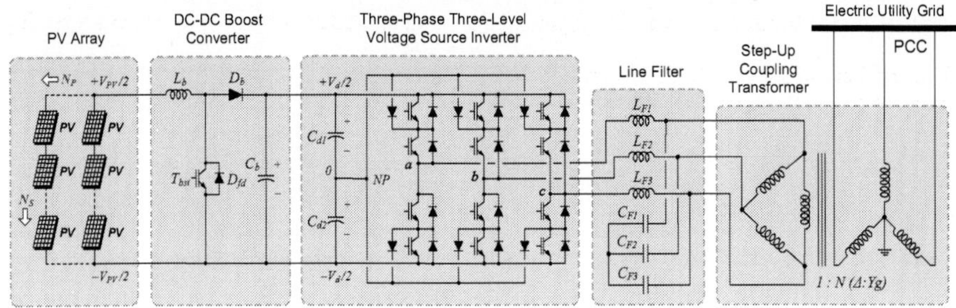

Figure 5. Detailed model of the proposed three-phase grid-connected photovoltaic system

The voltage source inverter presented in Fig. 5 consists of a multi-level DC-AC power inverter built with insulated-gate bipolar transistors (IGBTs) technology. This semiconductor device offers a cost-effective solution for distributed generation applications since it has lower conduction and switching losses with reduced size than other switching devices. Furthermore, as the power of the inverter is in the range of low to medium level for the proposed application, it can be efficiently driven by sinusoidal pulse width modulation (SPWM) techniques.

The VSI utilizes a diode-clamped multi-level (DCML) inverter topology, also commonly called neutral point clamped (NPC), instead of a standard inverter structure with two levels and six pulses. The three-level twelve-pulse VSI structure employed is very popular especially in high power and medium voltage applications. Each one of the three-phase outputs of the inverter shares a common DC bus voltage that has been divided into three levels over two DC bus capacitors. The middle point of the two capacitors constitute the neutral point of inverter and output voltages have three voltage states referring to this neutral point. The general concept of this multi-level inverter is to synthesize a sinusoidal voltage from several levels of voltages. Thus, the three-level structure attempts to address some restrictions of the standard two-level one by providing the flexibility of an extra level in the output voltage, which can be controlled in duration to vary the fundamental output voltage or to assist in the output waveform construction. This extra feature allows generating a more sinusoidal output voltage waveform than conventional structures without increasing the switching frequency. In this way, the voltage stress on the switching devices is reduced and the output harmonics distortion is minimized [19].

The connection of the inverter to the distribution network in the so-called point of common coupling (PCC) is made by means of a typical step-up Δ-Y power transformer with line sinusoidal filters. The design of this single three-phase coupling transformer employs a delta-connected windings on its primary and a wye/star connected windings with neutral wire on its secondary. The delta winding allows third-harmonic currents to be effectively absorbed in the winding and prevents from propagating them onto the power supply. In the same way, high frequency switching harmonics generated by the PWM control of the VSI are attenuated by providing second-order low-pass sine wave filters. Since there are two possibilities of fit ting the filters, i.e. placing them in the primary and in the secondary of the coupling trans‐ former, it is normally preferred the first option because it reduces notably the harmonics contents in the transformer windings, thus reducing losses as heat and avoiding its overrating.

The mathematical equations describing and representing the operation of the VSI can be derived from the detailed model shown in Fig. 5 by taking into account some assumptions with respect to the operating conditions of the inverter [20]. For this purpose, a simplified scheme of the VSI connected to the electric system is developed, also referred to as the averaged model, which is presented in Fig. 6. The power inverter operation under balanced conditions is considered as ideal, i.e. the VSI is seen as an ideal sinusoidal voltage source operating at fundamental frequency. This consideration is valid since the high-frequency harmonics produced by the inverter as result of the sinusoidal PWM control techniques are mostly filtered by the low pass sine wave filters and the net instantaneous output voltages at the point of common coupling resembles three sinusoidal waveforms spaced 120° apart. At the output terminals of the low pass filters, the voltage total harmonic distortion (VTHD) is reduced to as low as 1%, decreasing this quantity to even a half at the coupling transformer output terminals (or PCC).

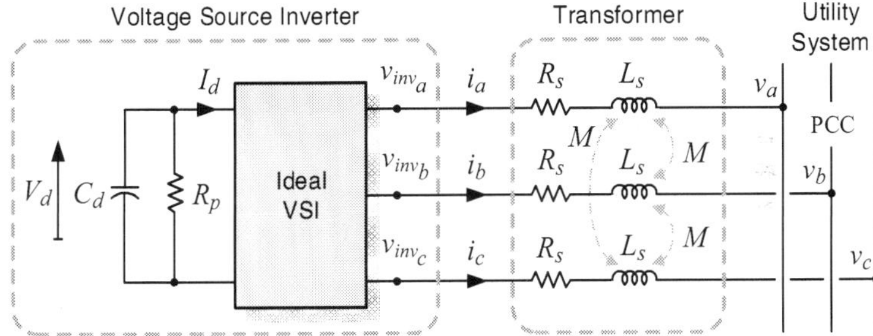

Figure 6. Equivalent circuit diagram of the VSI connected to the utility grid

The equivalent ideal inverter depicted in Fig. 6 is shunt-connected to the AC network through the inductance L_s, accounting for the equivalent leakage of the step-up coupling transformer and the series resistance R_s, representing the transformer winding resistance and VSI IGBTs conduction losses. The magnetizing inductance of the step-up transformer takes account of the mutual equivalent inductance M. On the DC side, the equivalent capacitance of the two DC bus capacitors, C_1 and C_2 (C_1=C_2), is described through C=C_1/2=C_2/2 whereas the switching losses of the VSI and power loss in the DC capacitors are represented by R_p.

The dynamic equations governing the instantaneous values of the three-phase output voltages on the AC side of the VSI and the current exchanged with the utility grid can be directly derived by applying Kirchhoff's voltage law (KVL) as follows:

$$\begin{bmatrix} v_{inv_a} \\ v_{inv_b} \\ v_{inv_c} \end{bmatrix} - \begin{bmatrix} v_a \\ v_b \\ v_c \end{bmatrix} = \left(R_s + s\, L_s \right) \begin{bmatrix} i_a \\ i_b \\ i_c \end{bmatrix}, \tag{9}$$

where:

s: Laplace variable, being $s = d/dt$ for $t > 0$

$$R_s = \begin{bmatrix} R_s & 0 & 0 \\ 0 & R_s & 0 \\ 0 & 0 & R_s \end{bmatrix} L_s = \begin{bmatrix} L_s & M & M \\ M & L_s & M \\ M & M & L_s \end{bmatrix} \tag{10}$$

Under the assumption that the system has no zero sequence components, all currents and voltages can be uniquely transformed into the synchronous-rotating orthogonal two-axes reference frame, in which each vector is described by means of its d and q components, instead of its three a, b, c components. Consequently, as depicted in Fig. 7, the d-axis always coincides with the instantaneous voltage vector and thus v_d equates $|v|$, while v_q is set at zero. Consequently, the d-axis current component contributes to the instantaneous active power and the q-axis current component to the instantaneous reactive power. This operation allows for a simpler and more accurate dynamic model of the VSI.

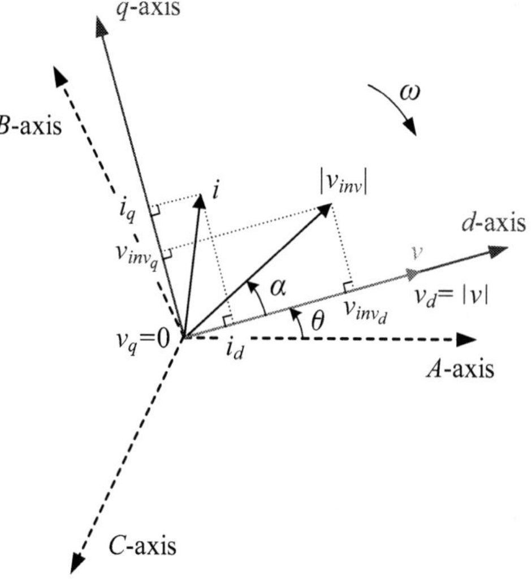

Figure 7. VSI vectors in the synchronous rotating d-q reference frame

By applying Park's transformation stated by Eq. (11), Eqs. 9 through 10 can be transformed into the synchronous rotating d-q reference frame as follows (Eqs. (12) through (14)):

$$K_s = \frac{2}{3} \begin{bmatrix} \cos\theta & \cos\left(\theta - \dfrac{2\pi}{3}\right) & \cos\left(\theta + \dfrac{2\pi}{3}\right) \\ -\sin\theta & -\sin\left(\theta - \dfrac{2\pi}{3}\right) & -\sin\left(\theta + \dfrac{2\pi}{3}\right) \\ \dfrac{1}{2} & \dfrac{1}{2} & \dfrac{1}{2} \end{bmatrix} \tag{11}$$

with:

$\theta = \int_0^t \omega(\xi)d\xi + \theta(0)$ angle between the d-axis and the reference phase axis, being ξ an integration variable

ω: synchronous angular speed of the network voltage at the fundamental system frequency f (50 Hz in this chapter)

Thus,

$$\begin{bmatrix} v_{inv_d} - v_d \\ v_{inv_q} - v_q \\ v_{inv_0} - v_0 \end{bmatrix} = K_s \begin{bmatrix} v_{inv_a} - v_a \\ v_{inv_b} - v_b \\ v_{inv_c} - v_c \end{bmatrix}, \quad \begin{bmatrix} i_d \\ i_q \\ i_0 \end{bmatrix} = K_s \begin{bmatrix} i_a \\ i_b \\ i_c \end{bmatrix}, \tag{12}$$

By neglecting the zero sequence components, Eqs. (13) and (14) are obtained.

$$\begin{bmatrix} v_{inv_d} \\ v_{inv_q} \end{bmatrix} - \begin{bmatrix} v_d \\ v_q \end{bmatrix} = R_s + s\, L'_s \begin{bmatrix} i_d \\ i_q \end{bmatrix} + \begin{bmatrix} -\omega & 0 \\ 0 & \omega \end{bmatrix} L'_s \begin{bmatrix} i_q \\ i_d \end{bmatrix}, \tag{13}$$

where:

$$R_s = \begin{bmatrix} R_s & 0 \\ 0 & R_s \end{bmatrix} L'_s = \begin{bmatrix} L'_s & 0 \\ 0 & L'_{ss} \end{bmatrix} = \begin{bmatrix} L_s - M & 0 \\ 0 & L_s - M \end{bmatrix} \tag{14}$$

It is to be noted that the coupling of phases abc through the term M in matrix L_s (Eq. (10)), was eliminated in the dq reference frame when the inverter transformer is magnetically symmetric, as is usually the case. This decoupling of phases in the synchronous-rotating system simplifies the control system design.

By rewriting Eq. (13), the state-space representation of the inverter is obtained as follows:

$$s\begin{bmatrix} i_d \\ i_q \end{bmatrix} = \begin{bmatrix} \dfrac{-R_s}{L'_s} & \omega \\ -\omega & \dfrac{-R_s}{L'_s} \end{bmatrix}\begin{bmatrix} i_d \\ i_q \end{bmatrix} + \dfrac{1}{L'_s}\begin{bmatrix} v_{inv_d} - |v| \\ v_{inv_q} \end{bmatrix}, \tag{15}$$

A further major issue of the dq transformation is its frequency dependence (ω). In this way, with appropriate synchronization to the network (through angle θ), the control variables in steady state are transformed into DC quantities. This feature is quite useful to develop an efficient decoupled control system of the two current components. Although the model is fundamental frequency-dependent, the instantaneous variables in the dq reference frame contain all the information concerning the three-phase variables, including steady-state unbalance, harmonic distortions and transient components.

The AC and DC sides of the VSI are related by the power balance between the input and the output on an instantaneous basis. In this way, the ac power should be equal to the sum of the DC resistance power and to the charging rate of the DC capacitors, as described by Eq. (16).

$$\frac{3}{2}\left(v_{inv_d} i_d + v_{inv_q} i_q \right) = -\frac{C_d}{2} V_d s V_d - \frac{V_d^2}{R_p}, \tag{16}$$

The VSI basically generates the AC voltage v_{inv} from the DC voltage V_d, in such a way that the connection between the DC-side voltage and the generated AC voltage can be described by using the average switching function matrix in the dq reference frame $S_{av,dq}$, as given by Eqs. (17) through (19). This relation assumes that the DC capacitors voltages are equal to $V_d/2$.

$$\begin{bmatrix} v_{inv_d} \\ v_{inv_q} \end{bmatrix} = S_{av,\,dq} V_d, \tag{17}$$

and the average switching function matrix in dq coordinates is computed as:

$$S_{av,\,dq} = \begin{bmatrix} S_{av,\,d} \\ S_{av,\,q} \end{bmatrix} = \frac{1}{2} m_i a \begin{bmatrix} S_d \\ S_q \end{bmatrix}, \tag{18}$$

where,

m_i: modulation index of inverter, $m_i \in [0, 1]$

α: phase-shift of the STATCOM output voltage from the reference position

$a = \dfrac{\sqrt{3}}{\sqrt{2}} \dfrac{n_2}{n_1}$: voltage ratio of the step-up coupling transformer turns ratio of the step-up Δ-Y coupling transformer and the average switching factor matrix for the dq reference frame,

$$\begin{bmatrix} S_d \\ S_q \end{bmatrix} = \begin{bmatrix} \cos \alpha \\ \sin \alpha \end{bmatrix},$$

(19)

with α being the phase-shift of the VSI output voltage from the reference position

Essentially, Eqs. (12) through (19) can be summarized in the state-space as stated by Eq. (20). This continuous state-space averaged mathematical model describes the steady-state dynamics of the VSI in the dq reference frame.

$$s\begin{bmatrix} i_d \\ i_q \\ V_d \end{bmatrix} = \begin{bmatrix} \dfrac{-R_s}{L'_s} & \omega & \dfrac{maS_d}{2L'_s} \\ -\omega & \dfrac{-R_s}{L'_s} & \dfrac{maS_q}{2L'_s} \\ -\dfrac{3}{2\,C_d}maS_d & -\dfrac{3}{2\,C_d}maS_q & -\dfrac{2}{R_p C_d} \end{bmatrix} \begin{bmatrix} i_d \\ i_q \\ V_d \end{bmatrix} - \begin{bmatrix} \dfrac{|v|}{L'_s} \\ 0 \\ 0 \end{bmatrix},$$

(20)

Inspection of Eq. (20) shows a cross-coupling of both components of the VSI output current through the term ω. This issue of the d-q reference frame modelling approach must be counteracted by the control system. Furthermore, it can be observed an additional coupling resulting from the DC capacitors voltage V_d. Moreover, average switching functions (S_d and S_q) introduce nonlinear responses in the inverter states i_d, i_q and V_d when α is regarded as an input variable. This difficulty demands to keep the DC bus voltage as constant as possible, in order to decrease the influence of the dynamics of V_d. There are two ways of dealing with this problem. One way is to have a large capacitance for the DC capacitors, since bigger capacitors value results in slower variation of the capacitors voltage. However, this solution makes the compensator larger and more expensive. Another way is to design a controller of the DC bus voltage. In this fashion, the capacitors can be kept relatively small. This last solution is employed here for the control scheme.

3.2. DC-DC boost converter

The intermediate DC-DC boost converter fitted between the PV array and the inverter acts as an interface between the output DC voltage of the PV modules and the DC link voltage at the input of the voltage source inverter. The voltage of the PV array is variable with unpredictable atmospheric factors, while the VSI DC bus voltage is controlled to be kept constant at all load conditions. In this way, in order to deliver the required output DC voltage to the VSI link, a standard unidirectional topology of a DC-DC boost converter (also known as step-up converter or chopper) is employed. This switching-mode power device contains basically two semiconductor switches (a rectifier diode and a power transistor) and two energy storage devices (an inductor and a smoothing capacitor) for producing an output DC voltage at a level greater than its input DC voltage. The basic structure of the DC-DC boost converter, using an IGBT as the main power switch, is shown in Fig. 5.

The converter produces a chopped output voltage through pulse-width modulation (PWM) control techniques in order to control the average DC voltage relation between its input and output. Thus, the chopper is capable of continuously matching the characteristic of the PV system to the equivalent impedance presented by the DC bus of the inverter. In this way, by varying the duty cycle of the DC-DC converter it is feasibly to operate the PV system near the MPP at any atmospheric conditions and load.

The operation of the converter in the continuous (current) conduction mode (CCM), i.e. with the current flowing continuously through the inductor during the entire switching cycle, facilitates the development of the state-space model. The reason for this is that only two switch states are possible during a switching cycle, namely, (i) the power switch T_b is on and the diode D_b is off, or (ii) T_b is off and D_b is on. In steady-state CCM operation and neglecting the parasitic components, the state-space equation that describes the dynamics of the DC-DC boost converter is given by Eq. (21) [21].

$$s\begin{bmatrix} I_A \\ V_d \end{bmatrix} = \begin{bmatrix} 0 & -\dfrac{1-S_{dc}}{L} \\ \dfrac{1-S_{dc}}{C} & 0 \end{bmatrix}\begin{bmatrix} I_A \\ V_d \end{bmatrix} + \begin{bmatrix} \dfrac{1}{L} & 0 \\ 0 & -\dfrac{1}{C} \end{bmatrix}\begin{bmatrix} V_A \\ I_d \end{bmatrix}, \tag{21}$$

where:

I_A: Chopper input current, matching the PV array output current, in A

V_A: Chopper input voltage, the same as the PV array output voltage, in V

V_d: Chopper output voltage, coinciding with the DC bus voltage, in V

I_d: Chopper output current, in A

S_{dc}: Switching function of the boost converter

The switching function is a two-level waveform characterizing the signal that drives the power switch T_b of the DC-DC boost converter, defined as follows:

$$S_{dc}\begin{cases} 0, \text{ for the switch} T_b \text{off} \\ 1, \text{for the switch} T_b \text{on} \end{cases} \tag{22}$$

If the switching frequency of T_b is significantly higher than the natural frequencies of the DC-DC boost converter, this discontinuous model can be approximated by a continuous state-space averaged (SSA) model, where a new variable D is introduced. In the [0, 1] subinterval, D is a continuous function and represents the duty cycle D of the DC-DC converter. It is defined as the ratio of time during which the power switch T_b is turned-on to the period of one complete switching cycle, T_s. This variable is used for replacing the switching function of the power converter in Eq. (21), yielding the following SSA expression:

$$s\begin{bmatrix} I_A \\ V_d \end{bmatrix} = \begin{bmatrix} 0 & -\dfrac{1-D}{L} \\ -\dfrac{1-D}{C} & 0 \end{bmatrix} \begin{bmatrix} I_A \\ V_d \end{bmatrix} + \begin{bmatrix} \dfrac{1}{L} & 0 \\ 0 & -\dfrac{1}{C} \end{bmatrix} \begin{bmatrix} V_A \\ I_d \end{bmatrix}, \tag{23}$$

The DC-DC converter produces a chopped output voltage for controlling the average DC voltage relation between its input and output. In this way, it is significant to derive the steady-state input-to-output conversion relationship of the boost converter in the CCM. Since in steady-state conditions the inductor current variation during *on* and *off* times of the switch T_b are essentially equal, and assuming a constant DC output voltage of the boost converter, the voltage conversion relationship can be easily derived. To this aim, the state-derivative vector in Eq. (23) is set to zero, yielding the following expression:

$$V_d = \frac{V_A}{(1-D)} \tag{24}$$

In the same way, by assuming analogous considerations, the current conversion relationship of the boost converter in the CCM is given by Eq. (25).

$$I_d = (1-D)I_A \tag{25}$$

4. PVG control strategy

The hierarchical control of the three-phase grid-connected PV generator consists of an external, middle and internal level, as depicted in Fig. 8 [21].

4.1. External level control

The external level control, which is outlined in the left part of Fig. 8 in a simplified form, is responsible for determining the active and reactive power exchange between the PV generator and the utility electric system. This control strategy is designed for performing two major control objectives, namely the voltage control mode (VCM) with only reactive power compensation capabilities and the active power control mode (APCM) for dynamic active power exchange between the PV array and the electric power system. To this aim, the instantaneous voltage at the PCC is computed by employing a synchronous-rotating reference frame. As a consequence, the instantaneous values of the three-phase AC bus voltages are transformed into d-q components, v_d and v_q respectively. Since the d-axis is always synchronized with the instantaneous voltage vector v_m, the d-axis current component of the VSI contributes to the instantaneous active power p while the q-axis current component represents the instantaneous reactive power q. Thus, to achieve a decoupled active and reactive power control, it is required to provide a decoupled control strategy for i_d and i_q. In this way, only v_d is used for computing the resultant current reference signals required for the desired PV output active and reactive power flows. Additionally, the instantaneous actual output currents of the PV system, i_d and i_q, are obtained and used in the middle level control.

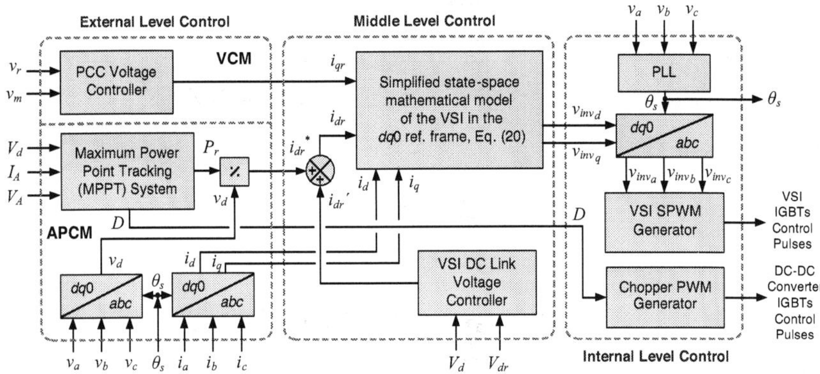

Figure 8. Multi-level control scheme for the three-phase grid-connected PV generator

In many modern electricity grids with high integration of intermittent renewable-based distributed generation, voltage regulation is becoming a necessary task at the distribution level. Since the inverter-interfaced sources are deployed to regulate the voltage at the point of common coupling of each inverter, the PVG can easily perform this control action and participate in the voltage control of the grid. To this aim, a control loop of the external level is the VCM, also called Automatic Voltage Regulation (AVR). It controls (supports and regulates) the voltage at the PCC through the modulation of the reactive component of the inverter output current, i_q. Since only reactive power is exchanged with the grid in this control mode, there is no need for the PV array or any other external energy source. In fact, this reactive power is locally generated just by the inverter and can be controlled simultaneously and independently of the active power generated by the PV array. The design of the control loop in the rotating frame employs a standard proportional-integral (PI) compensator including an anti-windup system. This control mode eliminates the steady-state voltage offset via the PI compensator. A voltage regulation droop is included in order to allow the terminal voltage of the PV inverter (PCC) to vary in proportion with the compensating reactive current. Thus, the PI controller with droop characteristics becomes a simple phase-lag compensator.

The main objective of the grid-connected solar photovoltaic generating system is to exchange with the electric utility grid the maximum available power for the given atmospheric conditions, independently of the reactive power generated by the inverter. In this way, the APCM allows dynamically controlling the active power flow by constantly matching the active power exchanged by the inverter with the maximum instant power generated by the PV array. This implies a continuous knowledge of not only the PV panel internal resistances but also the voltage generated by the PV array. This requirement is very difficult to meet in practice and would increase complexity and costs to the DG application. It would require additional sensing of the cell temperature and solar radiation jointly with precise data of its characteristic curve. Even more, PV parameters vary with time, making it difficult for real-time prediction.

Many MPPT methods have been reported in literature. These methods can be classified into three main categories: lookup table methods, computational methods (neural networks, fuzzy logic, etc.) and hill climbing methods [22-25]. These vary in the degree of sophistication, processing time and memory requirements. Among them, hill climbing methods are indirect methods with a good combination of flexibility, accuracy and simplicity. They are efficient and robust in tracking the MPP of solar photovoltaic systems and have the additional advantages of control flexibility and easiness of application with different types of PV arrays. The power efficiency of these techniques relies on the control algorithm that tracks the MPP by measuring some array quantities.

The simplest MPPT using climbing methods is the "Perturbation and Observation" (P&O) method. This MPPT strategy uses a simple structure and few measured variables for implementing an iterative method that permits matching the load with the output impedance of the PV array by continuously adjusting the DC-DC converter duty cycle. This MPPT algorithm operates by constantly perturbing, i.e. increasing or decreasing, the output voltage of the PV array via the DC-DC boost converter duty cycle D and comparing the actual output power with the previous perturbation sample, as depicted in Fig. 9. If the power is increasing, the perturbation will continue in the same direction in the following cycle, otherwise the perturbation direction will be inverted. This means that the PV output voltage is perturbed every MPPT iteration cycle k at sample intervals T_{trck}, while maintaining always constant the VSI DC bus voltage by means of the middle level control. Therefore, when the optimal power for the specific operating conditions is reached, the P&O algorithm will have tracked the MPP (the climb of the PV array output power curve) and then will settle at this point but oscillating slightly around it. The output power measured in every iteration step is employed as a reference power signal P_r and then converted to a direct current reference i_{dr} for the middle level control.

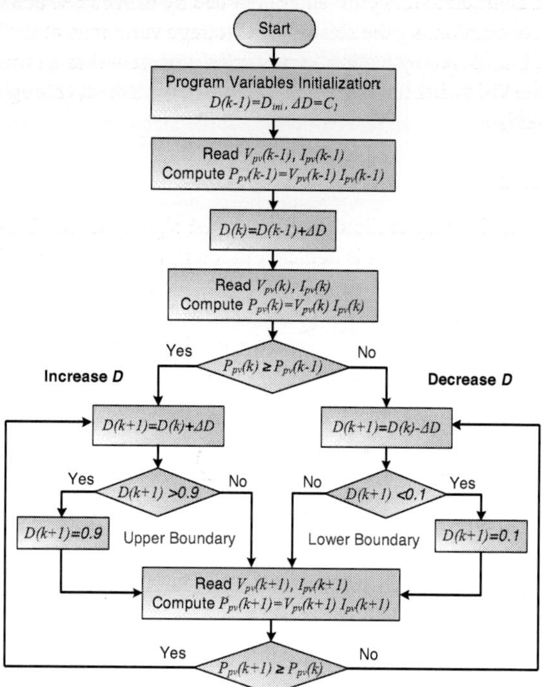

Figure 9. Flowchart for the P&O MPPT algorithm

4.2. Middle level control

The middle level control generates the expected output, particularly the actual active and reactive power exchange between the PV VSI and the AC system, to dynamically track the reference values set by the external level. This level control, which is depicted in the middle part of Fig. 8, is based on a linearization of the state-space mathematical model of the PV system in the d-q reference frame, described in Eq. (20).

In order to achieve a decoupled active and reactive power control, it is required to provide a decoupled current control strategy for i_d and i_q. Inspection of Eq. (20) shows a cross-coupling of both components of the PV VSI output current through ω. Therefore, appropriate control signals have to be generated. To this aim, it is proposed to use two control signals x_1 and x_2, which are derived from assumption of zero derivatives of currents in the upper part (AC side) of (20). In this way, using two conventional PI controllers with proper feedback of the VSI output current components allows eliminating the cross-coupling effect in steady state. Eq. (20) also shows an additional coupling of derivatives of i_d and i_q with respect to the DC voltage V_d. This issue requires maintaining the DC bus voltage constant in order to decrease the influence of V_q. The solution to this problem is obtained by using a DC bus voltage controller via a PI controller for eliminating the steady-state voltage variations at the DC bus. This DC bus voltage control is achieved by forcing a small active power exchange with the electric grid for compensating the VSI switching losses and the transformer ones, through the contribution of a corrective signal i_{dr*}.

4.3. Internal level control

The internal level provides dynamic control of input signals to the DC-DC and DC-AC converters. This control level, which is depicted in the right part of Fig. 8, is responsible for generating the switching signals for the twelve valves of the three-level VSI, according to the control mode (PWM) and types of valves used (IGBTs). This level is mainly composed of a three-phase three-level sinusoidal PWM generator for the VSI IGBTs, and a two-level PWM generator for the single IGBT of the boost DC-DC converter. Furthermore, it includes a line synchronization module, which consists mainly of a phase locked loop (PLL). This circuit is a feedback control system used to automatically synchronize the converter switching pulses with the positive sequence components of the AC voltage vector at the PCC. This is achieved by using the phase θ_s of the inverse coordinate transformation from dq to abc components.

5. PVG model and control implementation in MATLAB/Simulink

The complete detailed model and control scheme of the three-phase grid-connected PVG is implemented in the MATLAB/Simulink software environment using the SimPowerSystems (SPS) [8], as depicted in Figs. 10 to 11. SPS was designed to provide a modern design tool that allows scientists and engineers to rapidly and easily build models that simulate power systems. SimPowerSystems uses the Simulink environment, which is a tool based on a graphical user interface (GUI) that permits interactions between mechanical, thermal, control, and other disciplines. This is possible because all the electrical parts of the simulation interact with the extensive Simulink modelling library. These libraries contain models of typical power equipment such as transformers, lines, machines, and power electronics among others. As Simulink uses MATLAB as the computational engine, designers can also use MATLAB toolboxes and other Simulink blocksets.

Since the detailed model of the proposed PVG application contains many states and non-linear blocks such as power electronics switches, the discretization of the electrical system with fixed-step is required so as to allow much faster simulation than using variable time-step methods. Two sample times are employed in order to enhance the simulation, Ts_Power= 5 µs for the simulation of the power system, the VSI and the DC-DC converter, and Ts_Control= 100 µs for the simulation of the multi-level control blocks.

The three-phase grid-connected PV energy conversion system is implemented basically with the Three-Level Bridge block. The three-phase three-level Voltage Source Inverter makes uses of three arms of power switching devices, being IGBTs in this work. In the same way, the DC-DC converter is implemented through the Three-Level Bridge but using only one arm of IGBTs.

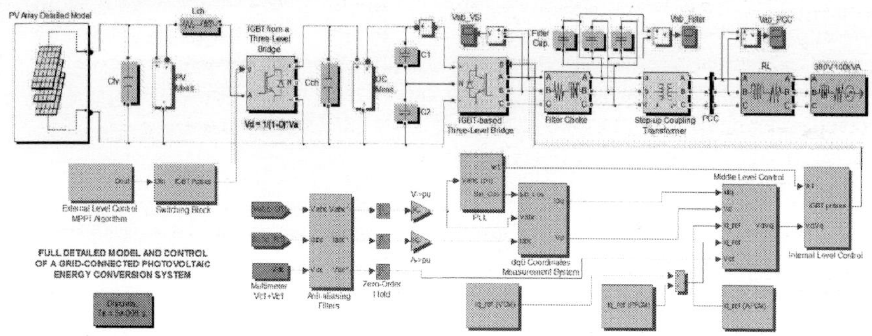

Figure 10. Detailed model and dynamic control of the grid-connected PVG in the MATLAB/Simulink environment

From the four power switching devices of each arm, just one device is activated for accomplishing the chopping function while the other three are kept off all the time. With this implementation approach, the turn-on and turn-off times (Fall time, Tail time) of the power switching device are not modelled, resulting in a faster simulation when compared to a single IGBT mask using an increased state-space model.

Fig. 11 shows the detailed model of the PV array in the MATLAB/Simulink environment, including the implementation of the equivalent circuit of the PV generator by using controlled current sources and resistances, and control blocks for implementing Eqs. 1 through 8 [26].

6. Simulation and experimental results

In order to analyze the effectiveness of the proposed models and control algorithms of the three-phase grid-connected PV system, time-discrete dynamic simulation tests have been performed in the MATLAB/Simulink environment. To this aim, the simulation of a 250 Wp (peak power) PVG has been compared with experimental data collected from a laboratory-scale prototype, which is presented in Fig. 12.

The PV array implemented consists of a single string of 5 high-efficiency polycrystalline PV modules (NS=5, NP=1) of 50 Wp (Solartec KS50T, built with Kyocera cells) [27]. This array makes up a peak installed power of 250 W and is linked to a 110 V DC bus of a three-phase three-level PWM voltage source inverter through a DC-DC boost converter. The resulting dual-stage converter is connected to a 380 V/50 Hz three-phase electric system using three 60 V/220 V step-up coupling transformers connected in a Δ-Yg configuration. The VSI has been rated at 1 KVA and designed to operate at 5 kHz with sinusoidal PWM. It is built with IGBTs with internal anti-parallel diodes and fast clamping diodes, and includes an output inductive-capacitive low pass filter. The DC-DC boost converter interfacing the PV string with the DC bus of the VSI is also built with IGBTs and fast diodes, and has been designed to operate at 5 kHz.

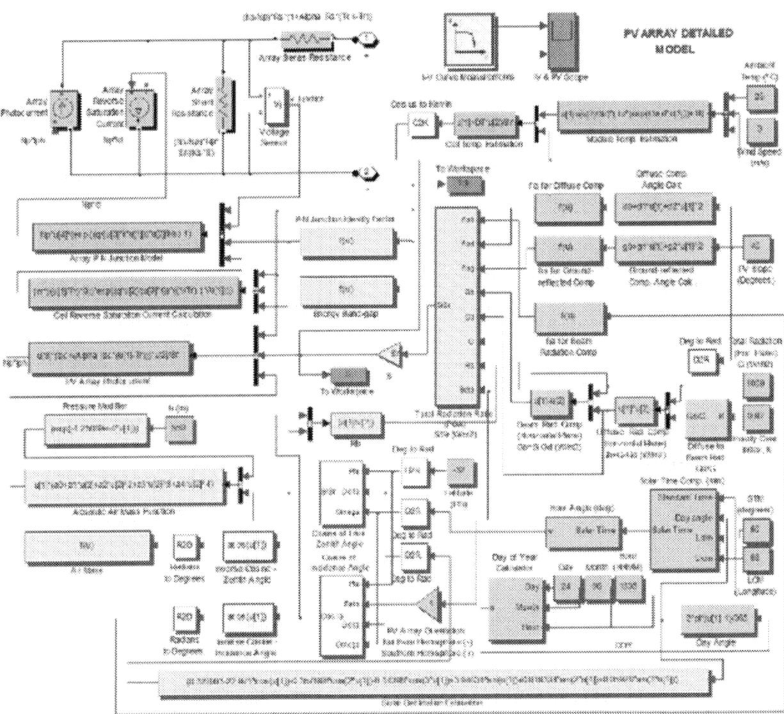

Figure 11. Detailed model of the PV array in the MATLAB/Simulink environment

Figure 12. Laboratory-scale prototype of the three-phase grid-connected PV system.

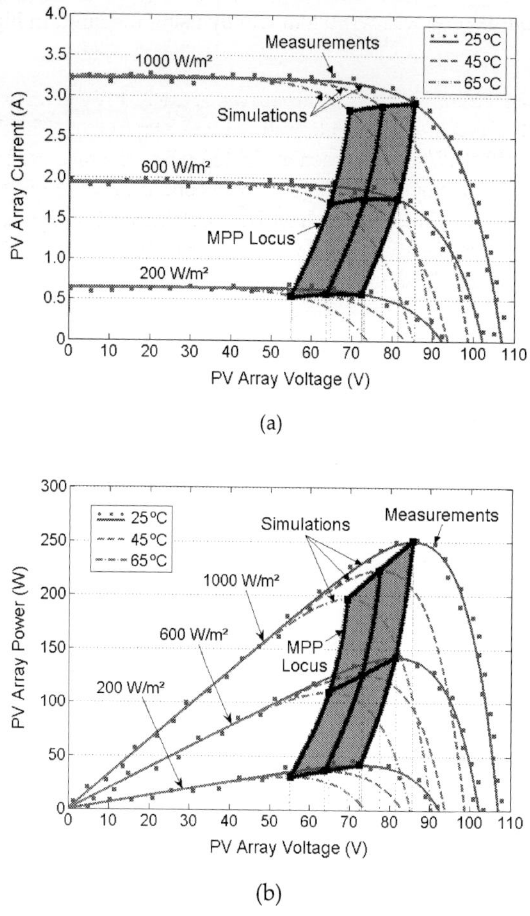

Figure 13. Simulated and measured characteristic curves of the test PV string for given climatic conditions: (a) *I-V* curve. (b) *P-V* curve.

The three-level control scheme was entirely implemented on a high-performance 32-bit fixed-point digital signal processor (DSP) operating at 150 MHz (Texas Instruments TMS320F2812) [28]. This processor includes an advanced 12-bit analog-to-digital converter with a fast conversion time which makes it possible real time sampling with high accuracy and real time *abc* to synchronous *dq* frame coordinate transformation. The DSP is operated with a selected sample rate of 160 ksps and low-pass filters were implemented using 5th order low-pass filters based on a Sallen & key designs. The control pulses for the VSI and the DC-DC chopper has been generated by employing two DSP integrated pattern generators (event managers). The gate driver board of the IGBTs has been designed to adapt the wide differences of voltage and current levels with the DSP and to provide digital and analog isolation using optically coupled isolators. All the source code was written in C++ by using the build-in highly efficient DSP compiler.

Fig. 13 depicts the *I-V* and *P-V* characteristic curves of the 250 W PV array for given climatic conditions, such as the level of solar radiation and the cell temperature. The characteristic curve at 25°C and 200/600/1000 W/m² have been evaluated using the proposed model (blue solid line) with the software developed and measurements obtained from the experimental set-up (blue dotted line). The experimental data have been obtained using a peak power measuring device and *I-V* curve tracer for PV modules and strings (PVE PVPM 1000C40) [29]. As can be observed, the proposed model of the PV array shows a very good agreement with measured data at all the given levels of solar radiation.

As can be derived from both characteristic curves of the PV system, there exist a specific point at which the generated power is maximized (i.e. MPP) and where the output *I-V* characteristic curve is divided into two parts: the left part is defined as the current source region in which the output current approximates to a constant, and the right part is the voltage source region in which the output voltage hardly changes. Since the MPP changes with variations in solar radiation and solar cell operating temperature, the PV array have to be continuously operated within the MPP locus (shaded region) for an optimized application of the system. In this way, a continuous adjustment of the array terminal voltage is required for providing maximum power to the electric grid.

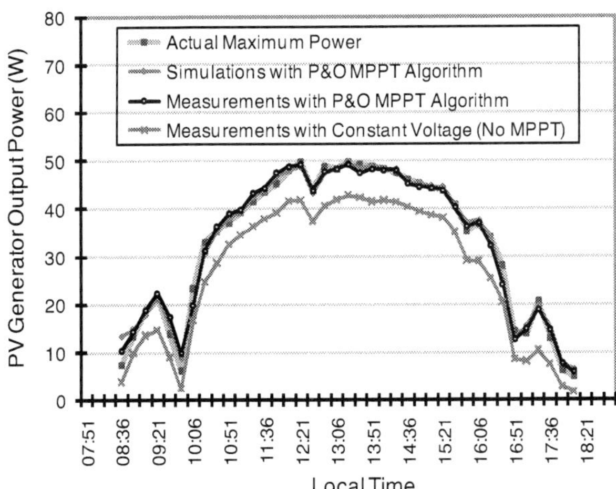

Figure 14. Comparison of actual, measured and simulated output power trajectory for the proposed 250 W PV system with and without the MPPT algorithm implementation.

Fig. 14 presents a comparison of actual, measured and simulated output power trajectory within a 10-hour period of analysis for a cloudy day with high fluctuations of solar radiation, for the proposed 250 W PV system with and without the implementation of the MPPT algorithm. The time data series shown in light blue solid line represents the actual maximum power available from the PV array for the specific climatic conditions, i.e. the MPP to be tracked at all times by the MPPT. Simulation results obtained with the MPPT algorithm are shown in red solid lines. In the same way, the two time data series shown in black and green solid lines, respectively represents the measurements obtained from the experimental setup with the control system with the MPPT activated (black) and with no MPPT (green). As can be observed, the MPPT algorithm (measurements and simulation) follows accurately the maximum power (actual available power) that is proportional to the solar radiation and temperature variations. In this sense, it can be noted from measurements and simulation a very precise MPP tracking when soft variations in the solar radiation take place, while differing slightly when these variations are very fast and of a certain magnitude. It can be also derived that there is a good correlation between the experimental and the simulation data. In addition, the deactivation of the MPPT control results in a constant voltage operation of the PV array output at about 60 V for the given prototype conditions. In this last case, a significant reduction of the installation efficiency is obtained, which is worsen with the increase of the solar radiation. This preceding feature validates the use of an efficient MPPT scheme for maximum exploitation of the PV system.

7. Conclusion

This chapter has presented a full detailed mathematical model of a three-phase grid-connected photovoltaic generator, including the PV array and the electronic power conditioning system. The model of the PV array uses theoretical and empirical equations together with data provided by manufacturers and meteorological data (solar radiation and cell temperature among others) in order to accurately predict the PV array characteristic curve. Moreover, it has presented the control scheme of the PVG with capabilities of simultaneously and inde-pendently regulating both active and reactive power exchange with the electric grid. The control algorithms incorporate a maximum power point tracker (MPPT) for dynamic active power generation jointly with reactive power compensation of distribution utility system. The model and control strategy of the PVG have been implemented using the MATLAB/Simulink environment and validated by experimental tests.

Acknowledgements

The author wishes to thank the CONICET (Argentinean National Council for Science and Technology Research), the UNSJ (National University of San Juan) and the ANPCyT (National Agency for Scientific and Technological Promotion) for the financial support of this work.

Author details

Marcelo Gustavo Molina[*]

Address all correspondence to: mgmolina@iee.unsj.edu.ar

Institute of Electrical Energy, National University of San Juan–CONICET, Argentina

References

[1] International Energy Agency. World Energy Outlook 2014 [Internet]. November 2014. Available from: http://www.iea.org/bookshop/477-World_Energy_Outlook_2014 [Accessed: January 2015]

[2] Razykov, T. M.; Ferekides, C. S.; Morel, D.; Stefanakos, E.; Ullal, H. S. and Upadhyaya, H. M. Solar Photovoltaic Electricity: Current Status and Future Prospects, Solar Energy, 2011; 85(8):1580-1608.

[3] El Chaar, L.; Lamont, L. A. and El Zein, N. Review of Photovoltaic Technologies, Renewable and Sustainable Energy Reviews, 2011; 15(5):2165-2175.

[4] Parida, B.; Iniyan, S. and Goic R. A Review of Solar Photovoltaic Technologies, Renewable and Sustainable Energy Reviews, 2011; 15(3):1625-1636.

[5] Molina, M.G. and Mercado, P.E. Modeling and Control of Grid-connected Photovoltaic Energy Conversion System used as a Dispersed Generator, 2008 IEEE/PES Transmission and Distribution Conference & Exposition Latin America, Bogotá, Colombia, August 2008.

[6] Ai, Q.; Wang, X. and He X. The Impact of large-scale Distributed Generation on Power Grid and Microgrids, Renewable Energy, 2014; 62(3):417-423.

[7] Ruiz-Romero, S.; Colmenar-Santos, A., Mur-Pérez, F. and López-Rey, A. Integration of Distributed Generation in The Power Distribution Network: The Need for Smart Grid Control Systems, Communication and Equipment for a Smart City - Use cases, Renewable and Sustainable Energy Reviews, 2014; 38(10):223-234.

[8] The MathWorks Inc. SimPowerSystems for Use with Simulink: User's Guide. Available from: www.mathworks.com [Accessed: August 2014].

[9] Molina, M.G. and Espejo, E.J. Modeling and Simulation of Grid-connected Photovoltaic Energy Conversion Systems, International Journal of Hydrogen Energy, 2014; 39(16):8702-8707.

[10] King, D.L.; Kratochvil, J.A.; Boyson, W.E. and Bower, WI. Field Experience with a New Performance Characterization Procedure for Photovoltaic Arrays. In: 2nd World Conference on Photovoltaic Solar Energy Conversion; 1998. P. 6-10.

[11] Duffie, J.A. and Beckman, W.A. Solar Engineering of Thermal Processes. Second ed. New York: John Wiley & Sons Inc.; 1991.

[12] Young, AT. Air mass and refraction. Applied Optics 1994; 33:1108-1110.

[13] Angrist, SW. Direct Energy Conversion, second ed. Boston: Allyn and Bacon; 1971.

[14] Scroeder, D.K. Semiconductor Material and Device Characterization, second ed. New York: John Wiley & Sons Inc.; 1998.

[15] Acharya, Y.B. Effect of Temperature Dependence of Band Gap and Device Constant on I-V Characteristics of Junction Diode. Solid-State Electronics 2001; 45:1115-1119.

[16] Malek, H. Control of Grid-Connected Photovoltaic Systems Using Fractional Order Operators, Thesis Dissertation, Utah State University, 2014. Available from: http://digitalcommons.usu.edu/etd/2157 [Accessed: December 2014].

[17] Teodorescu, R., Liserre, M. and Rodríguez, P. Introduction in Grid Converters for Photovoltaic and Wind Power Systems, John Wiley & Sons, Ltd, Chichester, UK; 2011.

[18] Carrasco, J.M., Franquelo, L.G., Bialasiewicz, J.T., Galván, E., Portillo-Guisado, R.C., Martín-Prats, M.A., León, J.I. and Moreno-Alfonso, N. Power Electronic Systems for the Grid Integration of Renewable Energy Sources: A Survey. IEEE Trans Industrial Electronics, 2006; 53(4):1002-1016.

[19] Pacas, J. M., Molina, M. G. and dos Santos Jr., E. C. Design of a Robust and Efficient Power Electronic Interface for the Grid Integration of Solar Photovoltaic Generation Systems", International Journal of Hydrogen Energy, 2012; 37(13):10076-10082.

[20] Molina, M. G. Emerging Advanced Energy Storage Systems: Dynamic Modeling, Control and Simulation. First ed. New York: Nova Science Pub., Inc.; 2013.

[21] Ahmed, A.; Ran, L.; Moon, S. and Park, J. H. A Fast PV Power Tracking Control Algorithm with Reduced Power Mode. IEEE Trans. Energy Conversion 2013; 28(3): 565-575.

[22] Chun-hua, Li; Xin-jian, Zhu; Guang-yi, Cao; Wan-qi, Hu; Sui, Sheng and Hu, Ming-ruo. A Maximum Power Point Tracker for Photovoltaic Energy Systems Based on Fuzzy Neural Networks. Journal of Zhejiang University - Science A, 2009; 10(2): 263-270.

[23] Molina, M.G., Pontoriero, D.H. and Mercado, P.E. An Efficient Maximum-power-point-tracking Controller for Grid-connected Photovoltaic Energy Conversion System. Brazilian Journal of Power Electronics, 2007; 12(2):147-154.

[24] Santos, J. L.; Antunes, F. and Cícero Cruz, A. C., "A maximum power point tracker for PV systems using a high performance boost converter", Solar Energy, vol. 80, pp. 772-778, 2006.

[25] Femia, N.; Petrone. G.; Spagnuolo, G. and Vitelli, M. Increasing the Efficiency of P&O MPPT by Converter Dynamic Matching, Proc. IEEE International Symposium on Industrial Electronics, pp. 1-8, 2004.

[26] Espejo, E. J.; Molina, M. G. and Gil, L. D. Desarrollo de Software para Análisis de Pérdidas de Productividad Debidas al Sombreamiento en el Sitio de Instalación de Parque Fotovoltaico Conectado a la Red, in Spanish, XXXVII workshop of the ASADES (Argentinean Association of Renewable Energy and Environment), and VI Latin-American Regional Conference of the International Solar Energy Society (ISES), Oberá, Misiones, Argentina, October 2014.

[27] Solartec S.A. KS50T - High Efficiency Polycrystalline Photovoltaic Module: User's Manual. Available from: http://www.solartec.com.ar/en/documentos/productos/3-25wp/SOLARTEC-KS50T-v0.pdf [Accessed: June 2014].

[28] Texas Instruments. TMS320F2812 - 32-Bit Digital Signal Controller with Flash: Technical documents. Available from: www.http://www.ti.com/product/TMS320F2812 [Accessed: January 2014].

[29] PV-Engineering GmbH. PVPM1000C40 - Peak Power Measuring Device and I-V Curve Tracer for Photovoltaic Modules up to 1000V and 40A DC: User's Manual. Available from: www.pv-engineering.de/en/products/pvpm1000c40.html [Accessed: December 2014].

Chapter 4

Method for Aligning Facets in Large Concentrators That Have Segmented Mirrors for Solar Thermal Applications

Sergio Vázquez y Montiel, Fermín Granados Agustín and
Lizbeth Castañeda Escobar

Additional information is available at the end of the chapter

http://dx.doi.org/10.5772/63472

Abstract

The high radiative flux solar furnaces already in operation, and the furnaces to be built in the future, have large concentrators consisting of multiple facets. It is, therefore, necessary to have an alignment procedure of the facets that guarantee the accuracy of alignment, allowing an alignment in a short time. This chapter presents a novel method for aligning the facets of a large concentrator, which uses a single optical system to achieve the required precision and the desired energy distribution in the focal plane. Simulation and experimental results obtained with our procedure show that the proposed alignment satisfactorily meets the specifications.

Keywords: Optical alignment, radiative flux, segmented concentrator, solar concentration, solar furnace

1. Introduction

Solar thermal energy (STE) is a form of solar energy to generate electrical energy or thermal energy for use in the residential and commercial sectors and in industry.

Concentrating solar power (CSP) plants use mirrors to concentrate the rays of sun and produce heat for electricity generation through a conventional thermodynamic cycle. These kinds of plants are candidates for providing the solar electricity needed within the next few decades.

According to thermodynamics and Planck's equation, the conversion of solar heat to mechanical work is limited by the Carnot efficiency, and then to achieve maximum conversion rates, the energy should be transferred to a thermal fluid or reactants at temperatures closer to that of the sun.

Solar radiation is a source of high temperature and energy. But for terrestrial use in average we have 1 kW/m^2. Therefore, it is an essential requisite for solar thermal power plants and high-temperature solar chemistry applications to make use of optical concentration devices that enable the thermal conversion to be carried out at high solar flux and with relatively little heat loss.

The high radiative flux solar furnace (HFSF) is the solar concentrator system with greater concentration ratio. This type of system uses large concentrators with dimensions of several meters, which cannot be constructed in one piece; they are built with multiple segments.

A key to these large concentrators operate efficiently is the alignment of the segments. This kind of concentrators has hundreds of segments that must be aligned with high precision and in a short time, until now; there is no known solution to this problem by the authors. The method that we propose in this chapter is a solution that was tested in the solar furnace high radiative flux of Mexico successfully.

The era of modern solar furnaces had started in the 1950s. Among the first furnaces built were the furnace of Arizona State College in the United States in 1956 [1] and the furnace of the Government Institute for Industrial Research in Japan in 1956 [2].

Several applications of the solar furnace focus on the study of properties such as expansion coefficients, emissivity, thermal conductivity, melting points [2], crystal growth, and purification of materials. Also began developing methods for the measurement of high temperatures in receivers [3] and the flux density of concentrated radiation [4]. The later have evolved calorimetric techniques [5, 6].

The combination of a stationary concentrator and a heliostat has come to be known as solar furnace. The advantage of this combination when compared with a parabolic dish directly tracking the sun is that the focus is stationary. Both concentrator and instrumentation can be placed indoors.

The estimated yearly average insolation in México is 5.5 kWh/m^2 per day higher than all other countries. For this reason, in Mexico a high radiative flux solar furnace, HFSF, was built to initiate the development of concentrating solar technologies for the production of solar fuels such as hydrogen. This development is part of a larger project known as "National Laboratory for Solar Concentration Systems and Solar Chemistry" [7].

The initial design considered an intercepted power of approximately 30 kW, with a target peak concentration of approximately 10,000 suns. A global standard deviation of the optical errors less than or equal to 4 mrad was chosen to reach such a goal. The optical design of the whole HFSF consists of a heliostat of 81 m^2 (9 m by 9 m), a shutter and a multifaceted concentrator of 409 spherical mirrors. Once the concentrator was built and aligned, it was verified that it generates a thermal power of 30 KW, with peaks of 18,000 suns (about 18,000 kW/m^2) and less than or equal to 10 cm in diameter sunspot.

The optical design of the concentrator of the new high radiative flux solar furnace is carried out through ray tracing simulation. In this large HSFS, the concentrator is a mirror with hundreds of facets and these facets must be aligned to achieve the desired energy distribution.

We consider the concentrator of HFSF formed with 409 mirror facets of hexagonal contour with a diameter of 40 cm each of them, each one being a spherical first surface mirror [8]. These facets are grouped in five sets of different focal length, placed on a spherical supporting structure, and with corrected orientations in order that the incident radiation on each facet is reflected to the same focus (see Figure 1 and Table 1) [8].

Group	Numbers of facets	Radius of curvature (mm)	Focal distance(mm)
A	85	7500	3750
B	104	8000	4000
C	130	8500	4250
D	64	9000	4500
E	16	9500	4750

Table 1. Number of facets in each mirror group, corresponding to different radii of curvature.

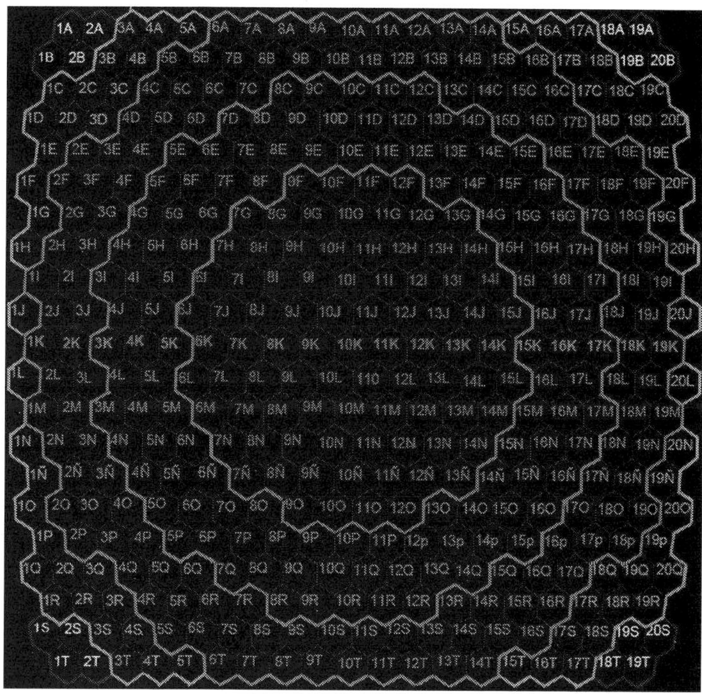

Figure 1. Arrangement of facets in the concentrator. Each color corresponds to a single focal length group.

To reduce manufacturing time and cost of the mirrors, they all have spherical surfaces. This means that the spatial and angular position of each mirror is calculated to compensate the aberrations and to reduce the spot size. Thus, the precise alignment of each mirror is essential for the proper operation of concentrator. Preliminary results of this method were reported by Vazquez et al [9].

In the optical design of the concentrator, the position of each mirror was calculated for collimated light. But it was not possible to use the sun as a light source for the alignment process due to the high temperature in the focal region (above 3000 K). The second reason for not using the sun in the alignment is that such a procedure would require the use of the HFSF heliostat to illuminate the concentrator. This would introduce an additional source of uncertainty, due to the imperfections in heliostat facet alignment and due to the sun tracking mechanism accuracy [9]. To decouple these sources of error from the process of concentrator facets alignment, a different procedure had to be proposed.

2. Alignment procedure facets

We proposed the following goals to be met by the alignment method:

- Use a light source other than the sun.
- The positioning of each facet can be achieved with the required accuracy.
- It is an easy method to implement.
- It works well in the environment where the concentrator is installed, independent of weather conditions.
- It does not require the use of the heliostat of the HFSF.

To fulfill those requirements, our method uses a quasi-point source, generated by a 25 mW HeNe laser (wavelength of 633 nm), together with a 60× microscope objective of 0.65 numerical aperture. This source is placed near the center of curvature of the set of mirrors to be aligned and the divergent light beam illuminates the concentrator. The light is reflected by each one of the mirror group. The reflection from each mirror forms a separate image at an observation screen. In the observation screen, the theoretical image produced by each mirror is calculated previously using the ray trace software. Alignment procedure is to match the actual spots with those calculated theoretically, by rotating the mirror around its supporting point. To this end, mirrors are attached to mechanisms enabling their movement in six degrees of freedom: displacement in three perpendicular directions and rotation around three axes. Each mirror is moved until the image matches the image generated theoretically. Figure 2 shows the optical arrangement used [9].

In Figure 3, a picture of the laser and the microscope objective used for generating the quasi-point light source is shown.

For each set of mirrors with the same radius of curvature, we place the point source generated by the optical arrangement of Figure 3 in a position near the center of curvature, then we need to find the correct position of the observation screen, this position is where the spots are clearly defined and the spots are not overlapping each other.

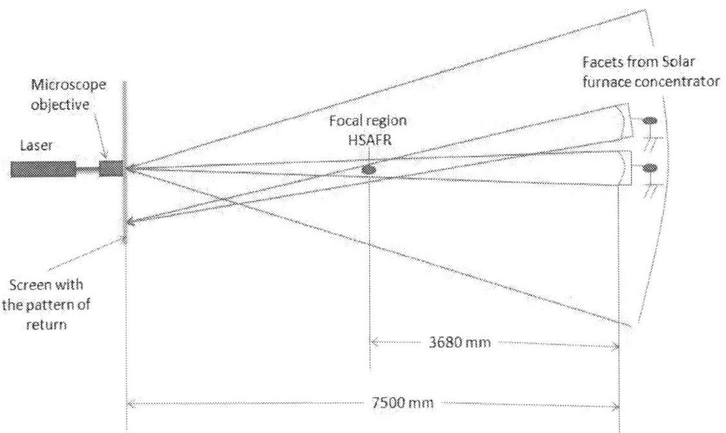

Figure 2. Scheme with the optical arrangement for the alignment of the HRFSF facets.

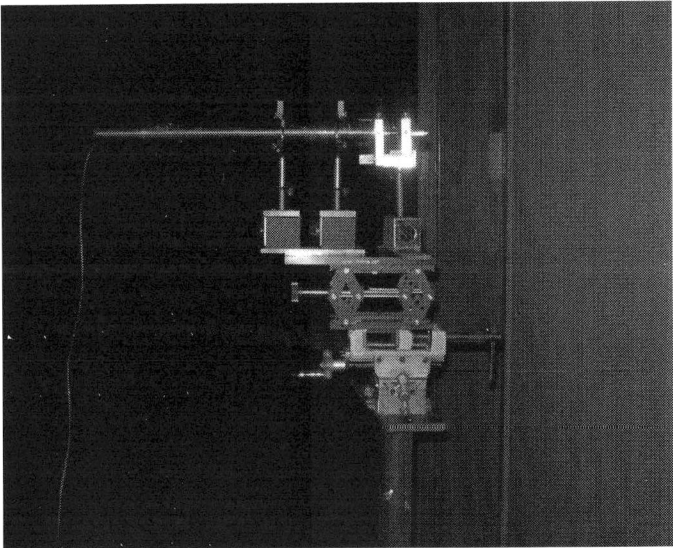

Figure 3. Optical components for generating the quasi-point light source.

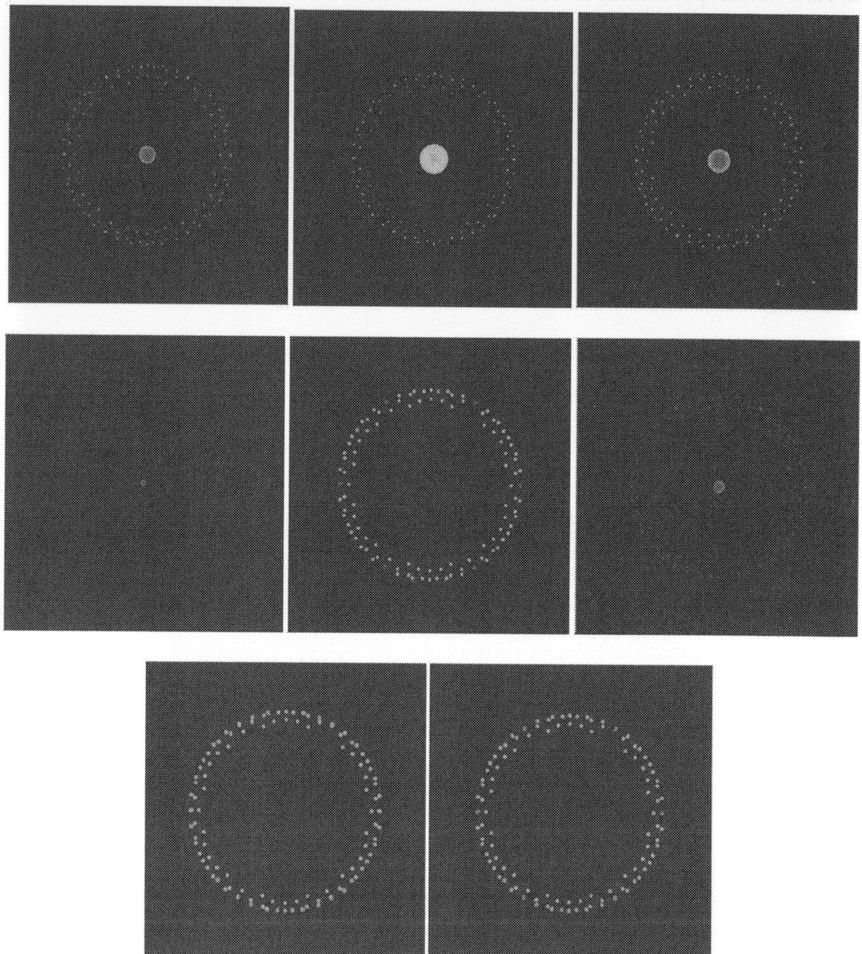

Figure 4. Simulation of images generated by mirrors with radius of curvature of 800 cm. Top left to bottom right, the images correspond to the following positions: –800 cm,– 802 cm,– 804 cm,– 806 cm, –808 cm,– 810 cm, –812 cm, and –814 cm.

To establish the position of the observation screen, we made a simulation with the software ZEMAX OpticStudio by varying the position of the screen and observing the generated image. In Figure 4, we show simulation images for the mirror set of radius of curvature of 800 cm, the light source is placed at –810 cm and the screen was placed at positions –800 cm, –802 cm, –804 cm, –806 cm, –808 cm, –810, –812 cm, and –814 cm. As shown, the size and position of the spots change with the position of the screen. If the screen is placed in front of the light source, there is a central spot causing noise during alignment and with this the intensity of the spots is reduced; however, if the screen is positioned behind the light source, the central spot disappears.

Figures 5 and 6 show the theoretically calculated alignment screens and mirrors corresponding to two sets of mirrors.

In Figure 7, a picture with the real spots on the alignment screen is shown. In the picture the spots generated by the mirrors with radius of curvature of 8500 mm and 9000 mm can be appreciated.

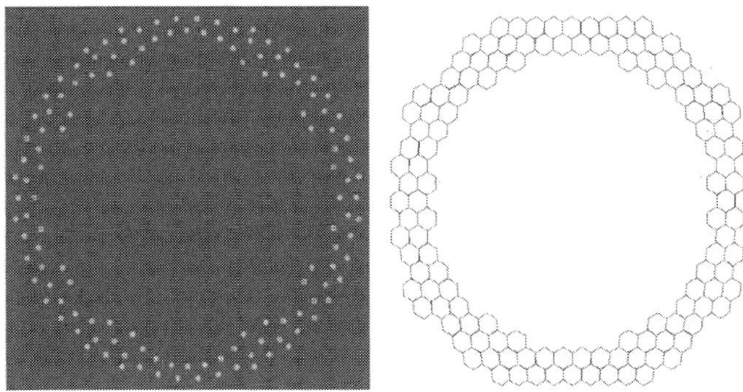

Figure 5. Screen and mirrors corresponding to 850 cm curvature radius.

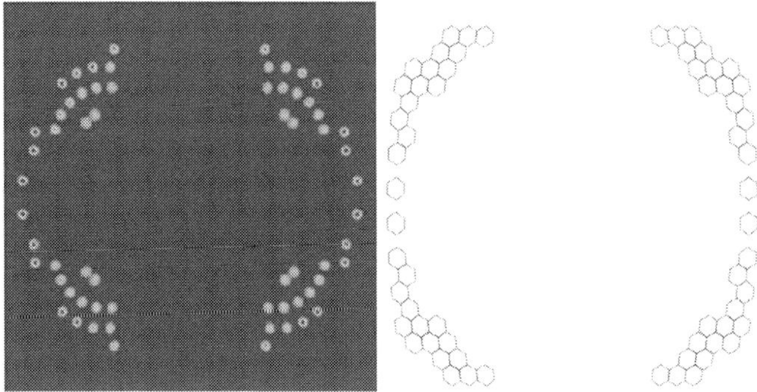

Figure 6. Screen and mirrors corresponding to 900 cm curvature radius.

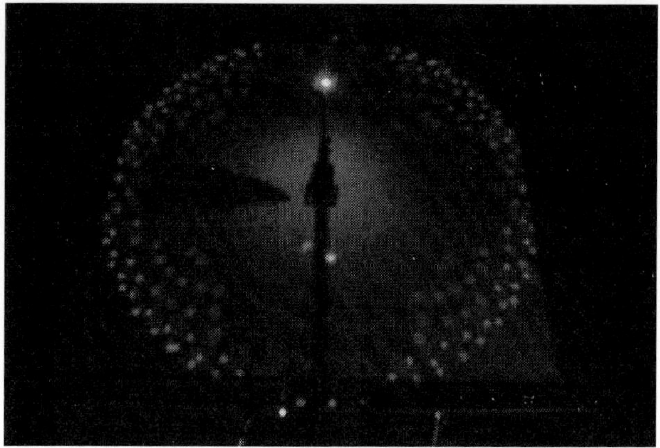

Figure 7. Picture with the real spots on the alignment screen [9].

3. Experimental results

After applying the alignment procedure for each of the concentrator mirrors, different tests were conducted to evaluate the accuracy of alignment. The first test was qualitative; the objective of this test was to observe the images of objects reflected by the concentrator [9]. As you can see, the images display excellent continuity and quality as a result of the alignment of the facets. It is a very good indication of the proper alignment of facets, see Figure 8.

Figure 8. Images reflected in the hexagonal facets after the alignment process [9].

For the second test, we use the 81 m² heliostat in conjunction with the concentrator, and the sun as a source of light. A water-cooled Lambertian target was positioned on the focal plane, and pictures of the solar image produced by the HFSF were taken with a charge-coupled device, CCD, camera [9], as can be seen in Figure 9.

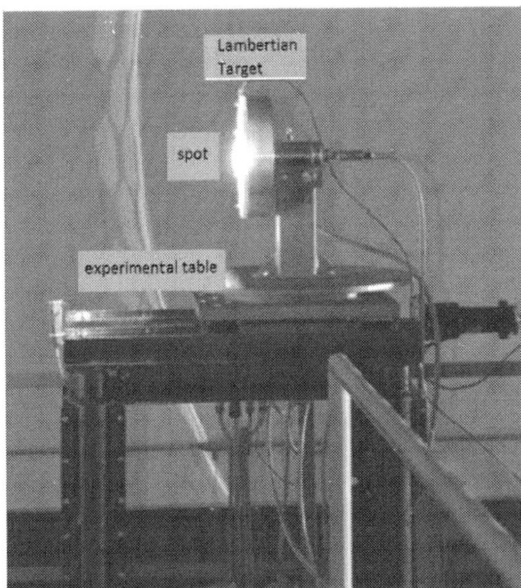

Figure 9. Water-cooled Lambertian target and example of solar image on the target.

To obtain the spot size and the distribution of irradiance we use MATLAB® software. With the results we estimate a spot diameter of 8 cm, and that 90% of the energy falls into a diameter of 6 cm. Comparing these results with ray tracing simulations, we conclude that our alignment procedure meets all requirements and that the standard deviation of the global optical error is of 2.7 ± 0.2 mrad.

Figure 10 shows an example of experimental image, the following steps to process the image and the resulting irradiance profile from the computer code.

The images obtained were compared with the results of the ray-tracing program called Tonalli [10]. The code calculates the irradiance flux on a focal plane using the convolution technique. We assume that the error distribution corresponds to a bidimensional Gaussian distribution characterized by a global standard deviation. The sun's model was taken from a standard solar radiation cone [10].

The images from CCD camera were analyzed, and we compared the profiles of irradiance in two orthogonal directions (horizontal and vertical) against the theoretical profile obtained for the global optical error, which gave the best fit with the experimental curve.

Image	Optical error, horizontal profile (mrad)	Optical error, vertical profile (mrad)
Image 1	2.8	2.8
Image 2	2.8	2.8
Image 3	2.8	2.4
Image 4	2.8	2.2

Table 2. Estimated optical error from different experimental images

Figure 11 shows an example of graphs obtained for the optimal adjustment in the horizontal and vertical image for first image as an example. Table 2 shows the summary of the results obtained with four images captured at different times. As you can see from the values in Table 2, the optical errors are consistent and repeatable; therefore, we can conclude that the global

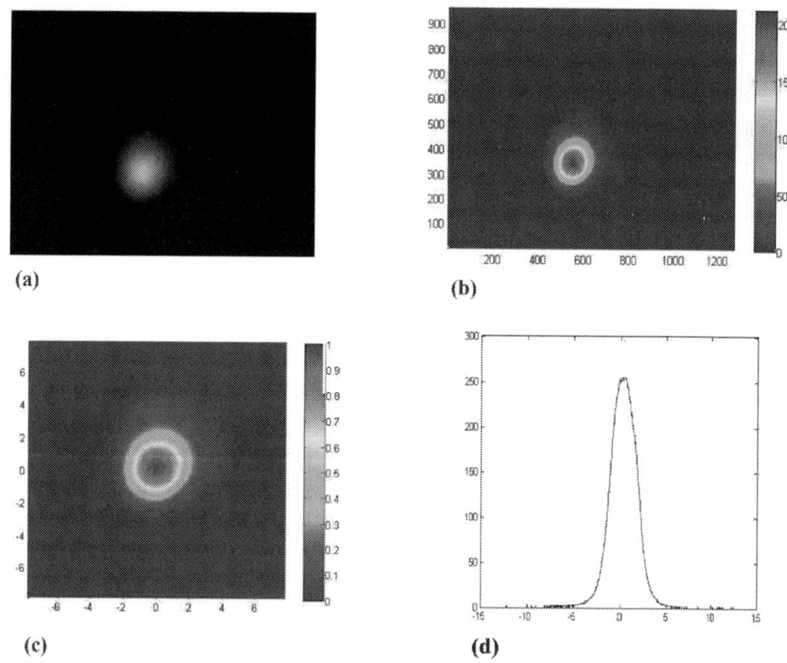

Figure 10. In (a) CCD picture from spot at the receiver at the focal point. (b) Picture translated to the computer for processing. The units of the image frame are mm. The values of the color scale must be multiplied by 100 W. (c) Image centered and processed. The units of the image frame are cm. The color scale is normalized. (d) Irradiance profile from the solar spot. The horizontal axis is cm and the vertical should be multiplied by 100 W.

optical error is approximately 2.7 ± 0.2 mrad. It is important to note that this error does not include the tracking error of the heliostat along the solar day.

Figure 11 only presents the adjustment in particular directions from the images, but they cannot show the symmetry of the image compared to those expected theoretically. Figure 12 shows theoretical and experimental comparison of the contours of the irradiance distributions at the receiver for image 1. To construct Figure 12, a particular level of irradiance is set in the experimental and the theoretical image, the curve formed by points with the same irradiance value is a contour. Each color in Figure 12 corresponds to a different level of irradiance in the image. There are two curves with the same color, corresponding to the experimental and theoretical images. The experimental and theoretical distributions were normalized with respect to the peak.

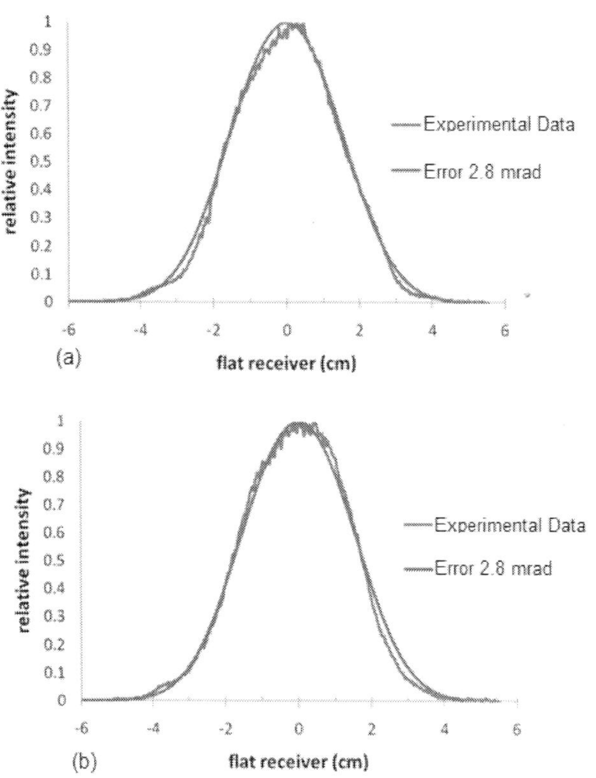

Figure 11. Comparison of experimental and theoretical irradiance profiles for image 1. (a) Vertical profile. (b) Horizontal profile.

Figure 12 shows a relative good correspondence with experimental results except in the lower contour in which there is a deviation from the circular shape of the image.

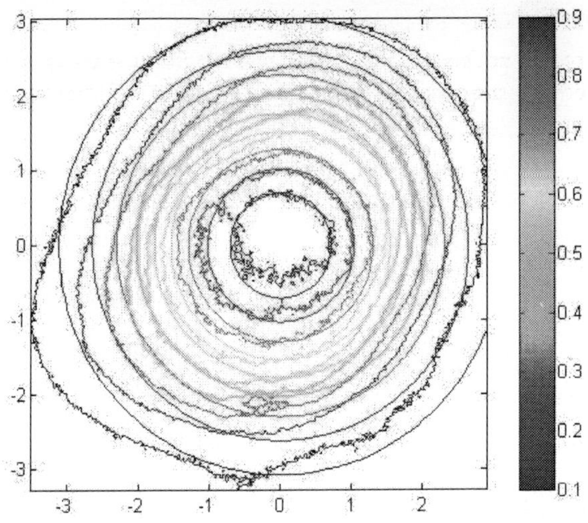

Figure 12. Comparison of experimental and theoretical contours, for image 1. The color scale units are Watts and is normalized.

4. Alignment tolerances of the method

Although the method has only geometric bases and their implementation requires fewer optical elements and sophisticated software is not required to process data as if it require other techniques, the results of the Section 3 shows that the alignment is within specified tolerances. To understand why the proposed method gives such good results, the following analysis was performed.

The optical concentrator design considers that the light source is at infinity and has the angular size of the sun, with these considerations the positions of each of the segments were defined for maximum concentration. However, in the alignment method the light source is near the center of curvature of each group of mirrors. Therefore, it is necessary to know as alignment errors are transformed into errors in the image.

In the mirror set of 850 cm of radius of curvature, we select one of them and will apply rotations around the X-axis in the range of −0.25 degrees to 0.25 degrees, and we calculate changes in image position for each angle. Figure 13 shows a graph of the results, the behavior is almost linear, and follows the following equation:

$$y = 22.737x + 0.116, \tag{1}$$

where the variable x represents the rotation angle of the mirror and variable y represents the displacement of the image position.

As the diameter of the spot generated by the mirror is 4 mm, the maximum possible error in the alignment process corresponds to a displacement of 4 mm and this corresponds to a mirror rotation 0.012 degrees.

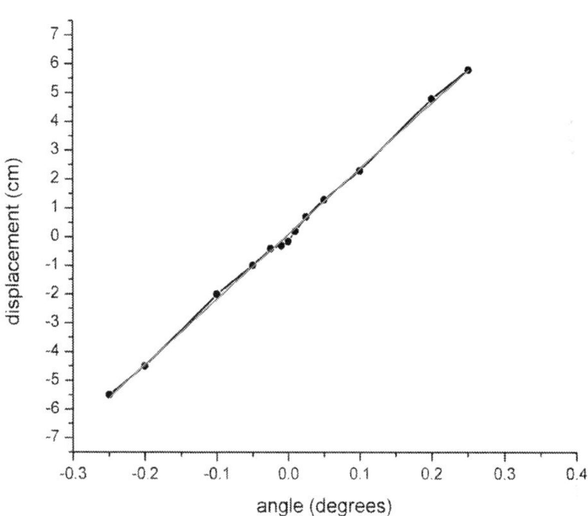

Figure 13. Image shift as a function of rotation of the mirror in the process of alignment. The dots (black) are the measured data and the solid line (red) corresponds to the fit of the data.

Figure 14 shows the ideal spot position without rotation and the spot position generated by applying a 0.01 degree rotation, during the alignment process. It is clear that this could be the maximum possible error.

Using the same mirror, but considering the sun as light source, we apply rotations of the mirror in a range of 0.01 degrees to 0.1 degrees, and we calculate the displacement of the image, the results are shown in the graph of Figure 15. As you can see, the behavior is almost linear and follows the following equation:

$$y = 32.757x - 0.0006, \tag{2}$$

where the variable x represents the rotation angle of the mirror and variable y represents the displacement of the image position.

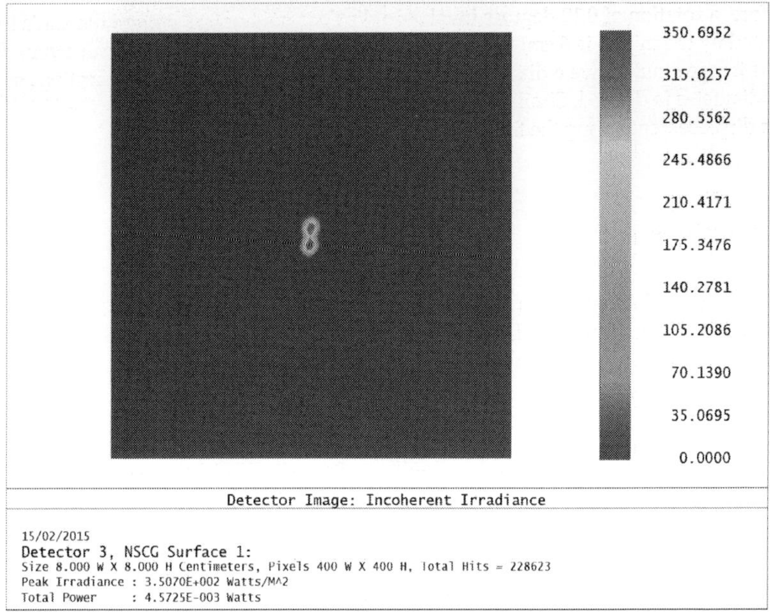

Figure 14. The ideal spot position and spot displaced by a rotation of 0.01 degrees.

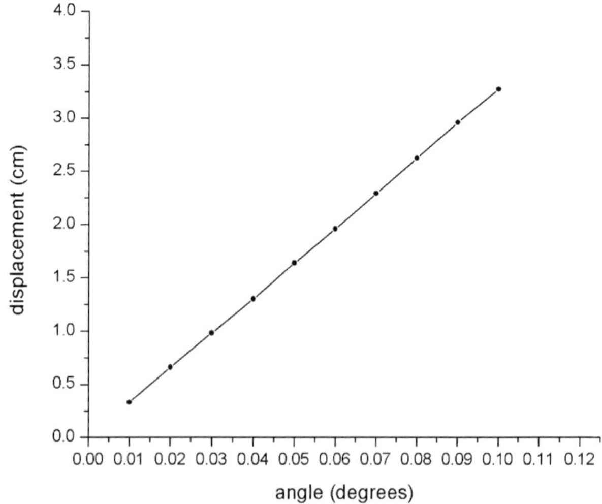

Figure 15. Image shift as a function of rotation of the mirror with the sun as a light source. The dots are the measured data and the solid line corresponds to the fit of the data.

Therefore, a rotation of 0.12 degrees involves a displacement of 0.39 cm. As the sun's image generated by the mirror is 6 cm, this displacement would generate a maximum error of 13% and the image would have a diameter of 6.78 cm corresponding to the specifications as an 8 cm in diameter is desired. Similar results are obtained with rotations about the Y-axis and mirror displacements along the three axes.

5. Conclusions

We have presented a novel method for aligning facets of segmented mirrors for applications of concentrating solar power. This is a method that requires only three optical components: it is cheap and easy to implement and can be used during day and night because the light source is a laser.

We showed that the proposed method allows the alignment of the segments and obtained the specifications with relative simplicity. We obtained the maximum errors that can occur in the alignment process and we find that the method generates errors tolerable without complicated computer programs.

After the alignment process, we used different tests to verify the actual operation of the concentrator and all tests showed a performance that exceeds expectations, demonstrating the viability of the proposed method.

Finally, if CCD cameras and image processing software are used, this method could be used to align the segmented mirror facets as used in large telescopes.

Author details

Sergio Vázquez y Montiel[1*], Fermín Granados Agustín[2] and Lizbeth Castañeda Escobar[3]

*Address all correspondence to: gatoangora2000@yahoo.com.mx

1 Polytechnic University of Tulancingo, México

2 National Institute of Astrophysics, Optics and Electronics, México

3 Instituto Tecnológico Superior de Xalapa, México

References

[1] Kevane, C.J., 1957. Construction and operation of the Arizona State College solar furnace. Solar Energy 1, 99–101.

[2] Hisada, T., Mii, H., Noguchi, C., Noguchi, T., Hukuo, N., Mizuno, M., 1957. Concentration of the solar radiation in a solar furnace. Solar Energy 1, 14–18.

[3] Brenden, B.B., Newkirk, H.W., S.H., 1958. A study of temperature measurement in a solar furnace. Solar Energy 2, 13–17.

[4] Loh, E., Hiester, N.K., Tietz, T.E., 1957. Heat flux measurements at the sun image of the California institute of technology lens-type solar furnace. Solar Energy 1, 23–26.

[5] Pérez-Rábago, C.A., Marcos, M.J., Romero, M., Estrada, C.A., 2006. Heat transfer in a conical cavity calorimeter for measuring thermal power of a point focus concentrator. Solar Energy 80, 1434–1442.

[6] Estrada, C.A., Jaramillo, O.A., Acosta, R., Arancibia-Bulnes, C.A., 2007. Heat transfer analysis in a calorimeter for concentrated solar radiation measurements. Solar Energy 81, 1306–1313.

[7] http://lacyqs.cie.unam.mx/es/index.php/instalaciones/horno-solar-de-alto-flujo-radiativo.

[8] Rivero-Rosas, D., Herrera-Vázquez, J., Pérez-Rabago, C. A., Arancibia-Bulnes, C.A., Vázquez-Montiel, S., Sanchez-González, M., Granados-Agustín, F., Jaramillo, O.A., Estrada, C.A. 2010. Optical design of a high radiative flux solar furnace for Mexico. Solar Energy 84, 792–800.

[9] Vázquez-Montiel, S., Pérez-Rabago C.A., Pérez-Enciso, R., Rivero-Rosas, D., Granados-Agustín, F., Arancibia-Bulnes, C.A., Estrada, C.A. 2011. Method for facets alignment for the high flux solar furnace at CIE-UNAM in Temixco, Mexico, First Stage. In: SolarPaces September 2011; September 2011; Granada Spain. 2011.

[10] Rivero-Rosas D., Pérez-Rabago C. A., Arancibia-Bulnes C. A., Pérez-Enciso R., Estrada C. A.. Concentration Images Profiles of the High-Flux Solar Furnace of CIE-UNAM in Temixco, Mexico, First Stage. In: SolarPaces September 2011 symposium; Granada Spain. 2011.

Chapter 5

Theoretical Analysis and Implementation of Photovoltaic Fault Diagnosis

Yihua Hu and Wenping Cao

Additional information is available at the end of the chapter

http://dx.doi.org/10.5772/62057

Abstract

The utilization of solar energy by photovoltaics (PVs) is seen in increase across the world since the technologies are getting mature and the material costs are being driven down. However, their operating costs are still very high, owing to their vulnerability to harsh outdoor environments they are working. Currently, the reliability of PV systems is the bottle-neck issue and is becoming a heated research topic. This chapter presents the state-of-the-art technologies for photovoltaic fault diagnosis, based on an intensive literature review and theoretical analysis. The chapter evaluates the fault mechanisms of photovoltaics at the cell, module, string and array levels. Analytical models are developed to understand the PV's terminal characteristics for diagnostic purposes. Offline and online fault diagnosis technologies are reviewed and compared based on the use of electrical sensors and thermal cameras. The aim of this chapter is to illustrate the PV faulty characteristics, to develop offline and online fault diagnosis, and to use the fault diagnosis information to achieve optimal operation (maximum power point tracking) under various PV faulty conditions, by using multi-disciplinary analytical, empirical and experimental methods.

Keywords: Fault diagnosis, MPPT, photovoltaics, solar energy, reliability

1. Introduction

Solar energy is the primary source of renewable energy. The generation of solar power using photovoltaic (PV) systems is gaining popularity in developed and developing countries across the globe because PV technologies are getting mature and the material costs are continuing to reduce. In this field, the operating costs are still high primarily due to the vulnerability of the PV panels at harsh operating environments. It is now a technical challenge to operate the PV system with a minimal interruption and a maximum power output. This chapter reviews and

develops the state-of-the-art technologies for PV fault diagnosis and maximum power point tracking (MPPT) by the intelligent control of direct current (DC)–DC converters. This chapter evaluates the fault mechanisms of photovoltaics at the cell, module, string, and array levels and reviews some of the fault diagnosis techniques. Analytical models are developed to understand the PV's terminal characteristics for fault diagnostic purposes. Offline and online fault diagnosis technologies are studied by using inexpensive electrical sensors and thermal cameras. The analysis of faulty characteristics and fault diagnosis of PV is conducted by using multidisciplinary analytical, empirical, and experimental methods. It is hoped to provide guidelines for developing economic solar power plants especially for the developing countries where solar energy utilization is a very cost-sensitive market.

2. PV faults

Photovoltaic (PV)-based solar power generation has proven to be a cost-effective and environmentally sustainable technology and, thus, it is under the rapid development over the last few decades. Nonetheless, it is presently hampered by relatively high operational costs, low system efficiency, and low effective service time (EST) of PV panels. A cost reduction roadmap for PV systems is illustrated in Fig. 1. In principle, there are two effective ways for achieving this. One is to increase the operating efficiency (and thus the yield) and the other is to extend the EST of the PV components. This chapter addresses both the issues by the intelligent control of the DC–DC converters for MPPT and the development of PV fault diagnosis technologies.

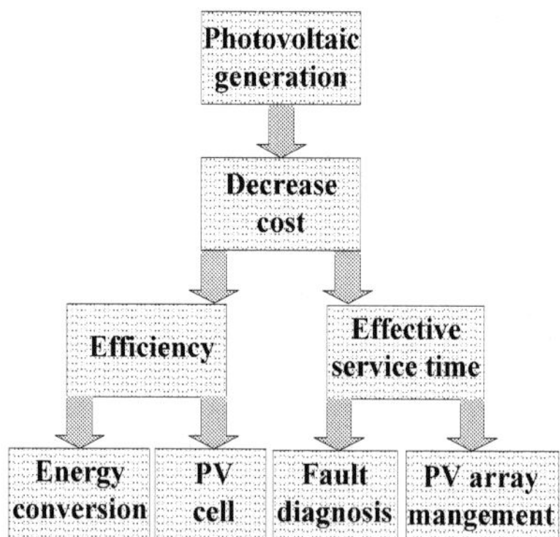

Figure 1. PV cost reduction roadmap.

PV system failures are typically caused by various types of faults or aging effect. In general, there are two types of faults in terms of their duration, temporary and permanent, as shown in Fig. 2. The temporary fault can be cleared automatically for a given time (e.g., shading) or by human intervention (e.g., dust), while the permanent fault is persistent or ongoing. Both the types can decrease the output of PV arrays. The temporary fault may be identified by human eyes, but the permanent fault usually cannot be identified without special equipment (e.g., thermal camera) before a severe damage is made.

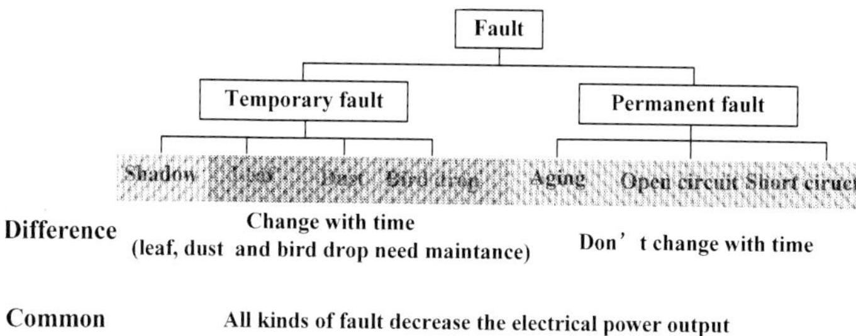

Figure 2. Types of PV faults.

Mismatch faults reduce the power output and cause potential damage to PV cells [1-10]. This section addresses both the issues by presenting a low-cost and efficient temperature-distribution analysis for identifying PV module mismatch faults using various technologies.

When a PV module is faulted or partial shading occurs, the PV system finds a non-uniform distribution of the generated electrical power and thermal profile, and the generation of multiple maximum power points (MPPs). If left untreated, this reduces the overall power generation and severe faults may propagate resulting in damage to the system [11-18]. In this chapter, any scenario that causes the output power of PV array to decrease is called a "fault."

3. PV fault characteristics

This section defines fault categories and mathematical models in terms of PV components. A PV module is composed of dozens of PV cells in series connections. A large number of PV modules connected in series form a PV string, which can be connected in parallel to form a PV array, as shown in Fig. 3. In order to restrict the hotspot phenomena of a PV module, a bypass diode is connected in parallel with PV cells, and the corresponding structure is named a cell-unit, which is composed of m PV cells. The PV module is connected in series by n cell-units to achieve the high-output voltage. The PV string is composed of s PV modules that are also connected in series.

3.1. Category of PV faults

When a fault occurs in the PV array, a temperature difference between the healthy and an unhealthy module is created, similar to partial shading observed from the terminal. Consequently, excessive heat and thermal stress can result in cell cracks. If the cell temperature exceeds its critical point, delamination of cell encapsulants may occur. If the reverse bias exceeds the cell's breakdown voltage, the cell will be damaged [19]. In terms of the severity of mismatch faults, this chapter defines three categories: minor, medium, and heavy faults. Their terminal characteristics are different in the following aspects:

i.   Under a minor fault, the faulted power unit in the PV panel can still operate to generate electricity. As illustrated by the single arrow in Fig. 4(a), the current still passes through the PV cell string to generate an output. In this case, the faulty cell becomes an electrical load, powered by the healthy ones.

ii.   Under a medium fault, PV cells in the string are characterized by the varying illumination levels. As presented in Fig. 4(b), the faulted cells can still operate as a source with a reduced power output. Because of the nonuniform illumination, the actual working point of the power unit is dictated by the operating point of the PV array.

iii. Under a heavy fault condition, the whole PV string is out of function while the bypass diode conducts to transmit the current, as indicated by the dotted arrow in Fig. 4(a). In essence, all PV cells in the string are open-circuited.

If there exists a meaningful temperature difference, hotspot suppression is needed to shift the system MPP and to minimize the impact of the mismatch fault [15].

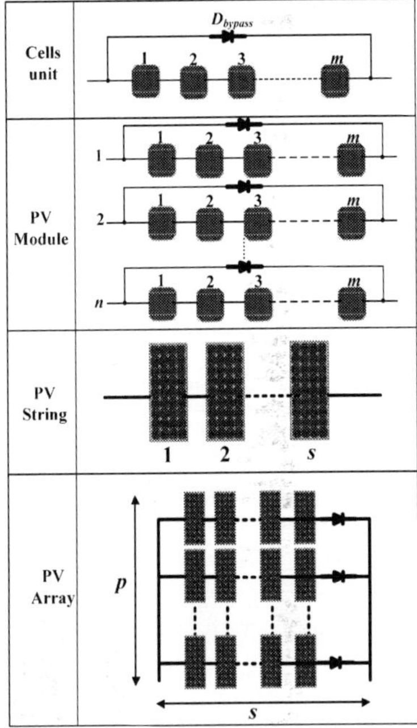

Figure 3. Component of PV arrays.

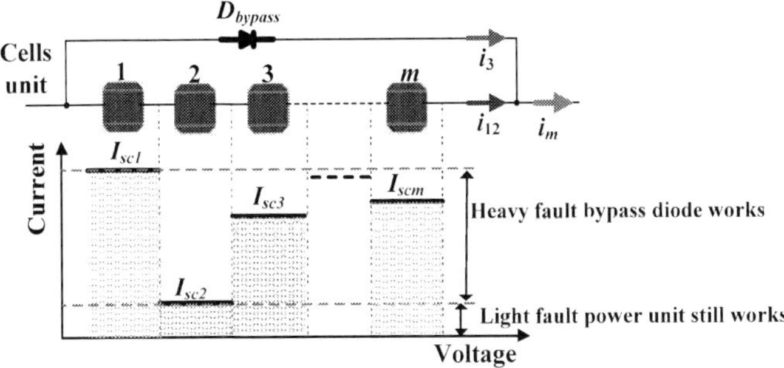

(a) Minor- and heavy-fault conditions

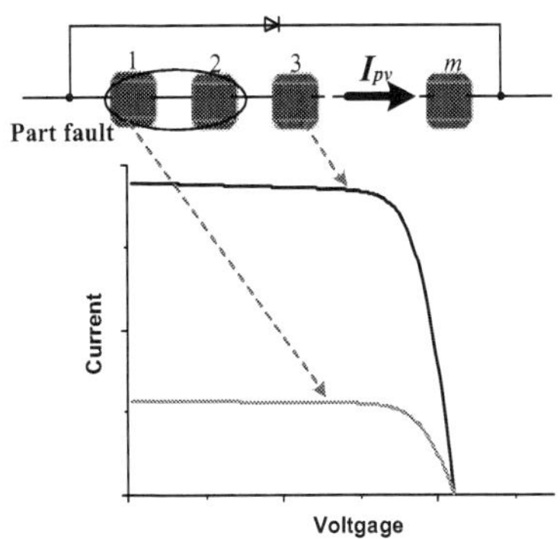

(b) Medium-fault condition

Figure 4. Three categories of mismatch faults defined for a PV system.

3.1.1. Analysis of minor faults

A temperature profile of the PV array under minor-fault conditions is presented in Fig. 5(a). The array consists of b rows and a columns of PV modules where Module 21 is faulted. I_{array} and V_{array} are the current and voltage of the PV array, respectively. I_H and I_f are the currents of healthy and faulty strings, respectively. V_H and V_f are the module voltages of healthy and faulty strings, respectively. T_H is the module temperature of a healthy string, $T_{H'}$ is the healthy module temperature within a faulted string, and T_f is the healthy cell temperature in a faulty power unit.

Under a PV minor fault, the faulty cell cannot generate electricity and becomes a resistive load (R_{eq}). Owing to the series of connection structure, the healthy cells supply power to the faulty

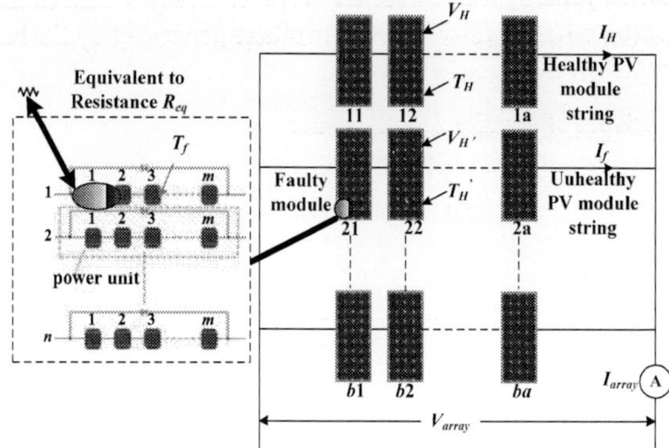

(a) Faulty PV array temperature distribution

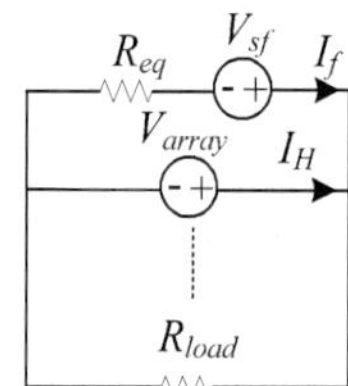

(b) Equivalent circuit upon a fault

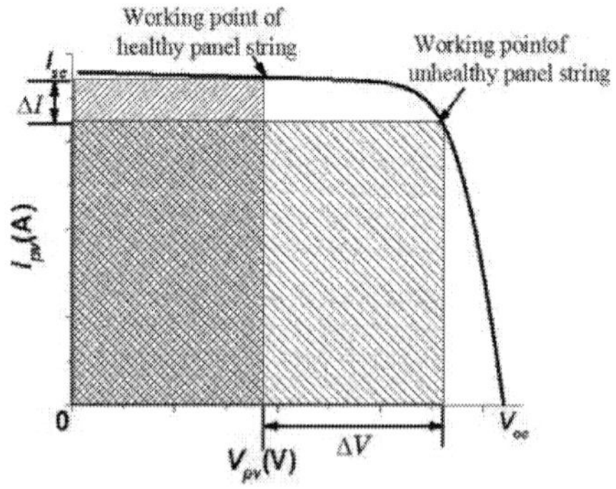

(c) Shift of working points

Figure 5. The PV system at a minor-fault condition [7].

PV cells (released as heat) and then create some hot spots. An equivalent circuit of the PV array is presented in Fig. 5(b), where V_{sf} stands for the voltage generated by the healthy PV cells in a faulty PV string and R_{load} is the load resistance.

The electric characteristics of a faulty PV string are:

$$V_{sf} - I_f R_{eq} = V_{array} \tag{1}$$

$$I_f = \frac{V_{sf}}{R_{eq} + R_{load}} \tag{2}$$

$$R_{eq} = \frac{V_{sf} - V_{array}}{I_f} \tag{3}$$

$$\Delta V = V_{H'} - V_{H} \tag{4}$$

$$\Delta I = I_{H} - I_{f} \tag{5}$$

$$I_f^2 \cdot R_{eq} < I_f (m - m_x) \frac{V_{H'}}{m \cdot n} \tag{6}$$

where ΔI is the current difference between the healthy and unhealthy strings, ΔV is the voltage difference between the healthy modules in healthy and unhealthy strings, and m_x is the number of faulty PV cells.

From Fig. 5(b), it can be seen the voltage of a PV cell in a healthy string is lower than that of a healthy cell in a faulty string; the current of a PV cell in a healthy string is higher than that of a healthy cell in a faulty string. Eqs. (4)–(6) express the mathematical relationship for faulty and healthy PV strings. Eq. (6) shows that when the output power of a faulted PV unit is higher than the I^2R power of its equivalent resistance, a minor fault is created and hot spots begin to form on the faulty cell.

Since the electrical power generated by healthy cells in the PV string supplies not only the load but also faulted cells (heating), the operating point in the current–voltage curve is effectively shifted. Fig. 5(c) demonstrates this in a PV system including healthy and unhealthy panel strings.

3.1.2. Analysis of heavy faults

Under a heavy-fault condition, the PV string containing the faulted cell/module loses production. Its operating points are illustrated in the output current–voltage curve in Fig. 6. Point A1 is the working point of the modules in the healthy string, A2 is the working point of the healthy modules in the faulty string, and A3 is the working point of the healthy cells in the faulty module.

Because the faulty power unit is short-circuited by a bypass diode, the healthy cells in the faulty string are effectively open-circuited. The relative positions of A1, A2, and A3 are determined by the PV array structure and its electrical characteristics. Due to the anti-parallel connection of the bypass diode, the faulty PV power unit is shorted by the diode. Therefore, its output voltage becomes zero. From Eq. (8), V_H is less than V_H'; I_H is greater than I_f, corresponding to the working points A1 and A2. T_H and $T_{H'}$ depend on the working points A1 and A2 in the curve. As the faulty power unit is shorted by a bypass diode, the PV cells are open-circuited, corresponding to point A3. The output power of the faulted power unit is lower than the needed power of the equivalent resistance upon a fault; the power unit is shorted by the bypass diode.

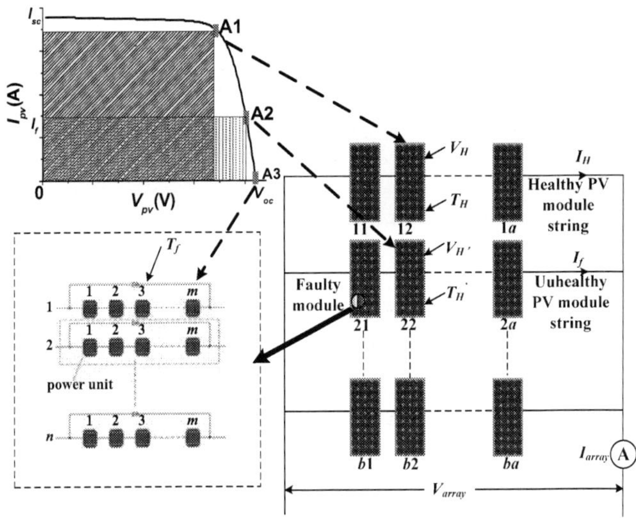

Figure 6. The PV system at a heavy-fault condition [7].

V_H and $V_{H'}$ are thus given by

$$V_H = \frac{V_{array}}{a} \tag{7}$$

$$V_{H'} = \frac{V_H \cdot a \cdot n}{a \cdot n - n_x} \tag{8}$$

where n_x is the number of faulty power units in the faulty PV panel string, which can be identified by thermal cameras.

3.1.3. Analysis of medium faults

The operating point of the PV array strongly affects the condition of the healthy PV modules in the healthy string and sometimes in the faulty string. Fig. 7(a) shows a 2×3 PV array under a medium fault, where module 21 is a faulted PV module, and the rest of the PV modules are healthy. Compared with other PV module (1000 W/m²), No. 21 has the lower illumination (300 W/m²). Fig. 7(b) and (c) presents the current–voltage and power–voltage curves, respectively, obtained from simulation.

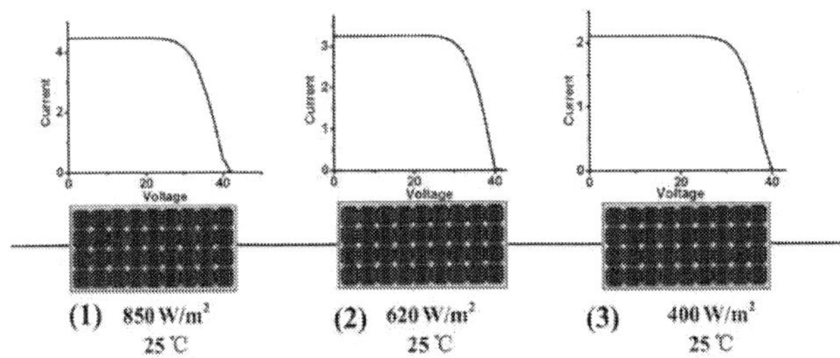

(a) PV string under medium-fault conditions

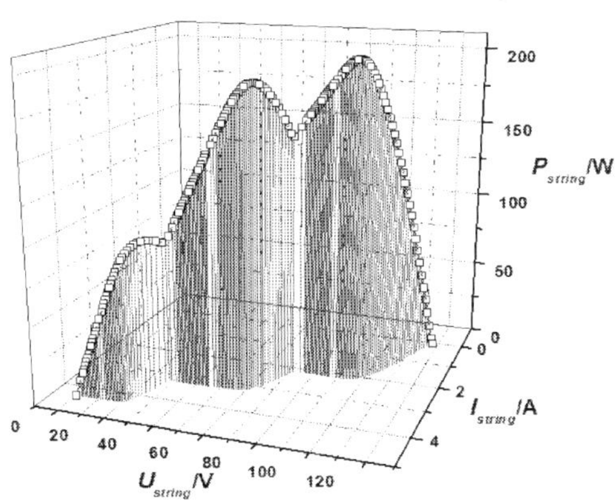

(b) Output characteristics of the medium faults

Figure 7. The PV system at a medium-fault condition.

Fig. 7 presents the typical output characteristics of the PV string under faulty conditions; the PV module simulation parameters are listed in Table 1. The string includes three modules with nonuniform illumination, with the corresponding environment parameters being 850 W/m^2, 25°C; 620 W/m^2, 25°C; and 400 W/m^2, 25°C, as shown in Fig. 7(a). Each module has uniform illumination. In this current–voltage–power waveform, there are three local maximum power points, corresponding to three working stages, as tabulated in Table 2. At stage 1, the PV string current is between 3.29 and 4.39 A; only PV module No. 1 can generate that level of current and PV modules 2 and 3 are shorted. As they are influenced by shadows and cannot generate electricity, the string output voltage is limited to 0–30 V. At stage 2, the operation string current is between 2.19 and 3.29 A; PV modules 1 and 2 can generate electricity; No. 3 is shorted by the bypass diode. The corresponding PV string output voltage is 30–70 V. At stage 3, the operation string current is 0–2.19 A; all the modules can generate electricity and the string output voltage area is 76–126 V. From this analysis, it can be found that:

i.   The multi-stage characteristics are caused by the differing output current ability of each module;

ii.   In the high-output current area, the faulty modules are short-circuited, and the terminal voltage of the corresponding faulty module is zero.

Parameter	Value
Open-circuit voltage	44.8 V
Short-circuit current	5.29 A
Power output	180 W
MPP current	5 A
MPP voltage	36 V
Current-temperature coefficient	0.037 %/K
Voltage-temperature coefficient	–0.34 %/K
Power-temperature coefficient	–0.48 %/K
Nominal operating cell temperature	46 ± 2°C

Table 1. Specifications of the PV module

Working stage	Voltage	Current	Working module
1	0–30 V	3.29–4.39 A	1
2	30–72 V	2.19–3.29 A	1 and 2
3	76–126 V	0–2.19 A	1, 2, and 3

Table 2. Analysis of different stages

4. Mathematical models of PV systems

Under different fault conditions, the mathematical models vary as follows.

4.1. Model of healthy PV cells

The electrical characteristics of PVs are influenced by both temperature and illumination. The electrical model of the PV cell is expressed by [3]:

$$I = I_L - I_o[\exp(\frac{\varepsilon \cdot V}{T_m}) - 1] \tag{9}$$

$$\varepsilon = \frac{q}{N_s \cdot K \cdot A} \tag{10}$$

$$I_L = \frac{G}{G_{ref}}[I_{Lref} + k_i(T_m - T_{ref})] \tag{11}$$

$$I_o = I_{oref}(\frac{T_m}{T_{ref}})^3 \exp[\frac{q \cdot E_{BG}}{N_s \cdot A \cdot K}(\frac{1}{T_{ref}} - \frac{1}{T_m})] \tag{12}$$

where I is the PV module output current, I_L is the photon current, q is the quantity of electric charge, A is the diode characteristic factor, K is the Boltzmann constant, I_o is the saturated current, T_m is the PV module temperature, G is the irradiance, V is the output voltage, G_{ref} is the reference irradiance level (1000 W/m²), I_{Lref}, I_{oref} are the reference values for I_L and I_o. k_i is the current–temperature coefficient provided by the PV manufacturer, T_{ref} is the reference temperature, N_s is the number of series-connected cells, T_m is the PV module temperature, and ε is a constant depending on q, Ns, K, A, and is calculated using the following equation:

$$I_{sc\_ref} - I_{mpp\_ref} = \frac{I_{sc\_ref}}{\exp(\frac{\varepsilon \cdot V_{oc\_ref}}{T_{ref}}) - 1}[\exp(\frac{\varepsilon \cdot V_{mpp\_ref}}{T_{ref}}) - 1] \tag{13}$$

where I_{mpp_ref}, I_{sc_ref}, V_{mpp_ref}, and V_{oc_ref} are the maximum power point (MPP) current, short-circuit current, MPP voltage, and open-circuit voltage at a reference condition, respectively, defined by the relevant standard.

4.1.1. Terminal characteristics of faulted cells

When a PV cell is subject to aging, a direct indication is its lower output power than normal. Due to the p–n junction characteristics of the PV cell, its open-circuit voltage only changes slightly while the short-circuit current changes dramatically. In this chapter, we use the short-circuit current to sense the aging condition of PV cells.

Fig. 8 presents a cell-unit with m nonuniformly aged PV cells, where I_{sc1}, I_{sc2}, I_{sc3}... I_{scm} are the short-circuit current for cells *1, 2, 3... m*, respectively. There are three ranges in the current-voltage output characteristics. In Range 1, the maximum current is the minimum of all cells' current (I_{sc1}, I_{sc2}, I_{sc3}... I_{scm}) and all the cells generate electricity. Minor fault is a transitional interval. Its equivalent circuit is presented in Fig. 8, and its terminal output voltage is presented in Eq. (14). Due to a voltage drop on R_e, the output voltage of the cell-unit is lower than a healthy cell-unit.

$$\sum_{1}^{m-1} V_{cell\_i} - i_{12} \cdot R_e = V_{cu} \tag{14}$$

where V_{cell} is the output voltage of the PV cell, R_e is the equivalent resistance of aged PV cell, and V_{cu} is the output voltage of the cell-unit.

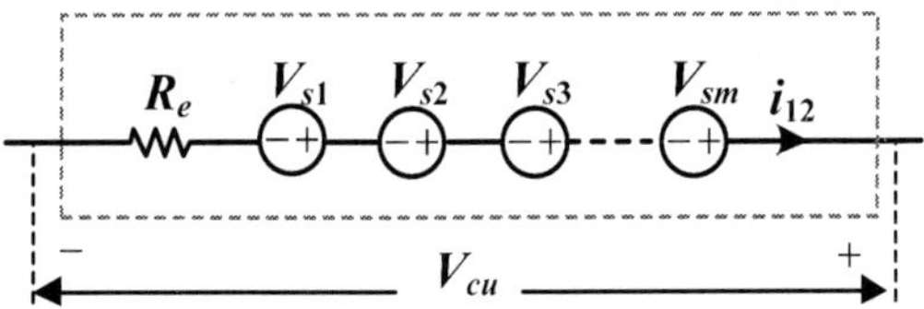

Figure 8. Equivalent circuit for the cell-unit at a minor fault.

As i_{12} increases, V_{cu} decreases to zero. The current switches from minor fault to heavy fault. In heavy fault, the cell-unit is bypassed by a diode, the corresponding terminal voltage is −0.5 V (i.e., diode voltage drop). In minor fault, the current passing the cell-unit i_j^q series-connected PV cells, the relationship between the output current i_{cu} and the terminal output voltage V_{cu}.depends on the operating points of the PV. To facilitate discussion on the three ranges, it is assumed that the magnitude of the short-circuit currents for m cells is

$$I_{sci_1} \leq I_{sci_2} \leq \cdots \leq I_{sci_m} \tag{15}$$

where i_{cell} is defined as the actual current passing the PV cells. When the current i_{cell} starts to increase from 0 to $Isci_1$, all the cells generate electricity. When exceeds $Isci_1$ but less than Ici_2, cell i_1 cannot generate electricity; it is either bypassed or turned into a resistor because of the bucket effect. As a result, the relationship of i_{cu} and V_{cu} is summarized as follows.

If $i_{cell} \leq Isci_1$, the unit-cell operates in normal condition:

$$i_{cu} = i_{cell} \leq I_{sci_1} \tag{16}$$

$$V_{cu} = mV_{cell} \tag{17}$$

where V_{cell} is equal to the voltage of every cell.

If $i_{cell} > Isci_1$, the cell-unit operates in heavy fault:

$$V_{cu} = -0.5V \tag{18}$$

$$i_{cell} = 0 \tag{19}$$

$$i_{cu} = i_{diode} \tag{20}$$

Where i_{diode} is the bypass current flowing through the diode.

The PV cells can work in minor fault if there exists an integer $k<m$. to satisfy the following conditions:

$$I_{sci_k} < i_{cell} \leq I_{sci_{k-1}}$$

$$(m-k)V_{cell} - i_{cell} \sum_{j=1}^{k} R_{ej} \geq 0 \tag{21}$$

When the cell-unit operates with a minor fault,

$$i_{cu} = i_{cell} \text{ and } V_{cu} = (m-k)V_{cell} - i_{cell} \sum_{j=1}^{k} R_{ej} \tag{22}$$

where R_{ej}.is the equivalent resistance of the jth cell.

Usually, normal operation and heavy fault are the steady-state operational conditions while minor fault is a short transitional range between the two and can often be ignored.

4.1.2. PV strings with nonuniformly faulty PV modules

A PV string consists of s V modules, with the terminal voltage V_{string}.and current i_{string}. Let the terminal voltage, current, and maximum current from the k PV module be $V_{module,k}$, $i_{module,k}$, and $i_{module,k}^{max}$ respectively. The following relationship can be established:

$$i_{string} = i_{module,1} = i_{module,2} = \cdots = i_{module,s} \tag{23}$$

$$V_{string} = V_{module,1} + V_{module,2} + \cdots + V_{module,s} \tag{24}$$

Similarly, the bucket effect indicates that the maximum current in the PV string is limited by the minimum $i_{module,k}^{max}$ f those non-bypassed modules. That is $i_{string} \leq i_{module,k}^{max}$, $1 \leq k \leq s$, and the kth module is not bypassed.

In practice, the cell-units within a PV module may be aged differently and thus have different maximum short-circuit currents. This case is called the "general non-uniform aging" in this chapter. A simpler case for nonuniformly aged PV modules is that all cell-units in the same PV module are aged uniformly so that the whole PV module can be characterized with a single maximum short-circuit current of any cell-unit. This is termed the simplified nonuniform aging in this chapter.

4.1.3. PV array with nonuniformly aged PV strings

A PV array consists of p parallel-connected PV strings; its terminal voltage and current are denoted by V_{array} and i_{array}, respectively. Let the terminal voltage and current for the jth PV string be $V_{array,j}$ and $i_{array,j}$, respectively. Therefore,

$$i_{array} = i_{array,1} + i_{array,2} + \cdots + i_{array,p} \tag{25}$$

$$V_{array} = V_{array,1} = V_{array,2} = \cdots = V_{array,p} \tag{26}$$

The power output from the PV array is the sum of p strings and is also limited by the bucket effect. That is, the maximum power output from the simplified nonuniform aging PV array can be written as $\sum_{j=1}^{p} \min\{P_{j,k}^{max} : 1 \leq k \leq s,$ and the (j, k) th module is un-bypassed $\}$ where $P_{j,k}^{max}$ is the maximum power output from the un-bypassed PV module at the position (j, k) (kth module in the jth string) of the PV array. $i_{module,j,k}$ is defined as the maximum short-circuit current in the (j,k). module; b and q as the number of PV modules in the jth string which generate electricity. Thus, $(s-q)$. PV modules are bypassed by diodes in the jth string. Then, the maximum power $P_{j,k}^{max}$ is calculated as

$$P_{j,k}^{max} = qV_{module}i_j^q \tag{27}$$

where V_{module}.is the MPP voltage supplied by a PV module, and i_j^q is the qth largest short-circuit current within the set $\{\ i_{module,\,j,1},\ i_{module,\,j,2},\ \cdots,\ i_{module,\,j,s}\ \}$. For a normal PV module consisting of three cell-units, $V_{module}=3V_{cu}$, and V_{cu} is the MPP voltage a PV cell-unit can provide.

5. PV array fault diagnosis and optimized operation

In this section, PV array fault diagnosis methods are reviewed and developed. Especially, PV fault diagnosis based on electrical sensors is proposed for small-scale PV array systems; thermal cameras are proposed for large-scale PV arrays; and optimized operation strategies under faulty conditions are also developed.

5.1. Existing fault diagnosis methods

Currently, thermal cameras, earth capacitance measurements (ECM), and time domain reflectometry (TDR) are three popular methods for PV fault diagnosis. In this section, the basic principles of those fault diagnosis are introduced [16, 20-34]. In the beginning of installation, ECM is an effective method to check the disconnection position of transmission line [23]. Due to the characteristics of PV modules, there is capacitance to ground existing in each module [35]. The capacitance to ground can be used to realize disconnection point location that is the theory of ECM. When the PV string has the disconnection point x_f, the disconnection module (x_f) from the positive node point to the fault point is calculated using the ratio of the earth capacitance values to the fault point and that of the whole line as shown in Fig. 9 and Eq. (28). Here, M is the total number of modules in the faulty string.

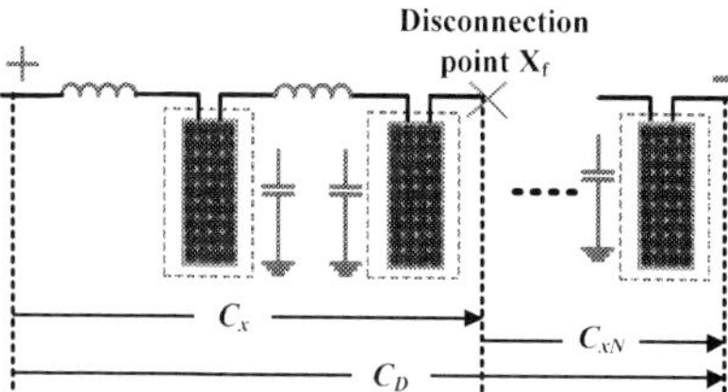

Figure 9. Fault diagnosis theory of ECM.

$$x_f = (C_x\,/\,C_D)M \tag{28}$$

For TDR, by injecting the signal into the transmission line, the signal will be distorted when mismatch is occurred [24]. Like a radar, the TDR method analyses the input signal and output signal, as shown in Fig. 10; the faulty point can be located and even aging condition can be estimated. But the illumination can influence the impedance of PV cell; therefore, TDR can only be used in the night.

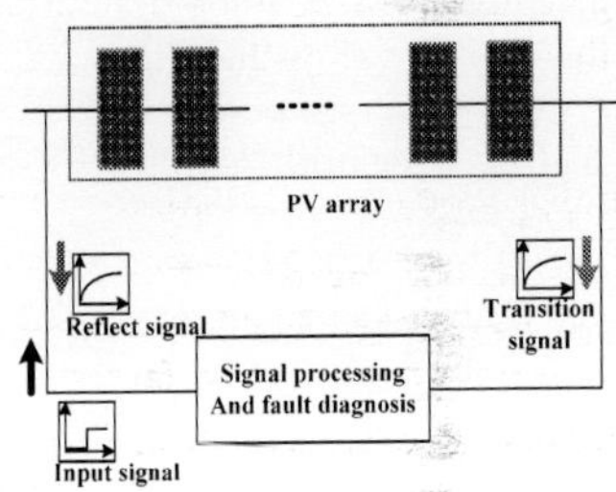

Figure 10. Theory of TDR.

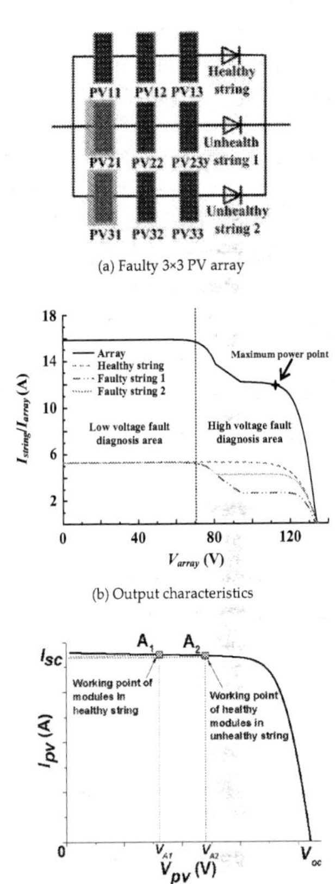

(a) Faulty 3×3 PV array

(b) Output characteristics

(c) Working points in the voltage fault diagnosis area

Figure 11. PV array under fault conditions.

The ECM can locate the disconnection of PV modules in the string while the TDR technology can predict the degradation of the PV array. Nonetheless, both ECM and TDR can only operate offline [23-24].

In practice, the online methods are highly demanded, which can take measurements while the tested device is in operation. To improve this, an automatic supervision and fault detection are proposed [25-26], based on power loss analysis. However, it requires surrounding environmental information and cannot identify the faulty module. An operating voltage window is then developed based on the PV string operation voltage and ambient temperature [27]. It can locate the open and short faults but still cannot identify the faulty module from the array. Currently, both offline and online fault diagnosis methods have been developed. Offline fault diagnosis cannot give real-time fault information that is the key factor for PV array optimization operation under fault condition. Currently, online fault diagnosis methods suffer from high costs or incapability of locating fault modules. A model-based reconfiguration algorithm is developed by [28] to realize the fault tolerance operation. But it needs a large number of electrical relays to reconfigure PV arrays. A similar technology, the *in situ* rearrangement strategy, can decrease the influence of shadow [29-32]. However, its success depends on three conditions: (i) a large number of relays need to be used; (ii) the health state of each PV module should be monitored; (iii) high-computing resource of the controller is required to calculate complex optimal arrangements. These increase the system cost and control complexity. An improved strategy is developed in [33], which not only needs the combination of power channels and relay matrix to combat the shadow influence but it also needs the healthy state of PV modules. A fingerprint curve of the PV array under shading conditions is proposed in [34] to find the key information (e.g., open-circuit and short-circuit points and MPP region) but it cannot locate the faulted modules.

5.2. PV fault diagnosis by using electrical sensors

It is clear that online fault diagnosis is important because: (i) it is the prerequisite for any array dynamical reconfiguration; (ii) it can provide crucial information for global MPPT; and (iii) it contains key state-of-health information useful for system maintenance. In this section, low-cost fault diagnosis for low-scale PV array and fault diagnosis strategy for large PV array fault diagnosis are proposed, respectively.

5.2.1. Fault diagnosis theory

When a PV array is faulted, the faulty module has lower illumination than healthy modules (e.g., a 3×3 array). Fig. 11(a) shows a multi-string faulty condition and Fig. 11(b) shows its output voltage–current characteristics. The output waveforms can be divided into two sections: the high-voltage fault diagnosis area and low-voltage fault diagnosis area (constant output current). In the latter area, the faulty module in the faulty string is shorted by bypass diodes where both healthy string and unhealthy string have the same current. PV string current sensors cannot distinguish the unhealthy string from healthy strings. Nevertheless, the healthy modules in the faulty string have a higher output voltage than the modules in the healthy string, illustrated as points A1 and A2 in Fig. 11(c). The voltage difference between the healthy module in the unhealthy string and module in the healthy string can be employed to locate the faulty module.

For a p row s column array, assume that there is x faulted modules in the unhealthy string. V_{A1} and V_{A2} can be expressed as Eqs. (29) and (30):

$$V_{A1} = \frac{V_{array}}{s} \tag{29}$$

$$V_{A2} = \frac{V_{array}}{s-x} \tag{30}$$

In Fig. 11(a), V_{A1} is the corresponding healthy string module voltage, such as No. PV11; V_{A2} is the normal module voltage in the unhealthy string, such as No. PV22.

5.2.2. Optimized sensor placement strategy

In order to achieve the PV array fault diagnosis, the reading of PV module voltage is needed. Due to the large number of PV modules employed, a large number of voltage sensors are also needed in the first instance.

There are three basic sensor placement methods, as shown in Fig. 12. In the PV array, in method 1, each module terminal voltage is measured by a voltage sensor; the total number of sensors is $p{\times}s$. In method 2, each voltage sensor measures the voltage between two nodes in the same column of adjacent strings; and $(p\text{-}1){\times}(s\text{-}1)$ voltage sensors are needed. In method 3, the electric potential difference of adjacent modules is measured; the corresponding number of sensors is $p{\times}(s\text{-}2)$. The large number of voltage sensors may increase system capital cost and information-processing burden. Therefore, the voltage placement method needs to be optimized.

Fig. 13 shows an equivalent PV matrix where a PV module is shown as a dot; the connection line of the adjacent module is represented by a node. The proposed voltage placement strategy is developed by the following steps:

i.   All the nodes should be covered by voltage sensors.

ii.   A sensor can only connect one node in a string.

iii.   Voltage sensor nodes cover different isoelectric points from different strings.

iv.   If p or s is an even number, each node is connected to and only to one sensor. If both p and s are odd, there is one and only one node to be connected to two different sensors, while each of the remaining nodes is connected to one sensor.

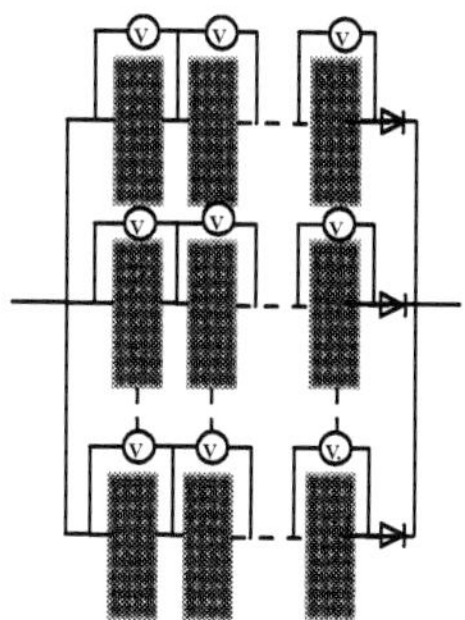

(a) Method 1

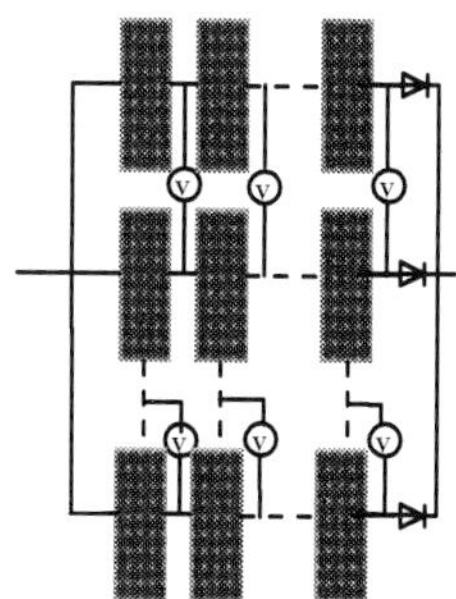

(b) Method 2

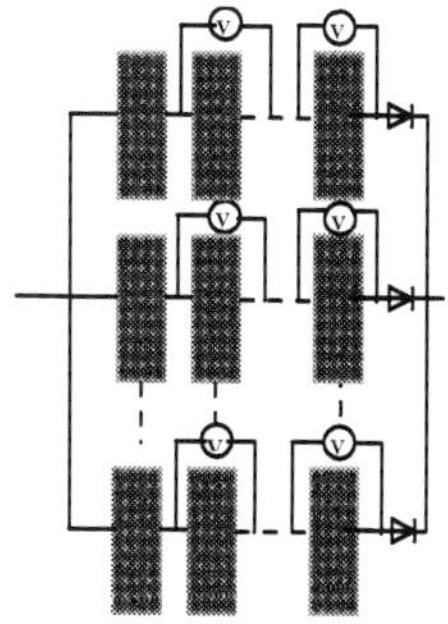

(c) Method 3

Figure 12. Sensor placement methods.

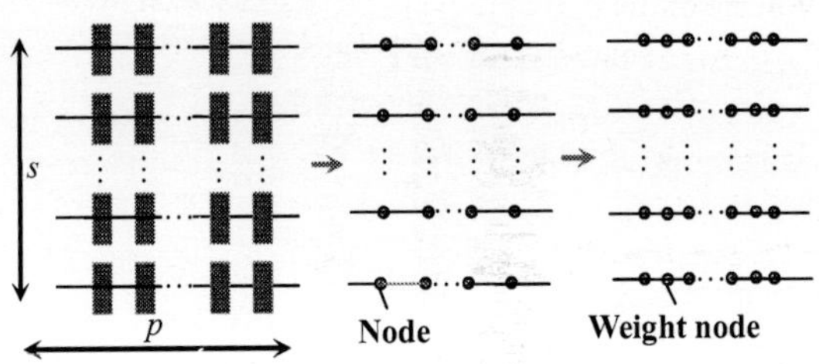

Figure 13. Equivalent matrix.

Fig. 14. presents an example of the 3×3 PV array. According to the proposed sensor placement strategy, only three voltage sensors are needed.

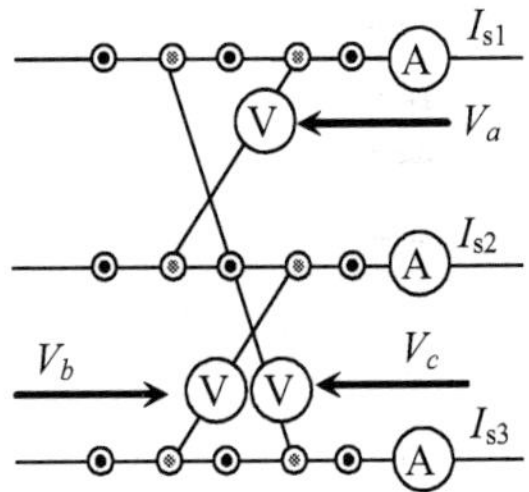

Figure 14. Simplified voltage sensor placement method for a 3×3 PV array.

A general principle is that the minimum number of sensors used to detect all possible faults should be $[p×(s-1)/2]$ where the notation $[a]$ is the ceiling function representing the smallest integer, which is not less than a. When a node is not connected to a sensor, the two adjacent PV modules of this node cannot be discriminated once a fault occurs at one of the two modules. The total number of nodes is equal to $[p×(s-1)/2]$. This is summarized in Table 3. It is clear that the proposed method can decrease the number of voltage sensors used compared with the other three methods.

Method	1	2	3	Proposed
No.	$p× s$	$(p-1)×(s-1)$	$p×(s-2)$	$p×(s-1)/2$

Table 3. Number of voltage sensors used by different methods

5.2.3. Mathematical model of proposed sensor placement strategy

The variable a_{ij} is defined as the healthy state of the PV module sitting at the i-th string and j-th module (denoted by (i, j)).in the $p×s$. array. If this module is healthy, then a_{ij}=1, otherwise

$a_{ij}=0$. The terminal voltage of the (i, j) module is denoted by u_{ij}, and the reading of a voltage sensor connecting the (i, j) module and another module sitting at the (r, k) position is denoted by $R_{i,j,r,k}$. Without loss of generality, consider the case that each string has at least one healthy module. According to the voltage division law and the fact that the number of healthy modules in the i-th string equalizes $a_{i1} + a_{i2} + \dots + a_{is}$, the terminal voltage u_{ij} of the (i, j) module is equal to a fraction of U_{array} and this fraction is 0 if $a_{ij}=0.$, and is $1/(a_{i1} + a_{i2} + \dots + a_{is})$ if $a_{ij}=1$ That is,

$$u_{ij} = \frac{a_{ij} U_{array}}{a_{i1} + a_{i2} + \dots + a_{is}} \tag{31}$$

Note that the total output voltage of the modules $(i, 1)$, $(i, 2)$,..., and (i, j) is the sum of the terminal voltage of each of these j modules (i.e.,). Similarly the total output voltage of the modules $(r, 1)$, $(r, 2)$,..., and (r, k) equalizes. Therefore, the voltage reading of connecting the (i, j) module and the (r, k) module is calculated as:

$$\begin{aligned}
R_{i,j,r,k} &= (u_{i1} + u_{i2} + \dots + u_{ij}) - (u_{r1} + u_{r2} + \dots + u_{rk}) \\
&= \frac{(a_{i1} + a_{i2} + \dots + a_{ij})U_{array}}{a_{i1} + a_{i2} + \dots + a_{is}} - \frac{(a_{r1} + a_{r2} + \dots + a_{rk})U_{array}}{a_{r1} + a_{r2} + \dots + a_{rs}}
\end{aligned} \tag{32}$$

These provide a solution to the unknowns a_{ij} When the working point of a PV string moves to the high-voltage area, the voltage output of the healthy modules increases until reaching the open-circuit output voltage ; faulted modules in the string will equally divide the remaining voltage Therefore, the following relations hold for a string including both healthy and unhealthy modules:

$$\begin{aligned}
u_{ij} &= a_{ij}U_{oc} + \frac{(1 - a_{ij})(U_{array} - (a_{i1} + a_{i2} + \dots + a_{is})U_{oc})}{s - (a_{i1} + a_{i2} + \dots + a_{is})} \\
&= \frac{(a_{ij}s - (a_{i1} + a_{i2} + \dots + a_{is}))U_{oc}}{s - (a_{i1} + a_{i2} + \dots + a_{is})} + \frac{(1 - a_{ij})U_{array}}{s - (a_{i1} + a_{i2} + \dots + a_{is})}
\end{aligned} \tag{33}$$

$$\begin{aligned}
R_{i,j,r,k} &= (u_{i1} + u_{i2} + \dots + u_{ij}) - (u_{r1} + u_{r2} + \dots + u_{rk}) \\
&= \frac{(s\sum_{l=1}^{j} a_{il} - j\sum_{l=1}^{s} a_{il})U_{oc}}{s - \sum_{l=1}^{s} a_{il}} + \frac{(j - \sum_{l=1}^{j} a_{il})U_{array}}{\sum_{l=1}^{s} a_{rl}} \\
&\quad - \frac{(s\sum_{l=1}^{k} a_{rl} - k\sum_{l=1}^{s} a_{rl})U_{oc}}{s - \sum_{l=1}^{s} a_{rl}} - \frac{(k - \sum_{l=1}^{k} a_{kl})U_{array}}{\sum_{l=1}^{s} a_{rl}}
\end{aligned} \tag{34}$$

The reading at the high-voltage status provides extra equations to solve the fault status variable That is, the system of Eqs (33) and (34) can be applied to determine the fault status variable.

It is observed that the placement of voltage sensors has two features. First, it is noted that the fault diagnosis problem is always solvable by placing voltage sensors only. This is because that if each PV module is installed with a voltage sensor, a PV fault can be detected from the terminal. Second, if there are always some methods to find the optimal sensor placement for any 2×s and 3×s PV arrays with the least number of sensors $(s-1)$ and $3 \times (s-1)/2$, respectively, these methods do not need the existence of any healthy strings. Then, there is a way to design the optimal sensor placement for any general p×s array with $p \times (s-1)/2$.sensors. The reason is explained as follows. When p is an even number, the p×s array can be divided into $\frac{p}{2}$ blocks of 2×s arrays. For each block of the 2×s array, it needs to apply the existing optimal sensor placement method to achieve the optimal $\frac{p}{2}$×s sensors. If p is odd, the p×s. array consists of one 3×s array and blocks of 2×s.arrays. It needs to apply the existing sensor placement method for these small blocks where the number of sensors is equal to $[3 \times (s-1)/2] + \frac{p-3}{2}(s-1)$.. By considering both even and odd cases, it can be found that:

$$\left[3 \times (s\text{-}1)/2\right] + \frac{p-3}{2}(s-1) = \left[p \times (s\text{-}1)/2\right] \tag{35}$$

Therefore, the optimal number of sensors can be obtained.

5.2.4. Locating the faulted PV module in uniform faulty PV modules

By the reading of current sensor in high-output voltage area, the unhealthy string can be located. After locating the unhealthy string, the next step is to find the faulty PV module. In the low-voltage fault diagnosis area, the faulty modules are shorted; the corresponding fault diagnosis eigenvalue of the mono-string faulty module is presented in Table 4, where the fully faulty module indicates that all cell-units in the module are faulty. No. 7 is the extreme case that all the modules in this string are faulty. Even though the PV array works in the low-voltage area, the modules are open-circuited when all modules are faulty. Table 5 shows the multi-string characteristic values, from which the faulty module can be identified easily.

PV31–PV33	V_a	V_b	V_c
100	$V_{array}/3$	$2\,V_{array}/3$	$V_{array}/6$
10	$V_{array}/3$	$V_{array}/6$	$V_{array}/6$
1	$V_{array}/3$	$V_{array}/6$	$2\,V_{array}/3$
110	$V_{array}/3$	$2V_{array}/3$	$-\,V_{array}/3$
11	$V_{array}/3$	$-\,V_{array}/3$	$2\,V_{array}/3$
101	$V_{array}/3$	$2\,V_{array}/3$	$2\,V_{array}/3$
111	$V_{array}/3$	$2\,V_{array}/3\text{-}V_{oc}$	$2V_{oc}-V_{array}/3$
0	$V_{array}/3$	$V_{array}/3$	$V_{array}/3$

Table 4. Characteristic values for the mono-string fully faulted modules (0: healthy, 1: faulty)

PV11–PV13/PV21–PV23	Va	Vb	Vc
100/100	$V_{array}/2$	$V_{array}/6$	$2V_{array}/3$
010/100	$V_{array}/2$	$V_{array}/6$	$V_{array}/6$
001/100	V_{array}	$V_{array}/6$	$V_{array}/6$
100/010	$V_{array}/6$	$V_{array}/6$	$2V_{array}/3$
010/010	$V_{array}/6$	$V_{array}/6$	$V_{array}/6$
001/010	$V_{array}/2$	$V_{array}/6$	$V_{array}/6$
100/001	0	$2 V_{array}/3$	$2V_{array}/3$
010/001	0	$2 V_{array}/3$	$V_{array}/6$

Table 5. Characteristic values for multi-string fully faulted PV modules

5.2.5. Locating faulty PV module in uniform faulty PV module

Both Tables 4 and 5 deal with the fully faulted module where all cell-units are faulted while a partially faulted module widely occurs where some cell-units are faulted. Usually, partial shadow is also accrued in one PV module. Due to the cell-unit structure of PV modules, even when only one cell is faulty (0 W/m^2), the whole cell-unit output power will decrease dramatically. Fig. 15(a) presents experimental results of the faulty cell-unit that is composed of 24 PV cells with only one faulty PV cell. The faulty cell-unit maximum output power is 4.75 W while that of healthy cell-unit is 48 W, making a loss of 90 %. From Fig. 15, we conclude that PV cell-unit cannot work when a cell fault occurs.

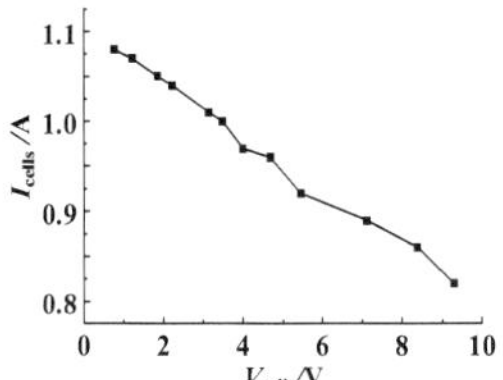

(a) PV cell-unit output under fault conditions

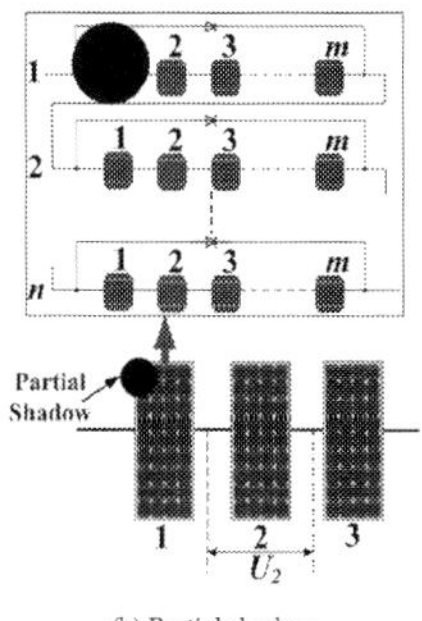

(b) Partial shadow

Figure 15. PV string under partial shadow conditions.

Therefore, when a PV module endures partial shadowing, its terminal output voltage is lower than the healthy module and higher than zero. In Fig. 15(b), PV module No. 1 loses one of the cell-unit, the PV module output voltage is reduced to $\frac{n-1}{n}$ of the healthy module output voltage.

Therefore, PV string fault diagnosis can be achieved by measuring the PV module voltage, which changes with the string working point. When the string works in the high-current output area, the faulty PV module can be located because its output voltage is zero (unified shading) or lower than the healthy module (partial shading). Table 6 gives the typical eigenvalue under nonuniform PV module faulty condition.

PV11–PV13	V_a	V_b	V_c	Comparison
100	$V_{array}/6<V_a<V_{array}/3$	$V_{array}/3$	$V_{array}/3<V_c<2V_{array}/3$	—
10	$V_{array}/6<V_a<V_{array}/3$	$V_{array}/3$	$V_{array}/6<V_c<V_{array}/3$	—
1	$V_{array}/3<V_a<2V_{array}/3$	$V_{array}/3$	$V_{array}/6<V_c<V_{array}/3$	—
110	$V_a<V_{array}/3$	$V_{array}/3$	$V_{array}/3<V_c<2V_{array}/3$	$2V_a<V_c<V_{array}$
11	$V_{array}/3<V_a<2V_{array}/3$	$V_{array}/3$	$V_c<V_{array}/3$	$V_a<2V_c<V_{array}$
101	$V_a<V_{array}/3$	$V_{array}/3$	$V_{array}/3<V_c<2V_{array}/3$	$V_a<2V_c''/>V_{array}$

Table 6. Characteristic values for the mono-string partially faulted modules

6. Experimental verification

A 3×3 PV array is built to verify the proposed fault diagnosis technique. The PV modules are the same as used for simulation, and the environment illumination is recorded by TS1333R.

6.1. Sensor-based PV fault diagnosis

In this experiment, typical fault scenarios are studied and the sensor readings are compared with eigenvalue in the high-voltage and low-voltage fault diagnosis areas to check the effectiveness of the proposed fault diagnosis technique.

Fig. 16 shows the mono-string, mono-module fault diagnosis. In the fault scenario 1 (see Fig. 16(a)), the illumination is 550 W/m² and temperature is 15℃. The P33 PV module is cast by shadow manually to emulate a fault. Fig. 16(b) shows the current–voltage output characteristics of faulty PV array. Due to the fault on module P33, string 3 cannot generate electricity in the output voltage range 82–120 V. Fig. 16(c) presents the sensor output curves. In the low-voltage area (10–70V), the sensor is V_a and the output voltage is $V_{array}/3$. This is a normal output voltage and the corresponding strings are healthy. That is, strings 1 and 2 connected by this sensor are healthy, which coincides with fault scenario 1 in Fig. 16(a). Fig. 16(e) illustrates the high-voltage and low-voltage fault diagnosis areas. In the low-voltage area, the reference eigenvalue is $V_{array}/6$; and in the high-voltage diagnosis area, the reference eigenvalue is $2V_{array}/3-V_{oc}$. The fact that the sensor V_b output is close to the reference value also verifies the proposed diagnosis method. The reference eigenvalue of V_c is $2V_{array}/3$; and the corresponding V_c sensor output also agrees with the reference eigenvalues. There is a slight deviation between V_a, V_b, and V_c and their reference values. This is caused by the diode voltage drop and the minor product irregularity between PV modules. From the sensor output results and information in Table 5, the fault type is classified as "001." The faulty module is P33 that also agrees with fault scenario 1 (Fig. 16(a)).

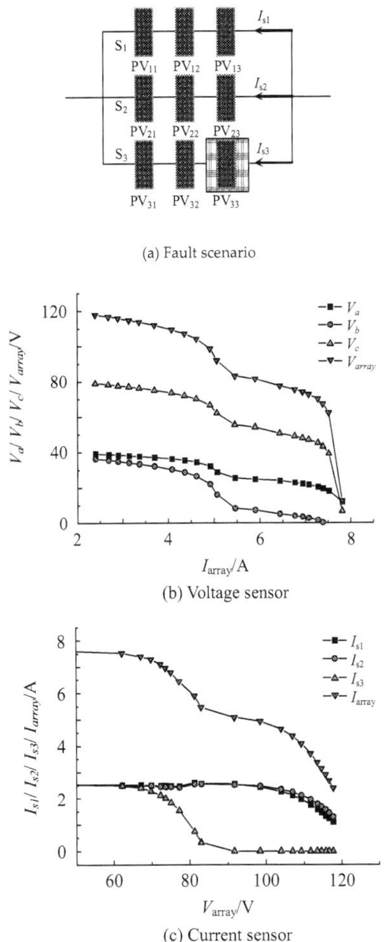

(a) Fault scenario

(b) Voltage sensor

(c) Current sensor

Figure 16. Mono-string mono-module fault diagnosis.

Fig. 17 shows the multi-string mono-module fault diagnosis. In fault scenario 2, the illumination is 580 W/m² with the temperature of 25℃. The P11 module in string 1 and P33 module in string 3 are cast by shadow manually to simulate faulty conditions. Fig. 17(b) presents the current–voltage output characteristics of the faulty PV array. Due to a fault that occurred in modules P11 and P33, strings 1 and 3 cannot generate electricity over the output voltage range 82–120 V. In the low-voltage area, the sensors a and b have the same output ($V_a=V_b=V_{array}/6$), as illustrated in Fig. 17(c). The voltage sensors a and b also satisfy the rule for locating healthy strings. Therefore, string 2 is diagnosed as being healthy, which coincides with the fault scenario in Fig. 17(a). Fig. 17(d) shows fault diagnosis progress. The faulty modules identified are P11 and P33.

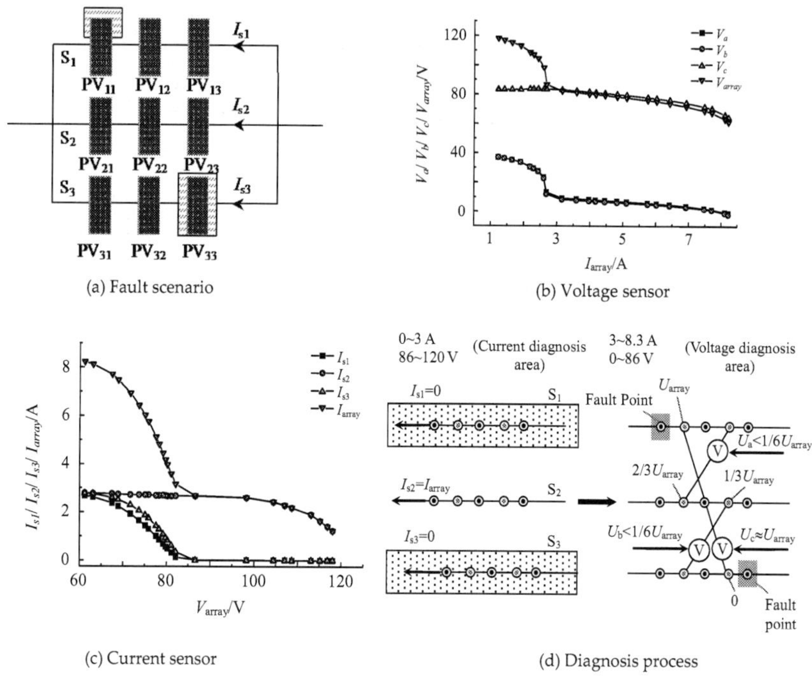

(a) Fault scenario

(b) Voltage sensor

(c) Current sensor

(d) Diagnosis process

Figure 17. Multibranches mono-module fault diagnosis.

In fault scenario 3 of Fig. 18(a), the illumination is 610 W/m² and the temperature is 30℃. P32 and P33 in string 3 are cast by partial shadow and full shadow, respectively. Fig. 18(b) demonstrates their output characteristics. String 3 can only generate electricity over the voltage range of 0–60 V. As presented in Fig. 18(a), $V_a=V_{array}/3$ in the whole output voltage range, indicating that strings 1 and 2 are both healthy. In the low-voltage area, V_c matches the reference value $2V_{array}/3$, verifying that P33 is faulty. In the high-voltage area, $V_c=2V_{oc}-V_{array}/3$. Therefore, either P31 or P32 is faulty. Fig. 18(d) presents the fault diagnosis progress.

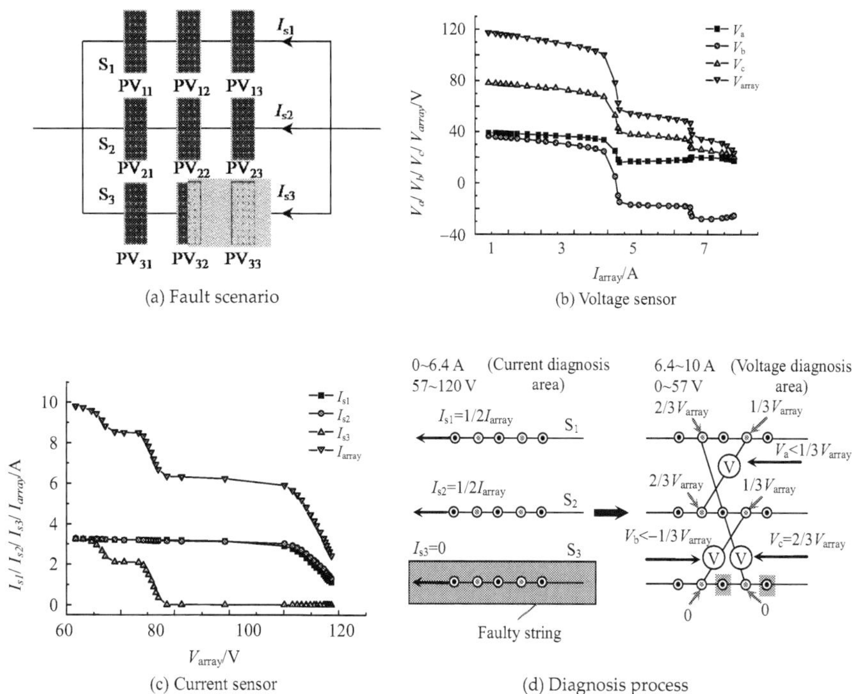

Figure 18. Mono-branch multi-module fault diagnosis.

6.2. Thermal camera-based PV fault diagnosis

Another fault diagnosis method is based on thermal images captured by thermal cameras, as shown in Fig. 19. The camera specifications are given in Table 7. Photovoltaic modules convert part of incident solar energy into electrical energy for commercial applications, with the rest being transferred to heat energy. The modeling of PV modules plays an important role in the fault diagnosis of a PV array. The objective of this section is to develop a parameter-based model of a PV module to interpret the thermal image.

In order to validate the proposed model, a scaled experimental setup was constructed as shown in Fig. 20(a). The mini-PV module used in the experiment and its respective datasheet are shown in Table 8. The type of the solar power meter used was TM-207.

Experimental tests are conducted at different load conditions, including MPP load, heavy load, light load, and partial-shading conditions. The MPP load (35 Ω) module was used to simulate the module in a healthy string working at MPP, and the 50 Ω load was used to simulate the module in fault string working at non-MPP. The sun illumination was measured at 560 W/m²

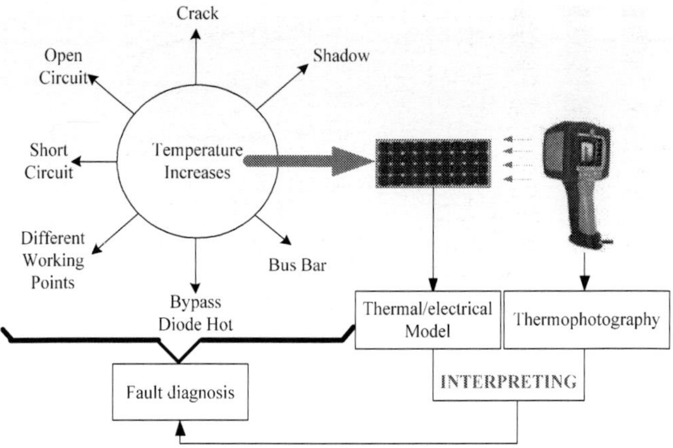

Figure 19. Schematic diagram of fault detection in PV systems.

Parameter	Value
Company	FLIR
Type	FLIR *i7*
IR resolution	140×140 pixels
Thermal sensitivity/noise equivalent temperature difference (NETD)	<0.1°C
Minimum focus distance	0.6 m
Spatial resolution (instantaneous field of view (IFOV)	3.7 mrad
Image frequency	9 Hz
Focus	Focus free

Table 7. Specifications of thermal camera

using the TM-207, and the voltage of MPP load and 50 Ω were 3.96 V and 4.37 V, respectively. The experiments were carried out at wind-speed conditions of less than 5 m/s. Fig. 20(b) is the thermography image obtained from thermal camera. Fig. 20(c) presents the Matlab processing results. The thermography produces colors by combining green, red, and blue light in varying intensities. The relationship between the constituent amounts of red, green, and blue (RGB) and the resulting color is not sensitivity. HSV method that stands for *hue, saturation,* and *value,* which is the common cylindrical-coordinate representation of points in an RGB color model, is introduced to process the thermography. By processing the image, the temperature difference can be clearly shown as illustrated in Fig. 20(c). Fig. 20(c) is the processing picture of Fig. 20(b).

(a) Experimental setup

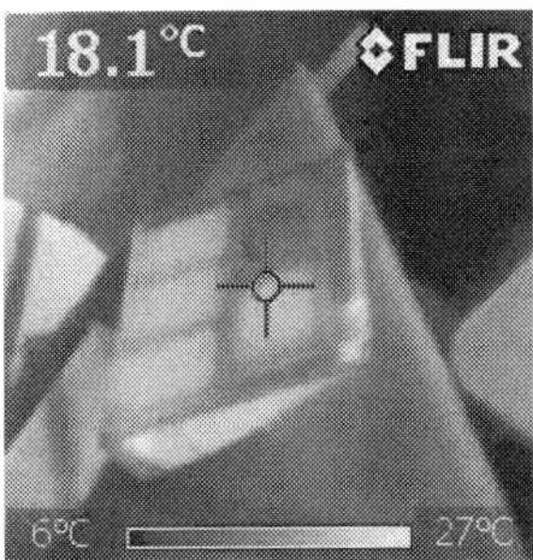

(b) Thermography

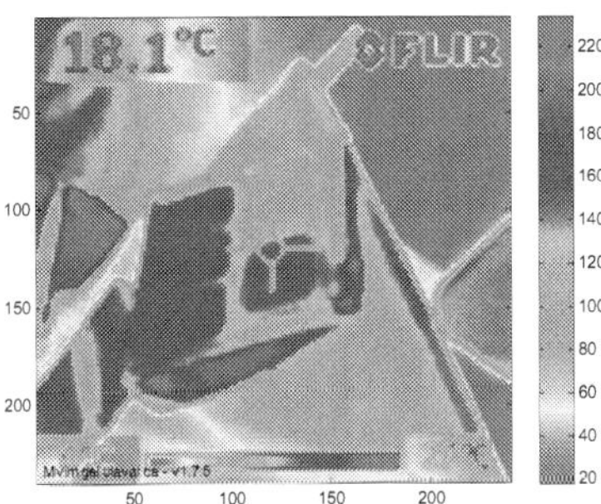

(c) Pretreatment of the thermography

Figure 20. Thermographical results at different load conditions.

Parameters	Value
Open voltage/V	4.8
Short current/A	0.23
MPP current/A	0.21
MPP voltage/V	3.85
Current temperature coefficient/°C	0.06%/K
Voltage temperature coefficient/°C	−0.36%/K
Power temperature coefficient /°C	−0.45%/K
Maximum power /W	0.8

Table 8. PV Module Parameter

7. Conclusion

This chapter has reviewed the fault mechanisms and model-based diagnosis techniques for PV systems. It can be concluded that: (i) the PV module electrical model and energy balance are coupled to establish a parameter-based model, (ii) the key parameters (S, U_{pv}, and T_a) of the PV model are calculated from two working points of the PV module and corresponding temperature, and (iii) fault diagnosis is realized by model-based methods. Notwithstanding the fact that a parameter-based model has been established and validated, other factors, such as wind and dust, which are not considered in this chapter, may contribute to the PV fault in reality. The vortex and dust distribution caused by wind can introduce non-uniform temperature distribution in the thermography. It is challenging to accurately estimate the temperature of PV modules. The pattern and features of historic information will be useful to for nonuniform illustration of PV systems.

Author details

Yihua Hu[1] and Wenping Cao[2*]

*Address all correspondence to: w.p.cao@aston.ac.uk

1 University of Strathclyde, Glasgow, UK

2 Aston University, Birmingham, UK

References

[1] Maki, A., Valkealahti, A., (2013). "Effect of photovoltaic generator components on the number of MPPs under partial shading conditions," *IEEE Transactions on Energy Conversion,* vol. 28, no. 4, pp. 1008-1017.

[2] El-Dein, M.Z.S., Kazerani, M., Salama, M.M.A., (2013). "Optimal photovoltaic array reconfiguration to reduce partial shading losses," *IEEE Transactions on Sustainable Energy,* vol. 4, no. 1, pp. 145-153.

[3] Mahmoud, Y.A., Xiao, W., Zeineldin, H.H., (2013). "A parameterization approach for enhancing PV model accuracy," *IEEE Transactions on Industrial Electronics,* vol. 60, no. 12, pp. 5708-5716.

[4] Alajmi, B.N., Ahmed, K.H., Finney, S.J., Williams, B.W., (2013). "A maximum power point tracking technique for partially shaded photovoltaic systems in microgrids," *IEEE Transactions on Industrial Electronics,* vol. 60, no. 4, pp. 1596-1606.

[5] Sullivan, C.R., Awerbuch, J.J., Latham, A.M., (2013). "Decrease in photovoltaic power output from ripple: simple general calculation and the effect of partial shading," *IEEE Transactions on Power Electronics,* vol. 28, no. 2, pp. 740-747.

[6] Patel, H., Agarwal, V., (2008). "Matlab-based modeling to study the effects of partial shading on PV array characteristics," *IEEE Transactions on Energy Conversion,* vol. 23, no. 1, pp. 302-310.

[7] Hu, Y., Cao, W., Finney, S., (2014). "Identifying PV module mismatch faults by a thermography-based temperature distribution analysis," *IEEE Transactions on Device and Materials Reliability,* vol. 14, no. 4, pp. 951-960.

[8] Hu, Y., Gao, B., Tian, G.Y., Song, X., Li, K., He, X., (2013). "Photovoltaic fault detection using a parameter based model," *Solar Energy,* vol. 96, pp. 96-102.

[9] Hu, Y., Cao, W., Ji, B., Song, X., (2015). "New multi-stage hysteresis control of DC-DC converters for grid-connected photovoltaic systems," *Renewable Energy,* vol. 74, pp. 247-254.

[10] Hu, Y., Cao, W., Wu, J., Ji, B., Holliday, D., (2014b). "Thermography-based virtual MPPT scheme for improving PV energy efficiency at partial shading conditions," *IEEE Transactions on Power Electronics,* vol. 29, no. 11, pp. 5667-5672.

[11] Tonui, J.K., Tripanagnostopoulos, Y., (2007). "Air-cooled PV/T solar collectors with low cost performance improvements," *Solar Energy,* vol. 81, no. 4, pp. 498-511.

[12] Ishaque, K., Salam, Z., Syafaruddin, A., (2011). "A comprehensive MATLAB Simulink PV system simulator with partial shading capability based on two-diode model," *Solar Energy,* vol. 85, no. 9, pp. 2217-2227.

[13] Munoz, M.A., Alonso-García, M.C., Vela, N., Chenlo, F., (2011). "Early degradation of silicon PV modules and guaranty conditions," *Solar Energy*, vol. 85, no. 9, pp. 2264-2274.

[14] Kaplani, E., (2012). "Detection of degradation effects in field-aged c-si solar cells through IR thermography and digital image processing," *International Journal of Photoenergy*, article ID: 396792, vol. 2012, pp. 11.

[15] Acciani, G., Simione, G.B., Vergura, S., (2010). "Thermographic analysis of photovoltaic panels," *International Conference on Renewable Energies and Power Quality (ICREPQ'10)*, Granada (Spain).

[16] Buerhopa, C.l., Schlegela, D., Niessb, M., Vodermayerb, C., Weißmanna, R., Brabeca, C.J., (2012). "Reliability of IR-imaging of PV-plants under operating conditions," *Solar Energy Materials and Solar Cells*, vol. 107, pp. 154-164.

[17] Parinya, P., Wiengmoon, B., Chenvidhya D., Jivacate, C., (2007). "Comparative study of solar cells characteristics by temperature measurement," *22nd European Photovoltaic Solar Energy Conference*, Milan, Italy.

[18] Krenzinger, A., Andrade, A.C., (2007). "Accurate outdoor glass thermographic thermometry applied to solar energy devices," *Solar Energy*, vol. 81, no. 8, pp. 1025-1034.

[19] Meyer, E.L., Ernest van Dyk, E., (2004). "Assessing the reliability and degradation of photovoltaic module performance parameters," *IEEE Transactions on Reliability*, vol. 53, no. 1, pp. 83-92.

[20] Simon, M., Meyer, E.L., (2010). "Detection and analysis of hot-spot formation in solar cells," *Solar Energy Materials and Solar Cells*, vol. 94, no. 2, pp. 106-113.

[21] Kurnik, J., Jankovec, M., Brecl, K., Topic, M., (2011). "Outdoor testing of PV module temperature and performance under different mounting and operational conditions," *Solar Energy Materials & Solar Cells*, vol. 95, no. 1, pp. 373-376.

[22] Zou, Z., Hu, Y., Gao, B., Woo, W.L., Zhao, X., (2014). "Study of the gradual change phenomenon in the infrared image when monitoring photovoltaic array," *Journal of Applied Physics*, vol. 115, no. 4, pp. 043522-1-11.

[23] Takashima, T., Yamaguchi, J., Otani, K., Oozeki, T., Kato, K., (2009). "Experimental studies of fault location in PV module strings," *Solar Energy Materials and Solar Cells*, vol. 93, nos. 6-7, pp. 1079-1082.

[24] Kumar, R.A., Suresh, M.S., Nagaraju, J., (2001). "Measurement of AC parameters of gallium arsenide (GaAs/Ge) solar cell by impedance spectroscopy," *IEEE Transactions on Electron Devices*, vol. 48, no. 9, pp. 2177-2179.

[25] Chouder, A., Silvestre, S., (2010). "Automatic supervision and fault detection of PV systems based on power losses analysis," *Energy Conversion and Management*, vol. 51, no. 10, pp. 1929-1937.

[26] Silvestre, S., Chouder, A., Karatepe, E., (2013). "Automatic fault detection in grid con-nected PV systems," *Solar Energy*, vol. 94, pp. 119-127.

[27] Gokmen, N., Karatepe, E., Silvestre, S., Celik, B., Ortega, P., (2013). "An efficient fault diagnosis method for PV systems based on operating voltage-window," *Energy Conversion and Management*, vol. 73, pp. 350-360.

[28] Lin, X., Wang, Y,. Zhu, D., Chang, N., Pedram, M., (2012). "Online fault detection and tolerance for photovoltaic energy harvesting systems," *The 2012 IEEE/ACM International Conference on Computer-Aided Design (ICCAD)*, San Jose, USA, pp. 1-6.

[29] Nguyen, D., Lehman, B., (2008). "An adaptive solar photovoltaic array using model-based reconfiguration algorithm," *IEEE Transactions on Industrial Electronics*, vol. 55, no. 7, pp. 2644-2654.

[30] Storey, J.P., Wilson, P.R., Bagnall, D., (2013). "Improved optimization strategy for ir-radiance equalization in dynamic photovoltaic arrays," *IEEE Transactions on Power Electronics*, vol. 28, no. 6, pp. 2946-2956.

[31] Velasco-Quesada, G., Guinjoan-Gispert, F., Pique-Lopez, R., Roman-Lumbreras, M., Conesa-Roca, A., (2009). "Electrical PV array reconfiguration strategy for energy extraction improvement in grid-connected PV systems," *IEEE Transactions on Industrial Electronics*, vol. 56, no. 11, pp. 4319-4331.

[32] Wang, Y., Lin, X., Kim, Y., Chang, N., Pedram, M., (2014). "Architecture and control algorithms for combating partial shading in photovoltaic systems," *IEEE Transactions on Computer-Aided Design of Integrated Circuits and Systems*, vol. 33, no. 6, pp. 917-929.

[33] Jonathan, S., Peter, R.W., Darren, B., (2014). "The optimized-string dynamic photovoltaic array," *IEEE Transactions on Power Electronics*, vol. 29, no. 4, pp. 1768-1776.

[34] Carotenuto, P.L., Manganiello, P., Petrone, G., Spagnuolo, G., (2014). "Online recording a PV module fingerprint," *IEEE Journal of Photovoltaics*, vol. 4, no. 2, pp. 659- 668.

[35] Lopez, O., Freijedo, F.D., Yepes, A.G., Fernandez-Comesana, P., Malvar, J., Teodorescu, R., Doval-Gandoy, J., (2010). "Eliminating ground current in a transformerless photovoltaic application," *IEEE Transactions on Energy Conversion*, vol. 25, no. 1, pp. 140-147.

Biomass and Other Sources

Chapter 6

Genset Optimization for Biomass Syngas Operation

Horizon Walker Gitano-Briggs and Koay Loke Kean

Additional information is available at the end of the chapter

http://dx.doi.org/10.5772/62727

Abstract

Although biomass is underrepresented in current methods for power generation, it has great potential to help meet the growing need for clean energy. This chapter details the modification of a gasoline-powered two-stroke genset for operation on syngas from a woodchip-powered gasifier. Generator and engine modifications along with a flexible air/fuel control system are described. Results from genset operation indicate a sustainable power output of 360 W with a biomass consumption rate of approximately 6 kg/hour. Optimum power production was achieved at an air/fuel ratio close to 1. After several hours of operation the engine was disassembled and inspected, revealing significant deposits on the piston and crank case parts, indicating that the engine would require weekly maintenance under such operating conditions.

Keywords: Biomass, Renewable, Gasifier, Syngas, Genset

1. Introduction

Today biomass is a neglected source of renewable energy, as most current efforts focus instead on sources such as wind, hydro, solar and even wave power. However, before the widespread use of fossil fuels (notably coal during the industrial revolution), biomass was the primary form of non-animal power for centuries. Burning biomass cooked food, heated dwellings in cold climates, and was involved in the transition out of the Stone Age as humans began smelting metals. Since the discovery of easily exploitable, apparently abundant fossil fuels, the relative importance of biomass has steadily fallen. Additionally with the increasing pressure of an exponentially growing human population, otherwise renewable power sources have often became non-renewable through over-exploitation. Consider that the hills around Katmandu Nepal have been completely deforested by the search for firewood. This reminder

emphasizes that *biomass* is not necessarily *sustainable*, and therefore can be considered non-renewable, depending on how it is exploited.

Currently, most consumed energy originates from the combustion of fossil fuels, such as petroleum, coal, and natural gas [1]. These fossil fuels formed over a period of millions of years within the Earth, and once the reserves of fossil fuels are depleted, civilization is likely to be literally out in the cold. Scientists estimate that the reserve of fossil fuels will be depleted within the next 50-120 years. Fortunately there are other viable fuel sources that can replace fossil fuels [2].

Accordingly, the use of renewable resources for energy production is becoming a strategic focus of many governmental institutions worldwide. Efforts are being made to increase the portion of renewable energies within many national energy supply structures [3]. A large number of renewable resources, including solar, wind, hydropower, geothermal, and biomass, are being examined for suitability as sources of energy for electrical power generation [4]. These energy sources are called "renewable" because it is assumed that they will be available as long as the planet remains habitable. Renewable energy offers the advantages that, in many cases, it can be produced locally and on a relatively small scale. These aspects provide a feasible solution for countries that do not have sufficient fossil fuel resources.

In most developed countries, the economy depends heavily on fossil fuels. For instance, the majority of road vehicles are powered by gasoline and diesel and most power plants use coal or other fossil fuels to produce electricity. Changing these systems can be expensive and time-consuming. Thus, recent research into renewable power has focused on developing new fuels for transportation and electricity supply [5].

One of the most promising renewable energy sources for electric power production is biomass, which is arguably the most important source of renewable energy today. If used correctly, biomass can supply power while maintaining environmental compatibility [3]. In fact, innumerable biomass waist products that could form the basis of a clean power generation system are currently causing disposal problems.

Biomass is defined as organic material that originates from plants and includes algae, trees and crops. Biomass is produced when green plants convert sunlight into chemical energy through photosynthesis. Thus, biomass can be considered as organic matter that stores the energy of sunlight in chemical bonds. Generally, biomass is categorized into four groups: woody plants, herbaceous plants and grasses, aquatic plants and manures [1].

One main thrust of biomass research is focused on developing liquid fuels or "biofuels" for transportation, including ethanol, biodiesel, and blended gasoline [6]. Ethanol is widely used as a biofuel in many countries, either as ethanol alone or blended with petroleum-based gasoline [7]. Aside from the use of biomass as fuel for transportation, many studies have been conducted on the use of biomass for electricity generation. These studies indicate that biomass gasification and subsequent combustion of the so-called "producer gas" or "synthetic gas" shows great promise for delivering a significant amount of today's electrical power require-ments [5, 8, 9].

In remote areas where grid electrical power does not exist, conventional fuels may also be scarce. Without a local source of electrical power, satellite communications, although available even in extremely remote locations, cannot be used. To address this issue, a "hybrid" photovoltaic and biomass genset power system was developed. The genset was powered with the combustible products of anaerobic gasification of biomass, or "syngas". This chapter describes the optimization of a small gasoline-powered genset for the production of electrical power using gasified woodchips. The genset was specified to be a two-stroke gasoline unit for ease of field servicing. This chapter details the conversion of the genset to syngas operation, optimization of the genset, development of a flexible air-fuel control system and performance measurements of the genset in syngas operation.

The desired specification was for a 2 kWh/day electrical output, ideally from a system operating less than 8 hours per day, 4 days per week, serving as backup power to a photovoltaic power system during extended periods of heavy clouding. As the largest readily available two-stroke genset is a 65 cc, 650 W unit (Yamaha EF950), it was agreed to use this as the base engine in this study.

2. Syngas

When organic material is heated with a limited air flow, it breaks down into simpler chemicals in the form of a hot gaseous exhaust. In a gasifier, organic material is burned with a restricted air flow to prevent complete combustion of the material, resulting in the production of a combustible mixture of gases known as "producer gas," "synthetic gas" or, in the case of woody biomass, "wood gas." Whereas individual components of the gas may vary based on the gasifier design, operating conditions and biomass, the primary energy carriers in the gas are generally carbon monoxide (CO), hydrogen (H_2) and methane (CH_4) [10]. As some air enters the system, allowing a small amount of combustion, the syngas also contains appreciable amounts of nitrogen (N_2), carbon dioxide (CO_2) and water (H_2O). Because these gases do not contribute to combustion, they act to dilute the fuel, reducing its heating value.

The gasifier used in this work was a small down-draft unit developed by the University Science Malaysia. This gasifier, shown schematically in Figure 1, is designed to burn wood pellets and delivers approximately 5 kW of combustible gas for direct combustion as well as for use in engines. Wood pellets are loaded into the gasifier from the top every 20-25 minutes of operation.

The ash settling tanks perform two tasks: cooling the gas and allowing any solid matter present to settle out. The condenser is responsible for condensing and removing water and the oil bath is used to clean the gas before introduction into an engine. The starting pump is used to establish air flow through the gasifier during starting. Once the gas begins flowing, it is burned at the flare column. When a flame can be sustained, it is then directed into the engine, and the starting pump is switched off.

A typical analysis of the resulting syngas is given in Table 1. The gas from this system generally produces a heating value of 4.5 MJ/kg.

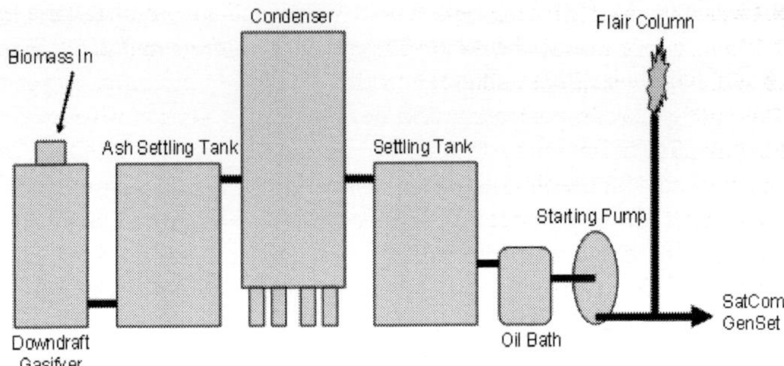

Figure 1. USM's 5 kW down-draft wood pellet gasifier system.

Component	Mole %
N_2	45
CO	20
CO_2	15
H_2	14
CH_4	4.0
O_2	0.2

Table 1. Syngas components [11]

3. Syngas combustion chemistry

The following chemical equation describes the stoichiometric combustion of the syngas in air [11, 12]:

$$0.45\ N_2 + 0.20\ CO + 0.15\ CO_2 + 0.14\ H_2 + 0.04\ CH_4 + 0.002\ O_2 + a\ (O_2 + 3.76\ N_2) = x\ CO_2 + y\ H_2O + z\ N_2$$

Carbon balance provides us with the following: $0.2 + 0.15 + 0.04 = x$ or $x = 0.39$. Hydrogen balance results in $2(0.14) + 4(0.04) = 2y$ or $y = 0.22$. Oxygen balance then gives $0.20 + 2(0.15) + 2(0.002) + 2a = 2x + y$ or $a = 0.248$. Thus, for stoichiometric combustion we have the fuel mass of

$$Mf = 0.45(28) + 0.20(28) + 0.15(44) + 0.14(2) + 0.04(16) + 0.002(32) = 25.8\ gm$$

and an air mass of

$$Ma = 0.25(32 + 3.76(28)) = 34.32\ gm$$

Together, these yield an air/fuel mass ratio of 34.32/25.8 = 1.33. In terms of moles (i.e., volume, assuming the two flows are at the same temperature), the air/fuel ratio is 0.248(4.76) / (0.45+0.2+0.15+0.14+0.04+0.002) or about 1.2 times as much as air consumed per unit fuel by volume. This result has two important ramifications for the conversion of the engine to syngas operation. First, the fuel induction system will have to be capable of delivering a large amount of fuel, commensurate with the air flow. For this reason the fueling control system will consist of two identical throttle valves, one for air and the other for fuel. Secondly, as the air intake is reduced to about half the normal flow (compared to gasoline operation), the power will also be reduced to about half of the gasoline rated power.

4. Engine modifications

The Yamaha 950 two-stroke, 240 VAC, 50 Hz, 3000 rpm, 650 W generator is shown in Figure 2. It was operated on gasoline to achieve the baseline fuel consumption of 682 g/hour and power production of 640 W, which is very similar to its rated power. Electric power–specific fuel consumption of the gasoline base model was 1066 gm/kWh, which is a rather high fuel consumption value typical of small crankcase scavenged two-stroke engines.

Figure 2. The Yamaha 950, 650 W genset, assembled by Kaba.

As the generator is an induction machine, a capacitor is placed in parallel with the AC output. This is necessary to cause the generator to self induce. In addition, the nominal value of the capacitor, 15 µF, is well matched to operation at 3000 rpm. As producer gas burns relatively slowly compared to gasoline, it was anticipated that the generator would be used at lower speeds. Lower speeds require a higher capacitance value for optimum output. Tests showed that increasing the generator's parallel capacitor from 15 µF to 35 µF (adding a 20 µF, 400 V capacitor in parallel), as shown in Figure 3, greatly improved the power production at lower speeds.

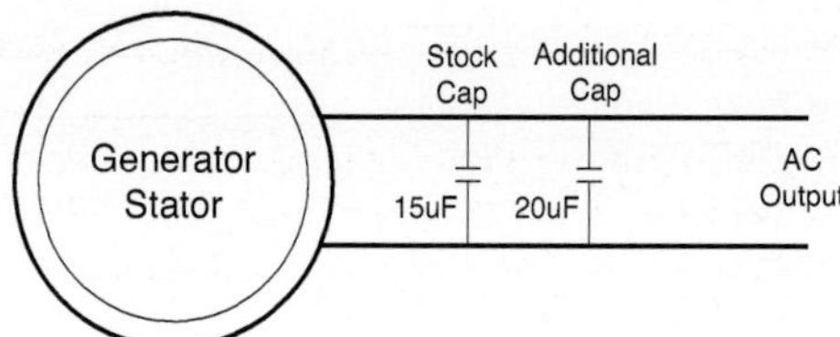

Figure 3. Output capacitance was increased from 15 to 35 μF.

The flame speed of syngas is significantly lower than gasoline. Thus, the spark timing was advanced by approximately 13° to improve power and efficiency when operating on syngas. This was accomplished by moving the flywheel/magneto keyway in such a way that the magneto (responsible for triggering the spark event) was advanced by 13° as shown in Figure 4.

Figure 4. Flywheel with mark indicating new keyway providing spark advance.

As mentioned above, the air and fuel flows are similar in terms of volume. Thus, to control air and fuel flows, an air/fuel control system was created using two identical servo throttles: one each for air and fuel. To simplify operation, an electronic controller was developed to provide two main control functions: the overall throttle as well as a separate air-fuel ratio (AFR) control. When the throttle control is opened, both the air and fuel throttle are opened proportionally. As the AFR control is adjusted, the air and fuel throttles are actuated in opposition; i.e., as the AFR is increased, the air valve is opened slightly and the fuel valve is closed by a similar amount. Reducing the AFR control opens the fuel valve and closes the air valve by a similar amount. This mechanism provides independent control over the throttle and AFR. The servo motors are pulse width modulated units, where position (and thus throttle opening) is proportional to the width of the pulse provided to the motor. Various throttle and AFR signals are shown in Figure 5.

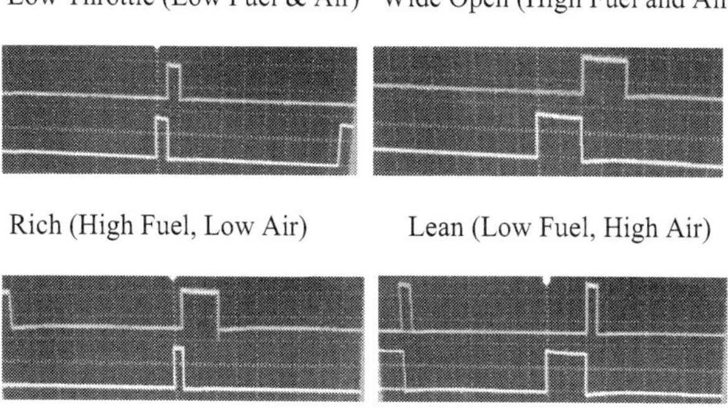

Figure 5. Fuel servo signal (blue, top) and air signal (yellow, bottom) at various throttle and AFR settings.

The digital display of the controller shows various operational parameters, including engine speed, temperature, throttle and AFR settings. The servomotor and digital display are shown in Figure 6. The servo motor is located below the throttle valve, and the pipe exiting to the right connects to the fuel throttle and fuel supply. Air and fuel are poorly mixed in the "T" junction, however, complete mixing occurs as the gases pass through the crank case of the engine before being transferred into the combustion chamber of the two-stroke engine. This provides good mixing of the air–fuel mixture.

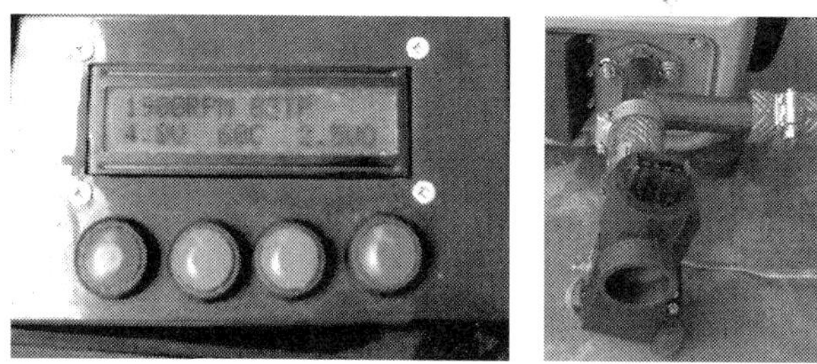

Figure 6. Digital display (left) and air servo valve (right).

5. Bottled syngas operation

Initial measurements were made using a bottled syngas, shown in Figure 7, consisting of 80% N_2 and 20% H_2. This allowed tuning and testing of the engine and instrumentation before the final runs using biomass-produced syngas. Fuel flow was measured via an instrumentation-grade "rotameter". Finally, as the engine is a two-stroke operating on gaseous fuel, provision for two-stroke oil had to be provided. This was accomplished via the addition of a two-stroke oil reservoir and electrical solenoid pump operated by the AFR controller. This allowed convenient adjustment of the two-stroke oil flow rate.

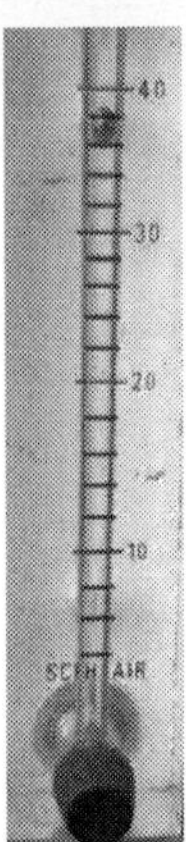

Figure 7. Bottled syngas (left) and fuel flow rotameter (right).

It was necessary to establish a starting procedure when using the bottled syngas for consistent starting. Fuel pressure was regulated down to near atmospheric pressure. However, in this configuration fuel flows even without engine operation, flooding the intake manifold. The throttle was set to approximately 10%, and the air fuel ratio was adjusted slightly lean to encourage air flow. Once the engine was started, the throttle could be opened and the air/fuel ratio adjusted for best operation. The genset running on bottled syngas with 13° spark advance and modified generator was able to exceed 3000 rpm and 300 V without an electrical load. Power production was evaluated with various loads and air/fuel ratios. The genset achieved a maximum electrical power output of 160 W (130 V at 1.15 A) with a fuel conversion efficiency of about 12.4%.

The power is significantly lower than the original gasoline configuration (which achieved 640 W), but the fuel conversion efficiency is better, as the gasoline engine achieved only 8%. The improvement in efficiency is likely a result of the "lean" combustion, as we have approximately 40% more dilution with additional N_2 when running the bottled syngas [13].

Low power production is common on syngas engines as approximately half of the incoming charge is fuel, reducing the available oxygen intake by about 50% [14]. Additionally, syngas tends to burn very slowly, as it is predominantly CO with abundant N_2 dilution. This results in a power de-rating of approximately 75% when run on syngas converted gasoline engines. Thus, the low power produced when running on the bottled syngas (160 W) was expected with the original gasoline production of 650 W.

6. Biomass syngas testing

Having verified the genset's functionality on bottled syngas, the engine was tested using syngas at the University Science Malaysia's Bio-Energy Laboratory. The gasifier is rated at approximately 5 kW (gas out) when burning wood pellets. Typically 4 kg is loaded at a time, and reloading occurs at 20 to 25 minute intervals (Figure 8). The main gasifier, measuring approximately 50 cm in diameter with a 125 cm height, feeds an ash-settling chamber approximately the size of a 200 liter barrel, a condensation unit measuring 50 cm × 50 cm × 150 cm tall, another settling barrel, an oil bath filter and a startup blower before being fed to the flare, sampling stack and the engine.

Figure 8. USM's 5 kW down-draft wood pellet gasifier.

Gasifier startup requires approximately 15 minutes before the gas can flare (Figure 9). Starting of the genset was performed on syngas, without gasoline assist. For each configuration tested, a specific startup procedure was required, and when followed, the genset started quite easily, generally with a single pull of the starter. To the surprise of the USM laboratory personnel, the engine would start and run on the biomass syngas even when the flare could not be sustained (typically a prerequisite for engine operation). This was attributed to a robust ignition system and careful air-fuel ratio control. Additionally, the genset was able to run and continuously produce power during re-fueling of the producer gas system.

Figure 9. Flare (left) and gas sampling for analysis (right).

Many different configurations of gasifier operation, engine settings and electrical load were evaluated. Initially, fuel flow was measured with the rotameter; however, this device quickly tarred up and became obscured with the deposits. Thus, flow measurements were made quickly, and the flow meter was subsequently bypassed. The final genset is shown in Figure 10. The electrical load (off camera to the left) consisted of various 1 kW, 62-ohm resistors configured as required.

Figure 10. Modified syngas genset. Two-stroke oil reservoir is at upper left. Controller, fuel meter and electrical loads are not shown.

During operation, the output voltage, current and frequency were measured using an Agilent true RMS digital multimeter. A typical run consisted of selecting an operating configuration, varying the load and varying or optimizing the engine controls. The engine was operated at the optimum operating point for extended periods, and the system was shut down briefly during configuration changes. The ancillary equipment and test environment can be seen in Figure 11.

Figure 11. AFR and throttle controller are to the left of the engine, as is the digital display. Note that the gas supply system has relatively large pipes to supply the unpressurized syngas.

7. Biomass syngas operation results and analysis

The first observation made was that the genset started on syngas surprisingly well. Starting and operational servo valve settings are strongly influenced by the configuration under test in expected ways: with greater restrictions in the fuel flow path (such as the fuel flow meter), the fuel valve had to be opened relative to the air valve to admit more fuel. When the startup blower was in operation, the fuel valve could be closed somewhat with respect to the air valve due to the higher pressure of the fuel.

During a single biomass load, the quality of the fuel varied from the beginning of the burn to the end, requiring slight adjustments to the AFR control to maintain optimum power [15]. It was also noticed that the fuel flow meter caused significant restriction to the fuel flow, which in turn resulted in a much lower optimized AFR controller setting. Once the fuel flow meter was removed the best AFR was near 1, as expected based on the chemistry presented earlier. Similarly, the optimum throttle depended on the specifics of the configuration under test. In all cases, the power output varied smoothly with changes in throttle and AFR, and the engine was very easy to control.

Low-power operation was not strongly dependent on the actual electrical load, but was restricted by the throttle and governor. The governor of the gasoline version controls operation to the rated speed of 3000 rpm. This was bypassed for most of the tests to investigate a wide range of operating speeds and outputs. Electric power output of 120 W was achieved on loads from 61 to 183 ohms with the fuel flow restricted by the flow meter. Fuel flow rates of 1.5 scfm (standard cubic foot per minute) were measured for the first 30 minutes before the flow meter became obscured by internal tarring. As the system was allowed more air and fuel, the power level rose to approximately 270 W for loads of 120–180 ohms. Further improvements in fuel flow and bypassing of the genset governor allowed full power operation with a power output of approximately 430 W on both the 122 and 183 ohm loads.

At these higher powers, the genset was running near the rated speed. Fuel consumption was estimated at approximately 2 scfm of gas (about 3.2 m^3/hour) or about one third of the gasifier's rated capacity.

Output	Load	Current	Power		Throttle		Fuel Flow	
V	Ohms	A	We	Hz	%	AFR	SCFM	Comment
122	122	1.0	122	30	44	0.54	1.1	Choked
86	61	1.4	121	26	62	0.50	1	Choked
150	183	0.8	123	33	72	0.50	1.1	Choked
220	183	1.2	264	41	44	0.50	1.5	With Flow Meter
184	122	1.5	278	38	18	0.50	1.4	With Flow Meter
245	183	1.3	328	45	52	2.08	~2	No Flow Meter
280	183	1.5	428	48	106	1.92	~2	Bypass Govenor
220	122	1.8	397	50	72	1.92	~2	Bypass Govenor
210	122	1.7	361	50	80	1.25	~2	Average over 1 Hour w/ Reloads
229	122	1.9	430	50	100	1.04	~2	Fully Optamized

Table 2. Summary of various loads, powers, and settings

From the above data, it can be observed that at very low power settings (choked operation), the power output is independent of applied load. Opening up the overall flow allows greater power production, with similar power recorded for the 122 and 183 ohm loads. The initial runs indicated that the system was not receiving sufficient fuel flow (as evidenced by the low AFR values). This led to the improvement of the fuel flow path, eventually necessitating the elimination of the fuel flow rotameter, which was becoming clogged with tar. As the engine operation approached the rated speed (50 Hz), the governor became the limiting factor. Once bypassed, the power rose further and approximately 430 W was produced with both the 122 and 183 ohm loads. During testing, steady runs of over 60 minutes were achieved in this configuration at powers of 360–400 W, even while refueling was taking place.

The gasifier requires a blower to start the gas flow. The gasifier's capacity is more than twice the genset's capacity, so the blower was kept on most of the time to insure good gas flow through the gasifier, and the extra gas was flared. In one test, the blower was shut down, and the full gas supply delivered to the genset. It was capable of sustained operation in this condition. However, over the course of 15 minutes, the gas quality began to degrade due to the lower flow rate of air entering the gasifier. The blower was placed back in operation and the whole system returned to steady state within several minutes. This underscores the need to carefully size the gasifier for the genset and power demand.

Power output was measured to be a smooth function of throttle, generally achieving maximum power at 50–100% throttle range. The data in Figure 12 indicates an optimum throttle position of about 60% with the flow meter in-line at a load of 183 ohms.

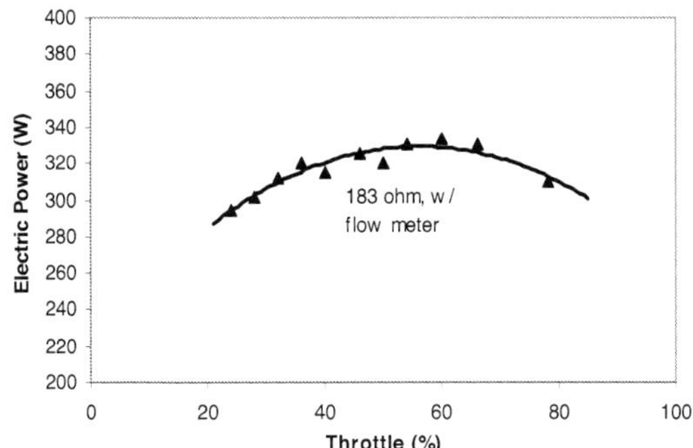

Figure 12. Variation in electrical power output as a function of throttle setting with flow meter in line.

For most runs, the AFR was optimum near stoichiometric with an AFR about 1.1–1.2 (Figure 13), in agreement with the estimated stoichiometric volumetric AFR of 1.2 as determined earlier. Again the power varied smoothly with AFR. Optimum AFR varied over the biomass burn cycle. Significantly, more power was produced when the flow meter was eliminated, as this was restricting the fuel delivery to the engine.

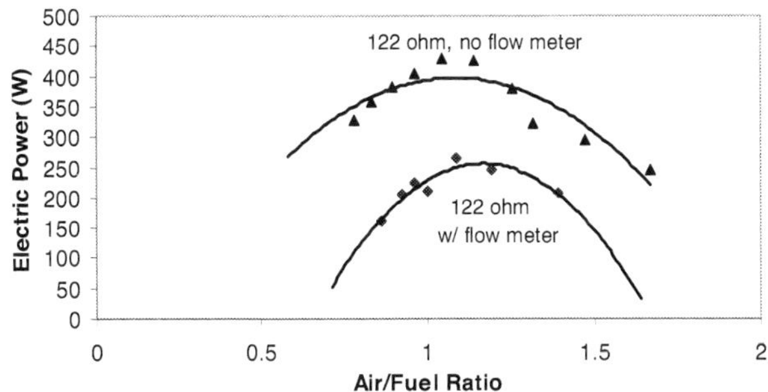

Figure 13. Variation in electric power output as a function of AFR. Peak power occurs near 1.1 AFR.

The genset was operated almost continuously for over 3.5 hours, with short shutdowns between configuration changes. Power output was approximately 360 W most of this time, resulting in a total energy production of 1.26 kWh (4.5 MJ). The total biomass fuel consumption during this time was 20 kg, producing a syngas with a heating value of approximately 4.5 MJ/

kg. As the genset is undersized for this gasifier, it was only capable of consuming about one third to one half of the syngas produced (the balance being flared). Thus, the overall energy consumption was determined to be approximately 7–10 kg or approximately 0.55 MJ of electric power per kg of biomass. The biomass is assumed to have a heating value of approximately 20 MJ/kg, resulting in an overall system efficiency of approximately 0.55/20 (2.75%). While this number appears quite low, it must be understood that this includes the efficiency of the gasification system, combustion, engine, and generator together. Looking at just the genset, it can be estimated that the fuel consumption was approximately 1.7 m^3/hour, which, assuming a heating value of 1.5 kWh/m^3, would give a power consumption of 2.55 kW (syngas), for a power production of 360 W, resulting in a genset efficiency of about 14%. This is a reasonable efficiency for a two-stroke induction generator and is in general agreement with the efficiency achieved using bottled syngas.

Higher powers require higher generator operating speeds. At higher speeds, however, the engine efficiency will drop. This is primarily a function of the speed of combustion of the fuel. As the engine speed is increased, there is insufficient time to completely burn the fuel before expansion of the mixture, resulting in burning occurring well into the expansion stroke or even into the blow-down phase and flame release into the exhaust system. This gives hot exhaust gases, as was observed at higher speeds, representing inefficient combustion. Thus, operation at lower speeds will improve the overall system efficiency, but will degrade the rated power production.

Finally, it was noticed that in contrast to gasoline engine operation, the exhaust emissions were perfectly clear. This is likely from better control of the two-stroke oil, as well as improved combustion of the oil, which is not mixed with a liquid fuel.

After 2 hours of operation, the spark plug was removed and inspected. Its light brown appearance indicated good operation, with no evidence of tarring or oil buildup that could cause fouling as shown in Figure 14. At the end of the run, the fuel valve was inspected and found to contain heavy deposits of tar. It was estimated that the fuel intake valve would require cleaning after approximately 10 hours of operation.

Figure 14. Spark plug at 2 hours of operation (left) and fuel valve after 3.5 hours (right).

8. Engine tare down inspection

As mentioned, the fuel flow meter became obscured by tar after about 30 minutes of operation. This occurred despite the presence of particulate filtering, condensation chamber, and oil bath filter. Tar buildup was expected, as tar tends to condense on cooler surfaces, and the flow meter was the coldest element prior to the engine.

The head showed (Figure 15) a thick buildup of soot, uniform except for a lighter color on the lower (hotter) side of the head. The piston crown similarly had a thick buildup of soot. The inside of the bore had a dark film of tar and oil, but still had decent tribological properties. The use of two-stroke oil was measured to be 15 ml for the 3.5-hour run, significantly less than the typical flow rate on gasoline, however, quite sufficient for syngas operation because the gas does not remove the oil from the target surfaces as gasoline does. Oil "wetting" marks can be seen on the edges of the piston in the photo below.

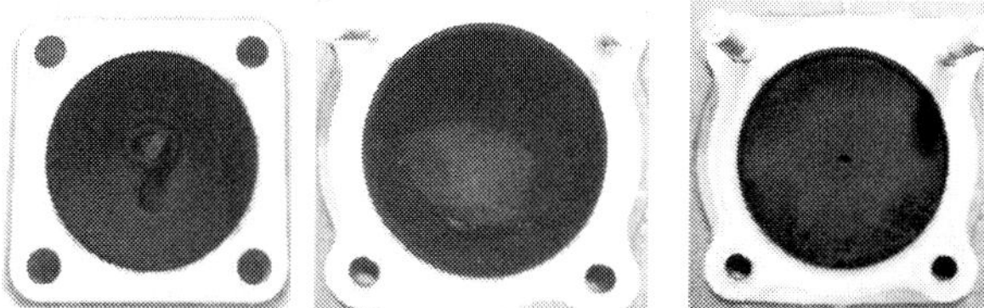

Figure 15. Cylinder head (left) bore (center) and piston crown (right) indicating significant tar or soot buildup, especially at the colder locations.

The crank case contained a smoky smelling oil and tar mixture. The buildup quantity was not enough to cause alarm, but the color was much darker than typically occurs with gasoline operation. The piston top land had significant carbon buildup, which extended down to the top of the skirt region (Figure 16). No abnormal wear was observed.

Figure 16. Crank case (left), and piston (right).

It is estimated that the engine should be capable of operating one shift per day, allowing approximately 7 hours of operation without any special attention. After 1 week of operation, however, it is believed that the carbon and tar buildup will be sufficient to warrant maintenance of the piston, bore and head. This maintenance consists of opening the head, cleaning the parts, reassembling and torquing the head bolts to specification. Properly done, this operation should take no more than 30 minutes. It should be noted that this rate of maintenance may adversely affect the life of the engine.

9. Conclusions

Several important conclusions resulting from this study, outlined below, will have ramifications on the commercial exploitation of this or similar biomass power generation systems.

First, the syngas genset was capable of sustained operation at over 360 W of electrical output on syngas, with a peak power of 430 W. This implies a de-rating of 45% from the original gasoline operation (expected from the fuel displacement of incoming air) with the described modifications. Power production efficiency is estimated at 0.55 MJ/kg of biomass, with an overall system efficiency of 2.75% and a genset efficiency of approximately 14%. Although the overall efficiency is rather low, it includes the efficiency of the gasifier, engine, and generator. In addition, a higher efficiency can be expected by better matching the size of the genset to the gasifier. It was observed that the engine was operating significantly faster than optimum for the combustion of the gas, as evidenced by the elevated exhaust gas temperatures. A larger displacement genset designed for lower speed operations would greatly improve thermal efficiency, with a commensurate improvement in the overall efficiency of the system. The air–fuel control system was capable of controlling the engine operation for easy starting and smooth operation over a wide range of circumstances. Two-stroke oil was delivered through an electronically controlled pump, at a consumption of 5 ml per hour of operation, which provided good lubrication. Assuming one shift of operation per day (7 hours of genset operation), it is estimated that the engine will require tear-down for cleaning every 30 hours of operation, or approximately weekly. The sparkplug remained clean during 2 hours of operation; however, the fuel flow meter became unusable after 30 minutes from tarring. The fuel valve showed significant tar deposits after 3.5 hours of operation, indicating that the fuel filtering should be improved or the fueling system would also require frequent cleaning.

10. Recommendation for future work

Based on the above data, a larger generator system is proposed to provide greater power and improved efficiency when compared to the two-stroke syngas genset described above. Such a system should have the following specifications:

Power production:	2k W (electrical)
	240 V, 8.3 A, 50 Hz
Displacement:	400 cc
Engine speed:	2000 rpm
Spark timing:	30° before TDC
De-carbonizing:	Monthly (or every 120 hours of operation)
Biomass consumption:	30 kg/hour(assuming clean, dry wood biomass)

Table 3. Optimized Genset Specifications

These specifications would likely require a non-standard engine and genset. For maximum efficiency, a permanent magnet generator could be used, allowing a somewhat smaller engine of perhaps 300 cc and proportionally reduced biomass consumption. Finally, the engine efficiency can be improved by increasing the compression ratio, as syngas has a much lower tendency to knock than gasoline. Four-stroke engines are somewhat mechanically more complex than the simple two-stroke engine investigated here; however, they are more common for genset operations and may require less servicing as the tar-laden syngas is not circulated through the crank case.

As we move forward in a world of increasing population, environmental concerns, and fuel prices, it is inevitable that the importance of biomass as a power source will also increase. The burden of developing clean, sustainable energy supplies depends on today's engineers and technicians. It is sincerely hoped that the information contained herein will be of benefit in this effort.

Author details

Horizon Walker Gitano-Briggs* and Koay Loke Kean

*Address all correspondence to: horizonusm@yahoo.com

University of Kuala Lumpur, Malaysian-Spanish Institute, Kulim, Malaysia

References

[1] P. Mckendry, "Energy Production from Biomass (Part 1): Overview of Biomass," *Bioresour. Technol.*, vol. 83, no., pp. 37–46, 2002.

[2] C. J. Diji, "Electricity Production from Biomass in Nigeria: Options, Prospects and Challenges," *Int. J. Eng. Appl. Sci.*, vol. 3, no. 4, pp. 84–98, 2013.

[3] M. Barz and M. K. Delivand, "Agricultural Residues as Promising Biofuels for Biomass Power Generation in Thailand," *J. Sustain. Energy Environ. Spec. Issue*, pp. 21–27, 2011.

[4] D. D. Guta, "Assessment of Biomass Fuel Resource Potential And Utilization in Ethiopia: Sourcing Strategies for Renewable Energies," *Int. J. Renew. Energy Res.*, vol. 2, no. 1, pp. 131–139, 2012.

[5] M. S. Islam and T. Mondal, "Potentiality of Biomass Energy for Electricity Generation in Bangladesh," *Asian J. Appl. Sci. Eng.*, vol. 2, no. 2, pp. 103–110, 2013.

[6] H. von Blottnitz and M. A. Curran, "A Review of Assessments Conducted on Bioethanol as a Transportation Fuel from a Net Energy, Greenhouse Gas, and Environmental Life Cycle Perspective," *J. Clean. Prod.*, vol. 15, no. 7, pp. 607–619, 2007.

[7] E. Gnansounou and A. Dauriat, "Ethanol Fuel from Biomass: A Review," *J. Sci. Ind. Res.*, vol. 64, pp. 809–821, 2005.

[8] N. S. Mamphweli and E. L. Meyer, "Evaluation of the Conversion Efficiency of the 180 Nm^3/h Johansson Biomass Gasifier," *Int. J. Energy Enviroment*, vol. 1, no. 1, pp. 113–120, 2010.

[9] A. Pirc, M. Sekavčnik, and M. Mori, "Universal Model of a Biomass Gasifier for Different Syngas Compositions," *J. Mech. Eng.*, vol. 58, no. 5, pp. 291–299, 2012.

[10] P. C. Munasinghe and S. K. Khanal, "Biomass-Derived Syngas Fermentation into Biofuels: Opportunities and Challenges." *Bioresour. Technol.*, vol. 101, no. 13, pp. 5013–5022, 2010.

[11] Y. Chhiti and M. Kemiha, "Thermal Conversion of Biomass, Pyrolysis and Gasification: A Review," *Int. J. Eng. Sci.*, vol. 2, no. 3, pp. 75–85, 2013.

[12] R. Barrera, C. Salazar, and J. F. Pérez, "Thermochemical Equilibrium Model of Synthetic Natural Gas Production from Coal Gasification Using Aspen Plus," *Int. J. Chem. Eng.*, vol. 2014, pp. 1–18, 2014.

[13] J. Singh Brar, K. Singh, J. Zondlo, and J. Wang, "Co-Gasification of Coal and Hardwood Pellets: A Case Study," *Am. J. Biomass Bioenergy*, vol. 2, no. 1, pp. 25–40, 2013.

[14] R. Dobson and T. Harms, "Simulation of a Syngas from A Coal Production Plant Coupled to A High Temperature Nuclear Reactor," *J. Energy South. Africa*, vol. 24, no. 2, pp. 37–45, 2013.

[15] L. D. Thi, Y. Zhang, J. Fu, Z. Huang, and Y. Zhang, "Study on ignition delay of multicomponent syngas using shock tube," *Can. J. Chem. Eng.*, vol. 92, no. 5, pp. 861–870, 2014.

Chapter 7

Radio Frequency Energy Harvesting - Sources and Techniques

M. Pareja Aparicio, A. Bakkali, J. Pelegri-Sebastia, T. Sogorb, V. Llario and A. Bou

Additional information is available at the end of the chapter

http://dx.doi.org/10.5772/61722

Abstract

Energy harvesting technology is attracting huge attention and holds a promising future for generating electrical power. This process offers various environmentally friendly alternative energy sources. Especially, radio frequency (RF) energy has interesting key attributes that make it very attractive for low-power consumer electronics and wireless sensor networks (WSNs). Ambient RF energy could be provided by commercial RF broadcasting stations such as TV, GSM, Wi-Fi, or radar. In this study, particular attention is given to radio frequency energy harvesting (RFEH) as a green technology, which is very suitable for overcoming problems related to wireless sensor nodes located in harsh environments or inaccessible places. The aim of this paper is to review the progress achievements, the current approaches, and the future directions in the field of RF harvesting energy. Therefore, our aim is to provide RF energy harvesting techniques that open the possibility to power directly electronics or recharge secondary batteries. As a result, this overview is expected to lead to relevant techniques for developing an efficient RF energy harvesting system.

Keywords: Energy Harvesting, Energy Source, Radio Frequency, RFID, Wireless Sensor Networks

1. Introduction

As the demand for wireless sensor networks (WSNs) increases, the need for external power supply drastically increases as well. Besides the problems of recharging and replacing, size and weight, batteries are an exhaustible source with an adverse environmental effect. For these reasons, it is highly desirable to find an alternative solution in order to overcome these power limitations.

The environment represents a relatively good source of available energy compared with the energy stored in batteries or supercapacitors. In this context, energy harvesting, also known as power harvesting and energy scavenging, is an alternative process for primary batteries, where energy is obtained from the ambient environment. An energy harvester typically captures, accumulates, stores, and manages ambient energy in order to convert it into useful electrical energy for autonomous wireless sensor networks. The use of energy scavenging minimizes maintenance and cost operation; therefore, batteries can be eventually removed in WSNs as well as in portable electronic devices.

Many potential ways to harvest energy from environment are available, including solar and wind powers, radio frequency energy and ocean waves, and thermal energy and mechanical vibrations [1–3]. The publications on this topic in the literature are rising to a great extent. Hence, many papers have been published on energy harvesting as a feasible alternative to batteries. Work by Sardini et al. [4] proposed an autonomous sensor powered by mechanical energy coming from airflow velocity. Therefore, the battery-less sensor uses the power harvested in order to provide measurements of air's temperature and velocity. A completely different approach is proposed by Tan et al. in Ref. [5]. The authors have explored a system for wind-powered sensor node. By measuring the equivalent electrical voltage or the frequency of a wind turbine generator output, the wind speed measurement can be indirectly obtained. Based on the sensed wind speed information, the fire control management system provides the spreading condition of a wildfire, so that the fire fighting experts can perform an adequate fire suppression action.

This paper focuses on the energy harvesting technology using electromagnetic energy captured from multiple available ambient RF energy sources, such as TV and radio transmitters, mobile base stations, and microwave radios. This technique is very useful for sensors located in harsh environments or remote places, where other energy sources, such as wind or solar sources, are impracticable. In this context, this work presents an overview of advances achieved in RF harvesting field. The main components of an RF energy harvesting system are discussed in Section 2. Section 3 provides different measurements of the ambient radio frequency energy obtained in published papers. An introduction to RF harvesting energy in radio-frequency identification (RFID) technology is presented in Section 4. Finally, conclusions are drawn in Section 5.

2. Overview of radio frequency energy harvesting system

The basic structure of a radio frequency energy harvesting system consists of a receiving antenna, matching circuit, peak detector, and voltage elevator. Where electromagnetic waves are captured by the antenna, voltage is amplified using the matching circuit, signal is converted to a voltage value thanks to the peak detector, and finally this voltage output is adjusted using the voltage elevator.

The whole system formed by receiving antenna, matching network, and rectifier is usually known as a rectenna or an RF/direct current (DC), which is able to harvest high-frequency energy in free space and convert it to DC power. The detail of each block is subsequently discussed in order to define specifications and limitations of the power conversion system.

Further, a block of power management and another for energy storage could be integrated into the energy harvesting system. The energy storage subsystem is responsible for storing all the captured energy and providing a constant output voltage.

Energy harvester is a promising power solution for WSNs. Instead of depending on centralized power sources for charging, sensor devices operate the existing energy in the environment. The DC voltage is stored in a holding capacitor or supercapacitor in order to power supply integrated circuits.

a.   Antenna

RF energy harvesting technique needs, as mentioned in the previous section, an efficient antenna with a circuit capable of converting alternating current (AC) voltage to direct current voltage. The front end is a key component to ensure the successful operation of RFEH system. It has the duty of capturing electromagnetic waves, which will be used later to power the integrated system.

Moreover, the antenna efficiency is related to the frequency: energy obtained from an antenna with small bandwidth, than a wideband receiver antenna used to capture signals from multiple sources. RF antenna can harvest energy from a variety of sources, including broadcast TV signal (ultrahigh frequency (UHF)), mobile phones (900–950 MHz), or Local Area Network (2.45 GHz/ 5.8 GHz).

In principle, power harvested from RF signals is enough to supply microelectronic devices gradually; however, this power can dramatically rise by using an array configuration. Therefore, the maximum possible power can be achieved by properly arranging similar antennas (with the same matching circuit and power management) [6, 7], or by using antennas operating at different frequencies [8]. The trend is to include the antenna, usually patch antenna, and the rectifier on the same printed circuit board [9].

The equivalent electrical model of an antenna is an AC voltage source (V_{ant}) with series impedance (Z_{ant}), as illustrated in Figure 1. Amplitude of the AC voltage source depends on the available power (P_{AV}) and the real impedance (R_{ant}). The average power received (P_{AV}) depends on the power density (S) and the antenna effective area (A_e), as expressed in Eq. (1):

$$P_{AV} = S \cdot A \tag{1}$$

Apparent power received (S) can be calculated using the Friis transmission equation (Eq. (2)). S is a function of several parameters: the transmitted power (P_{TX}), the transmitting antenna gain (G_{TX}), the received antenna gain (G_{RX}), the wavelength (λ), loss factor (L_C), and the distance between transmitter and antenna (r):

$$S = P_{TX} \cdot G_{TX} \cdot G_{RX} \cdot L_C \cdot \left(\lambda / 4 \cdot \pi \cdot r \right) \tag{2}$$

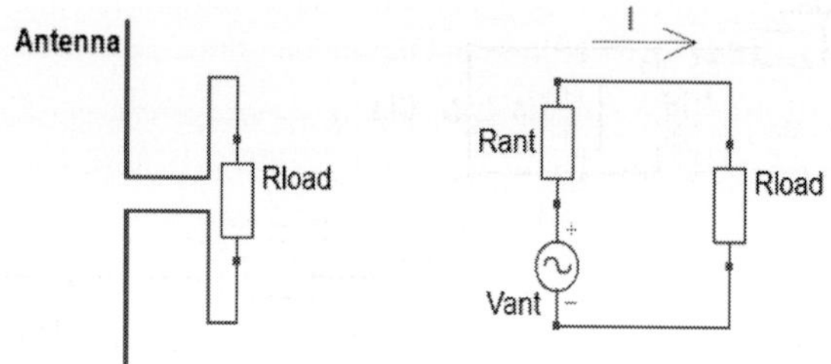

Figure 1. Antenna equivalent circuit.

The antenna impedance can be expressed by Eq. (3), where the real component is presented by two resistances: one is related to the material used (R_{loss}) and the other is due to the electromagnetic wave radiation (R_s). However, the imaginary component X_{ant} depends on the antenna structure, usually inductive for a loop antenna and capacitive for a patch antenna. Common Z_{ant} values are 300 Ω (closed dipole antenna), 75 Ω (open dipole antenna), and 50 Ω (wireless systems):

$$Z_{ant} = \left(R_{loss} + R_s \right) + jX_{ant} = R_{ant} + jX \tag{3}$$

Indeed, the concept of RF energy harvesting requires an efficient antenna with high performances. Hence, several researchers focused on highly efficient receivers for electromagnetic wave harvesting. Moon and Jung [10] proposed an interesting antenna design for RF energy harvesting system based on two radiators: the main one is a printed dipole radiator and the parasitic one with a loop structure. The parasitic radiator is suitable for receiving RF power in all directions from the main radiator. However, Xie et al. [11] designed a hexagonal microstrip patch antenna array that operates at 915 MHz, in order to achieve the maximum possible RF energy to convert into DC power for lighting light-emitting diodes (LED).

b. Matching circuit

Matching circuits are essentially used to match the antenna impedance to the rectifier circuit in order to achieve maximum power and improve efficiency, by using coils and capacitors [12, 13]. Several matching circuits are available; however, the main configurations that have been proposed are the transformer, the parallel coil, and the LC network, as shown in Figure 2.

For economic reasons, RFID tags and sensor networks use the shunt inductor and the LC network as matching networks instead of the transformer. Moreover, it is desirable for high-impedance antennas (e.g., dipole antenna) to use the parallel coil [12], whereas the LC network is used for small impedance antennas (e.g., Wi-Fi antenna) or when the available power P_{AV} is low [13].

As previously mentioned, the impedance matching circuit is designed to increase the voltage gain and reduce the transmission loss; this means that the impedance seen by the antenna is

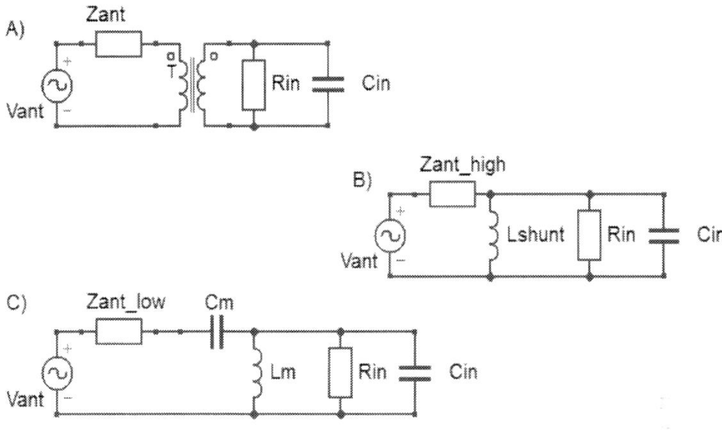

Figure 2. Matching network circuits: transformer (a), shunt inductor (b), and LC network (c) [12].

equal to the impedance of antenna [14]. The equivalent circuit and the normalized input voltage are shown in Figure 3. Therefore, Vin reaches its maximum level when α is equal to the unit, that is, when Rin and Rant are equal.

In radio frequency range, the impedance mismatch between the antenna and rectifier could be replaced by a tuning circuit, in order to adjust the receiver frequency [15, 16]. Multiband commercial antennas are typically equipped with filters [17]; however, the output power is lower than it should be [18]. An example of matching circuit impedance intended for television frequency band, formed by passive components and using the LC network, is discussed in Ref. [19].

Further, a shorted stub can be added to the matching circuit, which is represented by a wire with a length depending on wavelength and finishing on the ground plane. Therefore, the system performs as a tank circuit [9, 20]. However, in Ref. [21], the authors proposed an approximate method using a resistor in series with antenna. The current trend is to include antenna, impedance matching, and rectifier in a printed circuit board [19]. The RFEH system is designed on the same printed circuit board avoiding cable losses (cf. Figure 4).

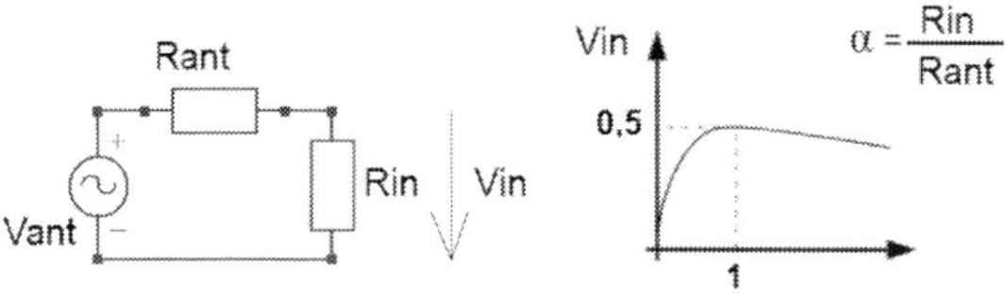

Figure 3. Transfer energy on matching circuit [14].

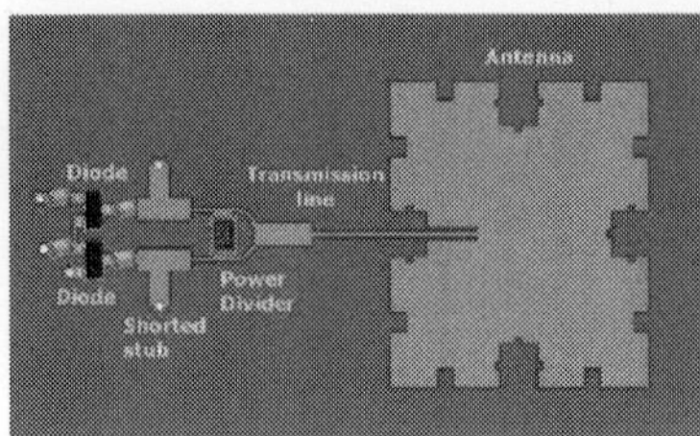

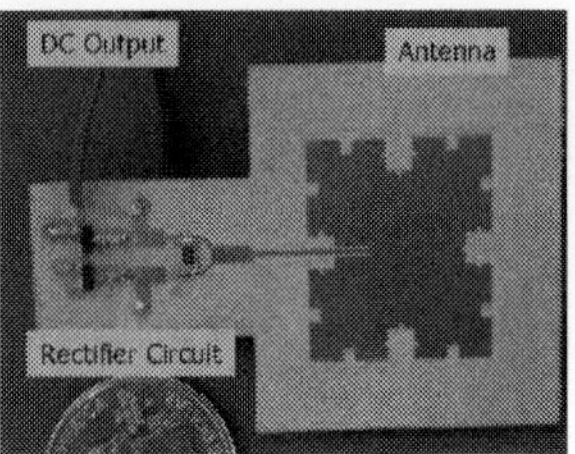

Figure 4. RFEH design circuit [9].

c.   Rectifier

Radio frequency signal captured by the antenna is an alternating current (AC) signal. In order to get a DC signal out of AC signal and improve the efficiency of the RF–DC power conversion system, a rectifier circuit is used. Rectification subsystem or peak detector, which has been already used on crystal radio, consists only of diodes and capacitors.

When the distance from the RF source is far and the received power is not high enough, the rectifier input needs to be amplified in order to power the circuit (sensor networks or RFID tags require at least 3.3 V). The most popular rectifier used is a modified Dickson multiplier, which has the function of rectifying the radio frequency signal and increases the DC voltage. Moreover, many works have used complementary metal–oxide–semiconductor (CMOS) technology to replace the diodes [13, 22]. Other different ways to rectify AC signals have been introduced, including Greinacher circuit or voltage doubler [23], Cockcorft–Walton circuit [20], multiplier resonant [24, 25], Villance multiplier [26], and boost converter [23, 27].

Choice of rectification circuits depends on the radio frequency signal and power received, since different values of DC voltage could be obtained with the same circuit and different radio frequency sources. The multiplier is usually formed using different stages; each stage includes two diodes and two capacitors. The voltage output is more important with a large number of stages. However, because diode loss increases with the stage number, the system efficiency is affected. The impact of rectifier stage number on the power received is presented in Figure 5. For the low received power (Pin < 0 dBm), the output voltage (V_{out}) is practically independent of the stage number, while efficiency is good for fewer stages. The high voltage range is achieved when the power received is around 0 dBm and the number of stages is large, whereas efficiency decreases when V_{out} reached its maximum. Therefore, it seems difficult to achieve a good design due to the received signal influence on the RFEH system.

Multiplier efficiency (η_{rect}) depends on the input and output powers (P_{in_rect} and P_{out_rect}, respectively), as expressed in Eq. (4). However, the efficiency of RFEH system (η_{RFEH}) depends on the power generated (P_{out_dc}) and the power received (P_{in_rf}). The η_{RFEH} can be calculated using Eq. (5):

$$\eta_{rect} = P_{out\_rect} / P_{in\_rect} \qquad (4)$$

$$\eta_{\text{RFEH}} = P_{\text{out\_dc}} / P_{\text{in\_rf}} \tag{5}$$

Diodes commonly used as rectification components are Schottky diodes, while Germanium diodes are also used for radio circuit of the peak detector. Performance analysis of some Schottky diodes is outlined in Table 1.

Device	I_S (A)	R_S (Ω)	C_{JO} (pF)	V_J (V)	B_V (V)	I_{BV} (A)
SMS7630	5E-6	20	0.14	0.34	2	1E-04
HSMS-282X	2.2E-8	6	0.7	0.65	15	1E-04
HSCH-9161	12E-6	50	0.03	0.26	10	10E-12

Table 1. Parameters of Schottky diodes.

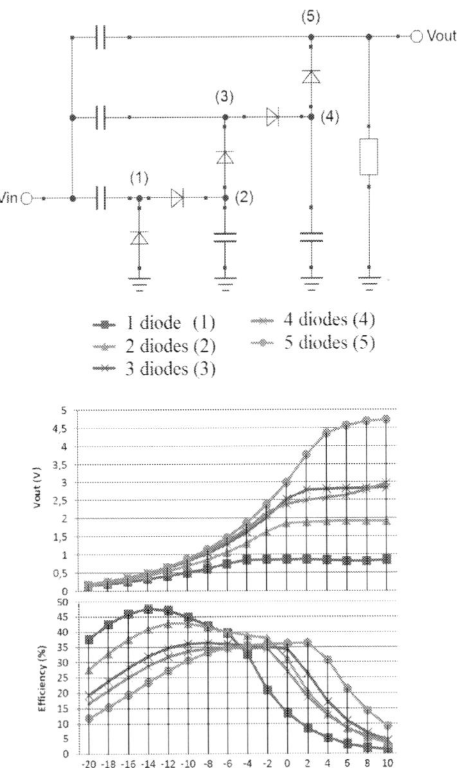

Figure 5. Stages multiplier versus output voltage and efficiency [28].

Rectifier equivalent circuit, as shown in Figure 6, is modeled by an input impedance $R_{in} \mid\mid C_{in}$, in addition to a current source depending on the input voltage, and a constant output resistor that presents the rectifier losses [14]. Output voltage value is determined by the stage number (N) of the multiplier.

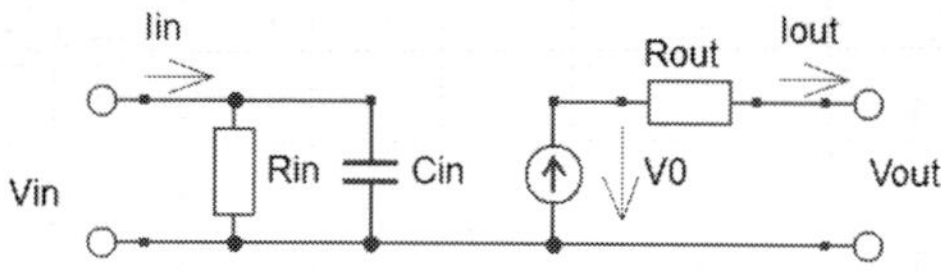

Figure 6. Multiplier equivalent circuit [14].

Further, multiplier equivalent circuit can also be obtained by using the mathematical equation [14], model simulation [12], or measurement [26].

3. Measurement of ambient radio frequency energy harvesting

As mentioned previously, an RFEH system is able to recover energy from available RF electromagnetic sources present in the ambient environment such as phone stations, radio, and television broadcasting. In Table 2, the main features of RFEH systems proposed in literature are summarized. As can be seen, the energy harvested is significantly very low that involves a decrease of the circuit performance.

Figure 7 shows the received power as a function of distance from RF power source at UHF. As it can be seen, for a free space distance of 40 m, the maximum theoretical power available for conversion is 1 µW and 7 µW for frequencies of 2.4 GHz and 900 MHz, respectively.

As mentioned above, many other sources of energy, including vibration, photovoltaic, and thermal, have been cleverly converted to useful energy using a variety of techniques. Table 2 presents some harvesting methods with their power generation capability.

Despite the fact that the power density of RFEH is lower than other sources, this powering method can be useful, especially for sensor nodes located in harsh environments, where other sources like wind or solar energies are not feasible.

Energy source	Power density (/cm²)
RF	0.01 to 0.1 µW
Vibration	4 to 100 µW
Photovoltaic	10 µW to 10 mW
Thermal	20 µW to 10 mW

Table 2. Comparison of energy harvesting sources [29].

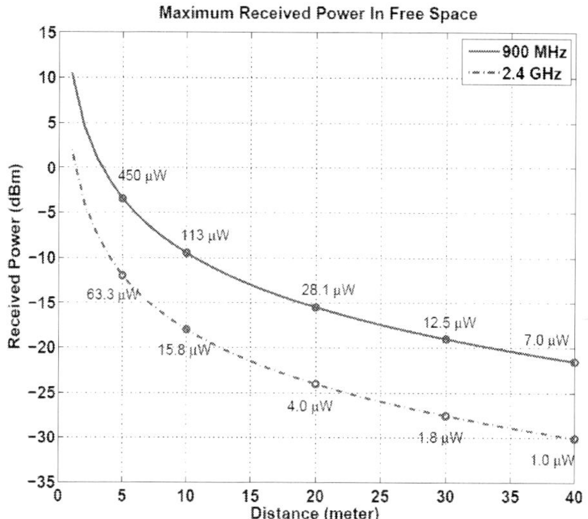

Figure 7. Received power versus distance [13].

Therefore, RFEH is a promising technology and an alternative source of energy to power the sensor nodes. As a result, these devices do not require any battery since they can use the power harvested from the ambient RF energy. Since battery replacement or its recharging is impracticable, autonomous WSNs need to exploit the RF ambient energy harvesting especially for long-duration applications.

It is important to note that the power density available depends on the radio frequency source and the distance. Values of this power are presented in Table 3 for different RF energy sources.

Source	Distance	Density power available
50kW AM radio station	5/10 [km]	159/40 [µW/m²]
100W GSM base station	100/500/1000 [m]	800/32/8 [µW/m²]
0,5 mobile phone	1/5/10 [m]	40/1.6/0.4 [mW/m²]
1W Wi-Fi router	1/5/10 [m]	80/3.2/0.84 [mW/m²]

Table 3. Power density on RFEH with different sources [30].

Table 4 provides a summary of results obtained from various studies in the RFEH field, with a brief description of the significant components used: RF source, antenna, matching circuit, and rectifier circuit.

Reference	Description	Frequency	Maximum Voltage	Maximum Performance	Power/Energy
BHA2006 [6]	Patch antenna array is used (4×4). Maximum power received is −10 dBm	2.4 GHz	n/a	n/a	373.248 μW
MIK2011a [8]	Using the same antenna for different frequency band, TV signal (74% to 42.6%) and RFID reader	470–770 MHz 950–956 MHz	n/a	74% (Pin=0 dBm) 54% (Pin=-20 dBm) 2% (Pin= −40 dBm)	0.74 mW (Pin=0 dBm) 5.4 μW (Pin=-20 dBm) 2 nW (Pin=−40 dBm)
UR2010 [15]	FM radio signals with loop antenna, tuned circuit, and Dickson charge pump 6 stages. AA supercapacitor is used to store energy	945 kHz	520 mV	n/a	60.4 μJ
BOU2010 [18]	Matching circuit using limited filter and rectifier with 1 stage. Maximum power received is −42 dBm (63 nW)	2.4 GHz	n/a	0.60%	400 pW
MIK2011b [31]	Patch antenna, matching circuit, and rectifier with 1 stage and boost converter	500–700 MHz	134 mV (Pin=−15 dBm)	18.2% (Pin=−20 dBm)	n/a
MIK2011c [32]	Microwave tooth antenna with filter, to matching circuit. A supercapacitor is used to store energy	470–505 MHz 520–560 MHz	3.7 V	> 50% (Pin=−5 dBm)	30 mW
AMA2011 [33]	Commercial UHF antenna and Dickson multiplier with 4 stages	UHF band	6 V	n/a	n/a

Table 4. Review of measurement of RFEH.

A comparison of the commercial requirements for sensor network nodes is presented in Table 5. Therefore, the use of RFEH for WSNs depends especially on the application, the distance from the base station, the radio-frequency band, distance between nodes, etc.

The results deduced from Tables 4 and 5 indicate that the RFEH is insufficient as a primary power source. Thus, it can be combined with other energy harvesting sources. As an example, for outdoor applications, when the base station is away from the sensor nodes, RFEH can be combined with photovoltaic energy. In a similar way, for human body sensors, this energy can be combined with thermal or vibration energy.

However, when the WSN is near to the base station, it is possible to use only RFEH as the power supply; in this case, the antenna and matching circuit must be compatible with the base station frequency. This system cannot be used for generic applications.

Operation Conditions	Crossbow MICAz	Waspmote	Intel IMote2	Tmote Sky (Telosb)
Radio standard	IEEE 802.15.4/Zig Bee	IEEE 802.15.4/Zig Bee	IEEE 802.15.4	IEEE 802.15.4
Typical range	100 m outdoor 30 m indoor	500 m	30 m	125 m outdoor 50 m indoor
Data rate (Kbps)	250	250	250	250
Sleep mode	15 µA	62 µA	390 µA	2.6 µA
Processor consumption	8 mA	9 mA	31–53 mA	500 µA
RX	19.7 mA	49.56 mA	44 mA	18.8 mA
TX	17.4 mA	50.26 mA	44 mA	17.4 mA
Supply voltage (min)	2.7 V	3.3 V	3.2 V	2.1 V

Table 5. Comparison of power consumption of sensor network nodes [34, 35].

Further, the energy-harvested design for powering sensor networks depends on different modes: sleep, transmission, reception, and minimal supply voltage required to run, (cf. Table 5), that is, it depends on the application.

4. RFEH and RFID

The RF harvesting technique is certainly a viable option for wide-range applications, including the passive RFID tags, where the signal used for communication is also used for powering [36, 37]. Therefore, RFID tags typically use the radio signal from a dedicated interrogator for power and communication. The antenna used could be designed for power harvesting and communication.

Table 6 shows the results of various studies that have been focused on RFID system powered by RFEH.

Reference	Frequency Band	Standard	Power Transmission	Antenna Gain	Distance	Efficient	Voltage	Consumption
OLG2010[9]	2.45 GHz	4 W EIRP	n/a	n/a	3.1–2.1 m	70%	1.6 V	1.6 V LED
KIT2005[20]	900 MHz	4 W EIRP	100 mW	7.5 dBi	3–3.5 m	n/a	0.6 V	2 µA
KIT2004[30]	2.45 GHz	RCD STD-1	300 mW	20 dBi	10 m	40%	>1 V	30 µW

Table 6. Review of RFID system using RFEH power.

It is well known that RFID systems generate and radiate electromagnetic waves; thus, they are justifiably classified as radio systems. However, they are not considered as RFEH systems, since they get their energy from readers. Hence, an RFID system uses the radio frequency signal in order to power and activate the tag, whereas in RFEH system, the energy source is usually not controlled by the reader. The identification process is presented in Figure 8. The energy is sent by using a radio frequency signal in order to receive the information from tags. Furthermore, the passive tags, as no battery, are smaller and lighter than the active and semi-passive tags.

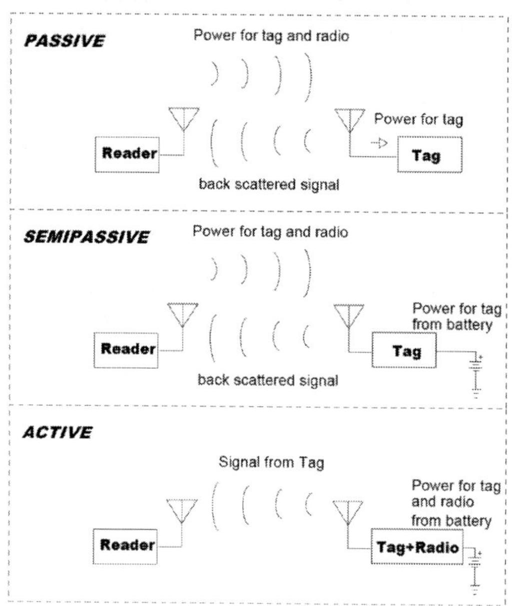

Figure 8. RFID systems [37].

Regarding RFID frequencies, there are four main frequency bands available for RFID systems:

- Low frequency (LF: 125–134 KHz).

- High frequency (HF: 13.56 MHz).

- Ultrahigh frequency (UHF: 956 MHz in USA and 866 MHz in Europe).

- Microwave band (2.45–5.8 GHz).

Despite the excellent progress made in the RFID technology, several issues still need to be addressed appropriately related to reliability, security, speed of communications, and evolution to a global standard. Therefore, it is highly suitable to develop compact transponders applicable for a long reading range, with a low price and a long life.

5. Conclusion

RF energy harvesters open up new exciting possibilities in wireless communication and networking by enabling energy self-sufficient, environmentally friendly operation with practically infinite lifetimes, and synergistic distribution of information and energy in networks. The energy is harvested from commercial RF broadcasting stations, especially for powering wireless sensor networks or other applications that require only a small amount of energy (10^{-3} to 10^{-6} W). Further, RFID sensors can be powered by scavenging ambient power from radio frequency signals in order to prolong the lifetime to several decades and reduce maintenance costs.

This study is expected to provide a survey that offers a holistic view of RF energy harvesting process. Therefore, this paper covers various approaches of RF energy harvesting in order to meet the future demand for self-powered devices. All the subsystems of an RF harvester are discussed, including the receiving antenna, the matching circuit, and the rectifier. Hence, several research groups have proposed RF harvesters in order to achieve optimum power density and ensure a permanent power supply. Finally, RF energy harvesting is an emerging and active research area where more advancement is required to harvest energy efficiently.

Future works can be made to design antennas operating at several frequencies including at 2.3 GHz (Wimax), 2.4 GHz (WLAN), 2.6 GHz (LTE/4G), as well as 5.2 GHz (WLAN). Furthermore, the DC voltage of the rectenna needs to be improved in order to ensure that the optimum power transfer can be delivered.

Acknowledgements

This study was supported in part by the EMMAG Program, 2014, funded by the European Commission.

Author details

M. Pareja Aparicio[1], A. Bakkali[1,2], J. Pelegri-Sebastia[1*], T. Sogorb[1], V. Llario[1] and A. Bou[1]

*Address all correspondence to: jpelegri@upv.es

1 Research Institute for Integrated Management of Coastal Areas, Universitat Politècniva de Valencia, Spain

2 Laboratory of Innovative Technologies, Abdelmalek Essaadi University, Morocco

References

[1] S. Sudevalayam, P. Kulkarni, Energy Harvesting Sensor Nodes: Survey and Implications, IEEE Communications Surveys & Tutorials, Vol. 13, No. 3,pp.443-461, 2011.

[2] M. T. Penella Lopez, Methods and Circuits for the Efficient Management of Power and Energy in Autonomous Sensors, Doctorate Thesis, University Polytechnic of Catalunya, 2010.

[3] S. Chalasani, J. M. Conrad, A Survey of Energy Harvesting Sources for Embedded Systems, IEEE Southeast Conference, IEEE, Huntsville, AL, 2008.

[4] E. Sardini, M. Serpelloni, Self-Powered Wireless Sensor for Air Temperature and Velocity Measurements with Energy Harvesting Capability, IEEE Transactions on Instrumentation and Measurement, Vol. 60, No. 5, pp. 1838-1844, 2011.

[5] Y. K. Tan, S. K. Panda, Self-Autonomous Wireless Sensor Nodes with Wind Energy Harvesting for Remote Sensing of Wind-Driven Wildfire Spread, IEEE Transactions on Instrumentation and Measurement, Vol. 60, No. 4, pp. 1367-1377, 2011.

[6] S. A. Bhalerao, A. V. Chaudhary, R. B. Deshmukh, R. M. Patrikar, Powering Wireless Sensor Nodes using Ambient RF Energy, IEEE International Conference on Systems, Man and Cybernetics, Taiwan, 2006.

[7] P. Nintanavongsa, U. Muncuk, D. R. Lewis, K. R. Chowdhury, Design Optimization and Implementation for RF Energy Harvesting Circuit, IEEE Journal on Emerging and Selected Topics in Circuits and Systems, Vol. 2, Issue 1, pp. 24-33, 2012.

[8] C. Mikeka, H. Arai, Dual Band RF Energy Harvesting Circuit for Range Enhancement in Passive Tags, Proceeding of the 5th European Conference on Antennas and Propagation (EUCAP'11), IEEE, Rome, 2011.

[9] U. Olgun, C.C. Chen, J.L. Volakis, Wireless Power Harvesting with Planar Rectennas for 2.45 GHz RFIDs, International Symposium on Electromagnetic Theory, IEEE, Berlin, 2010.

[10] J.I. Moon, Y.B. Jung, Novel Energy Harvesting Antenna Design Using a Parasitic Radiator, Progress In Electromagnetics Research, Vol. 142, pp. 545–557, 2013.

[11] F. Xie, G. M. Yang, W. Geyi, Optimal Design of an Antenna Array for Energy Harvesting, IEEE Antennas and Wireless Propagation Letters, Vol. 12, pp.155-158, 2013.

[12] M.T. Penella, M. Gasulla, Powering Autonomous Sensors, Springer Netherlands, Springer Science + Business Media, DOI:10.1007/978-94-007-1573-8, 2011.

[13] T.T. Le, Efficient Power Conversion Interface Circuits for Energy Harvesting Applications, Doctorate Thesis, Oregon State University, 2008.

[14] J.P. Curty, N. Joehl, F. Krummeanacher, C. Dehollain, M.J. Declercq, A model for u-Power Rectifier Analysis and Design, IEEE Transactions on Circuits and systems I, 2005.

[15] M. Muramatsu, H. Koizumi, An Experimental Result using RF Energy Harvesting Circuit with Dickson Charge Pump, IEEE Conference on Sustainable Energy Technologies (ICSET), Sri Lanka, 2010.

[16] T. Sogorb, J. V. Llario, J. Pelegrí, R. Lajara, J. Alberola, Studying the Feasibility of Energy Harvesting from Broadcast RF Station for WSN, IEEE International Instruments and Measurement Technology Conference, Canada, 2008.

[17] R.H. Chen, Y. C. Lee, J. S. Sun, Design and Experiment of a Rectifying Antenna for 900 MHz Wireless Power Transmission, IEEE Asian-Pacific Microwave Conference, IEEE, Macau, 2008

[18] D. Bouchouicha, F. Dupont, M. Latrach, L. Ventura, Ambient RF Energy Harvesting, International Conference on Renewable Energies and Power Quality (ICREPQ'10), Spain, 2010.

[19] U. Olgun, C. C. Chen, J. L. Volakis, Investigation of Rectenna Array Configurations for Enhanced RF Power Harvesting, IEEE Antennas and Wireless Propagation Letters, Vol. 10, pp. 262-265, 2011.

[20] H. Kitayoshi, K. Sawaya, Long Range Passive RFID-Tag for Temperature Monitor System, Proceeding of International Symposium of Antennas and Propagation (ISAP'06), Korea, 2006.

[21] M. Arrawatia, M. S. Baghini, G. Kumar, RF Energy Harvesting System at 2.67 and 5.8 GHz, Proceeding of Asia-Pacific Microwave Conference, IEEE, Yokohama, 2010.

[22] F. Yuan, CMOS Circuits for Passive Wireless Microsystems, Springer-Verlag New York, Springer Science + Business Media, Doi. 10.1007/978-1-4419-7680-2, 2011.

[23] K. Avuri, K.A. Devi, S. Sadasivam, N.M. Din, C.K. Chakrabarthy, Design of a 377 Ω Patch Antenna for Ambient RF Energy Harvesting at Downlink Frequency of GSM 900, 17th Asian Pacific Conference on Communications, IEEE, Sabah, 2011.

[24] M. Pinuela, P.D. Mitchenson, S. Lucyszyn, Analysis of Scalable Rectenna Configurations for Harvesting High Frequency Electromagnetic Ambient Radiation, International Conference on Micro and Nanotechnology for Power Generation and Energy Conversion Applications, Belgium, 2010.

[25] H. Jabbar; Y. S. Song, T. T. Jeong, RF Energy Harvesting System and Circuits for Charging of Mobile Devices, IEEE Transactions on Consumer Electronics, Vol.56, No. 1, 2010, pp. 247–253.

[26] Nimo, D. Grgic, L. M. Reindl, Impedance Optimization of Wireless Electromagnetic Energy Harvester for Maximum Output Efficiency at uW Input Power, Active and

Passive Smart Structure and Integrated Systems Conference, San Diego, California, 2012.

[27] F. Kocer, M. P. Flynn, An RF-Powered, Wireless CMOS Temperature Sensor, IEEE Sensor Journal, Vol. 6, No. 3, 2006.

[28] O. A. Escala, Study of the Efficiency of Rectifying Antenna Systems for Energy Harvesting, Thesis for the degree of Engineer, UPC-Barcelona, Spain, 2010.

[29] Mikeka, H. Arai, Chapter 10: Design Issues in Radio Frequency Energy Harvesting System, Sustainable Energy Harvesting Technologies – Past, Present and Future, In-Tech, Croatia, 2011.

[30] A. M. Zungeru, Li-Minn Ang, S. R. S Prabaharan, K, P. Seng, Radio Frequency Energy Harvesting and Management for Wireless Sensor Networks, Green Mobile Devices and Networks: Energy Optimization and Scavenging Techniques, CRC Press Taylor and Francis Group, pp. 341–368, DOI: 10.1201/b10081-16, Oxon, Uk, 2011.

[31] Mikeka, H. Arai, A. Georgiadis, A. Collado, DTV Band Micropower RF Energy-Harvesting Circuit Architecture and Performance Analysis, IEEE International Conference on RFID-Technologies and Applications, Sitges, Spain, 2011.

[32] C. Mikeka, H. Arai, Microwave Tooth for Sensor Power Supply in Battery free Applications, Proceeding of the Asia-Pacific Microwave Conference, IEEE, Melbourne, VIC, 2011.

[33] N. Amaro, S. Valchev, An UHF Wireless Power Harvesting System Analysis and Design, International Journal of Emerging Sciences, Vol. 1, No. 4, pp. 625-634, 2011.

[34] A.Pavone, A. Buonanno, F. D.Corte, Design Considerations for Radio Frequency Energy Harvesting Devices, Progress in Electromagnetic Research B, Vol. 45, pp. 19–35, 2012.

[35] R. Lajara, J.Pelegrí, J. J. Perez, Power Consumption Analysis of Operating Systems for Wireless Sensor Networks. Sensors 2010, Vol. 10, pp. 5809–5826.

[36] H. Kitayoshi, K. Sawaya, Development of a passive RFID-Tag with 10-m Reading Distance under RCR STD-1 Specification, Proceeding of International Symposium of Antennas and Propagation (ISAP'04), Japan, 2004.

[37] M. Dobkin, The RF in RFID: Passive UHF RFID in Practice, Elsevier, Oxford, UK, 2008.

System Integration

Chapter 8

Renewable Energy, Emissions, and Health

María del P. Pablo-Romero, Rocío Román, Antonio Sánchez-Braza and Rocío Yñiguez

Additional information is available at the end of the chapter

http://dx.doi.org/10.5772/61717

Abstract

The deployment of renewable energy sources is reviewed in this research showing the importance that they have reached in most countries. The International Energy Agency has insisted on the importance of their promotion all over the world, considering that at least one renewable energy source is available in all countries. The aim of this chapter is to show that although renewables are an effective alternative to the use of fossil fuels, there are other important positive externalities. As the fossil fuels are the main source of greenhouse emissions and other air pollutants, the negative effects that they have on human health, such as respiratory and cardiovascular diseases, have been recently shown in many studies. When renewables contribute to reducing the use of fossil fuels and associated air pollutant emissions, they have a positive effect on human health. Therefore, policy makers have to take into consideration all these positive externalities of renewable sources, when evaluating the possibility of their promotion. However, this evaluation should also take into consideration that not all renewable energy sources have equivalent positive effects. Our final conclusion is that governments should be supported by recent research when deciding the most appropriate energy mix for a country.

Keywords: Renewable energy, environmental health, air emissions, environmental policy

1. Introduction

Since 2006, when the Stern report was published [1], commissioned by the UK Treasury, new research has revealed the negative impacts that climate change may have on economic growth and health. The impacts of economic change are based on the assumption that greenhouse gases will raise the average temperature in the next 50 years by between 2°C and 3°C, which represents a threat to the basic elements of human life in different parts of world: access to water, food production, health, land use, and environment. Furthermore, as empirical

evidence has shown, these harmful effects resulting from climate change will be accelerated to the extent that the world will warm more, with poor countries being mainly affected [2].

Weather conditions affect the patterns of diseases such as diarrhea, malaria, malnutrition, etc., most of which affect children and youth in low-income countries [3]. Some of these effects have already taken place [4, 5], not by generating new diseases, but rather by exacerbating existing illnesses, the effects concerning respiratory diseases being particularly important.

According to the Global Energy Assessment Report 2012 [6], the direct effects of climate change, which increase health issues, are manifested in the increase in heat waves and rising sea levels. Climate change favors the increase of heat waves and, these heat waves generate an increase in the number of deaths due to heart attacks, when temperatures rise above 30 degrees and certain level of humidity is reached. Also, higher temperatures increase the formation of ozone, which in turn has negative effects on respiratory and cardiovascular health. Furthermore, scientists expect that the sea level may rise by a meter or more this century [7, 8]. This increase in sea level will have negative effects on the health of the population and crops. Additionally, climate change will also affect health indirectly, through changes in nutrition and changes in the development of infectious diseases that often increase when temperatures are rising.

Within the field of combating climate change is set the target of reducing the use of polluting energy. Two policies are combined to obtain the reduction: energy efficiency and the substitution of polluting energy by alternative sources such as nonpolluting energy sources or renewable energies.

Renewable energy production has its basis in reducing emissions to the environment, either by reducing CO_2 emissions or other air pollutants such as $PM_{2.5}$ associated with transportation, which also have a direct negative effect on health.

At the global level, various policies are intended to replace polluting fuels with renewable energies in different percentages in the coming years. The ultimate goal of such substitution is also the reduction of emissions by certain percentages. For example, this is the case of the EU target for 2020, in which the need to reduce emissions by 20% is established, along with the target of 20% of renewable energy participation in energy consumption.

The development of these clean energy sources is encouraging the study of their effects on health, either by mitigating climate change or by improving air quality. In this regard, several studies have been developed to assess the impact on the health cost of implementing renewable energy. Previous analysis gives an overall perspective for Denmark [9] and gives a more specific analysis about the impact that the mitigation of certain particles has on health, that is, the impact of renewable energy on health by reducing $PM_{2.5}$ [10].

The novelty of our chapter is to highlight that one of the most important positive externalities of using renewable energies is that they contribute to reduce CO_2 and air emissions and improve human health.

The key question is that renewable energies are becoming a solution to the abatement of CO_2 in most developed countries as substitute fossil fuels that are pollutants, they contribute to reducing CO_2 and air pollutant emissions.

Additionally, the research studies link the effects of climate change, or more accurately, the effects of CO_2 emissions on health are becoming more frequent. Of course, the epidemiological studies are those that provide information about the effects that the air pollutants cause on human health.

The final aim of this chapter is to combine this information in order to improve the research studies on this important issue, the link among renewable energies, emissions, and health.

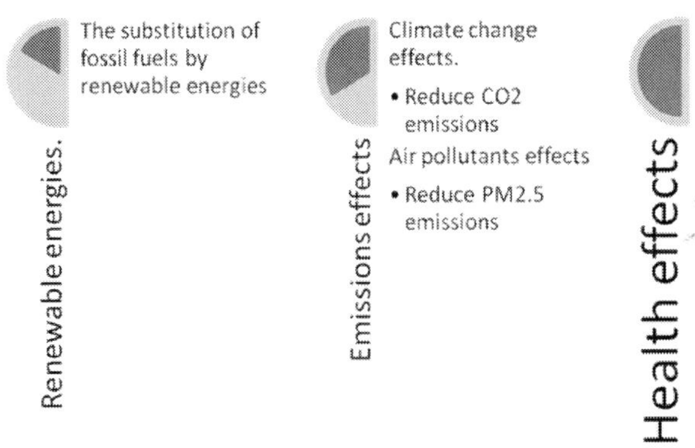

Figure 1. Renewable energies, emissions, and health.

This book chapter aims to establish a recent literature review on studies linking the establishment of renewable energy systems with their effect on health, distinguishing between the studies that are focused on the link between climate change and health with a global perspective, and those studies that link the effects of some specific pollutant on health.

Therefore, after an initial introduction about the health effects of climate change and environmental pollution, Section 2 is devoted to the current situation and development of renewable energy, and the main proposals for increased production of these energies replacing polluting energies. Section 3 gives a literature review of analysis linking renewable energy and impact on health. A discussion of energy policies in relation to the results of previous studies is made in Section 4. Finally, Section 5 summarizes the conclusions.

2. The development of renewable energy

2.1. CO_2 emissions intensity

REmap 2030, made by the International Renewable Energy Agency (IRENA), shows that under current policies and national plans (the "business as usual" case), average carbon dioxide (CO_2) emissions will only drop to 498 g/kWh by 2030. That is not enough to keep atmospheric CO_2 levels below 450 parts per million (ppm), and, what is more, severe climate change is expected to take place. A doubling in the share of renewable energies could help check climate change by reducing the global average emissions of CO_2 to 349 g/kWh, which is equivalent to a 40% intensity reduction compared to the 1990 levels (see Figure 2) [11, 12].

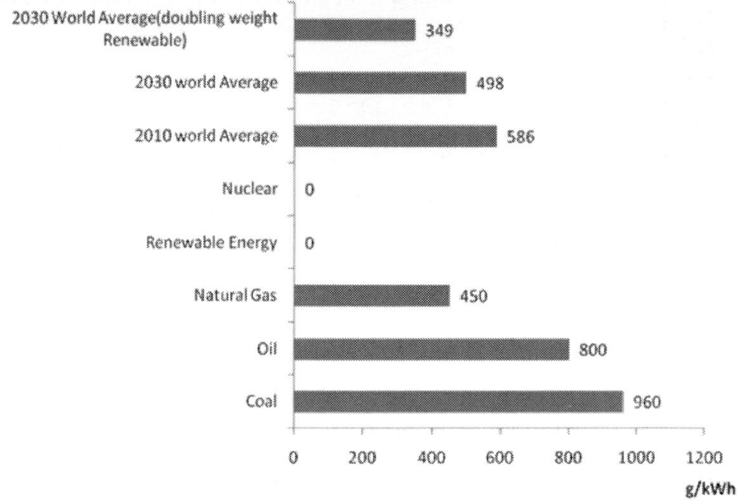

Figure 2. CO_2 emissions intensity

2.2. Renewable energy capacity

Each year since 2011 there has been, on a global basis, an addition of more than100 GW of new renewable energy capacity. This figure is the same as Brazil's total installed generation capacity, or double Saudi Arabia's. Renewables are more than 50% of the net capacity which has been added to the global power sector since 2011. This means that more new renewables capacity is being set up than that in fossil and nuclear power put together. This led to the share of renewables in total electricity production surpassing a record 22% by 2013; 16.4% of this was hydro and 3.6% was solar photovoltaics (PV) and wind. The renewable energy industry employed about 6.5 million people in 2013, an increase of 14% over 2012 [11].

We can see an increasing use of up-to-date renewable energy in four different markets: transport fuels, power generation, heating and cooling, and rural/off-grid energy services. When we break down modern renewables, taken as a share of total final energy use in 2012, we get the following: hydropower generated about 3.8%; other renewable power sources were 1.2%; heat energy was around 4.2%; and transport biofuels supplied more or less 0.8% [13]. According to REN21 [13], from 2009 to 2013, the installed capacity, as well as the output of most renewable energy technologies, increased at rapid rates, especially in the power sector (see Figure 3).

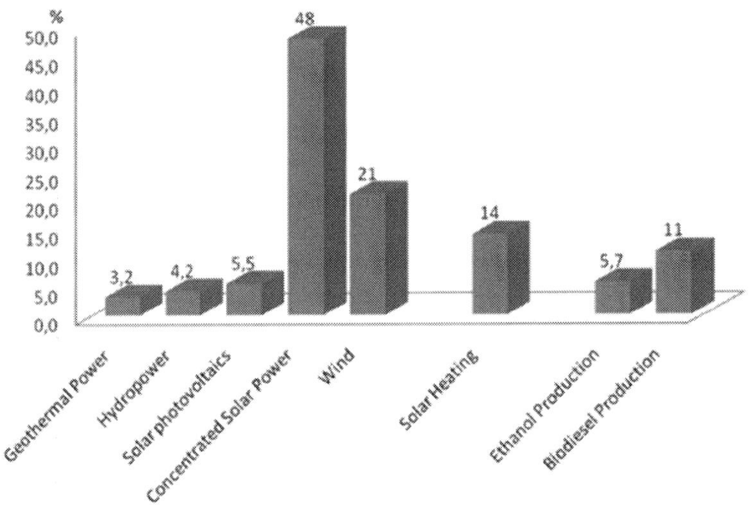

Figure 3. Average annual growth rates of renewable energy capacity and biofuels production, 2009–2013

However, a number of European countries that is on the increase have decreased, at times ex post facto, their financial backing for renewables in excess of the reduction of in technology costs. This has been brought on, partly, by the continuing economic crisis in certain and by increasing competition with fossil fuels. Doubts about policy have raised capital costs, hindering the financing of projects and lowering investment. During 2013, Europe continued to see a notable loss of start-up companies, especially solar PV (in the PV market, capacity additions dropped for a second year in a row, from 17GW in 2012 to 10GW in 2013).

Furthermore, renewables work in an uneven area where the prices of energy are not completely in line with external factors. Worldwide subsidies for fossil fuels and nuclear power remain high despite talk about phasing them out, thereby fostering energy use which is inefficient and at the same time also impeding investment in renewables. Estimates for the global cost of fossil fuel subsidies range from USD 544 billion to USD 1.9 trillion, figures which are notably above those for renewable energy [13].

2.3. Renewable energy by regions

In general, despite some exceptions in Europe and the United States, there were some significant and positive developments for renewables in 2013. Wind power advanced more steadily into Africa and Latin America; concentrating solar thermal power moved its focus further to the Middle East and to North and South Africa; renewable process heat fueled industries from Chile to Europe to India; and solar PV maintained its spread across the globe, with most capacity on-grid but also notable increases in off-grid markets in developing countries [13].

In the United States, the percentage of renewable generation increased to almost 12.9% (12.2% in 2012). This was in spite of a decrease in hydropower output and competition from cheap natural gas coming from shale. Meanwhile, between 2008 and 2013 the quantity of net coal-based electricity generation went down by nearly 19%.

China had not before had more new renewable power capacity than new fossil and nuclear capacity. Altogether, renewables were more than 20% (> 1,000 TWh) of its electricity generation. Seventy-two percent of new electric capacity was due to renewable power installations in the European Union. This is 70% more than in 2012. A decade before, in the EU-27 plus Norway and Switzerland the situation was completely different, with conventional fossil generation being approximately 80% of new capacity.

Spain was the first country to generate more electricity from wind power than from any other source, with 20.9% of the total.

There was an increase of over 4 GWin India's renewable capacity, up to approximately 70.5 GW. Though hydropower was most of the total (62%), solar PV and wind was close to 70% of 2013 renewable additions. All the same, India's power capacity is increasing fast, and renewables did not reach 17% of total additions from all sources in 2013.

Due to its competitiveness compared to the rest of the sources of energy generation, wind power was not allowed to be part of an auction in Brazil. The year ended with Brazil having 3.5 MW of wind power capacity, and there were contracts for another 10 GW. Although worldwide investment in solar PV decreased by almost 22% compared to 2012, new capacity installations went up by over 32%. By the beginning of 2013, 18 countries used non-hydro renewable resources to generate over 10% of their electricity, an increase – in 2010 only 8 countries did so. Austria, Kenya El Salvador, Lithuania, and Denmark were among these countries.

Looking at communities and regions from around the world, we see many that have targeted, or have already completed a successful transition to 100% renewable electricity. Some examples are Djibouti, Scotland, and the small-island state of Tuvalu. All of them plan to use renewable sources to get all their electricity by the end of the decade [13].

On a worldwide basis, homes and work places are choosing "green" instead of from both traditional utilities and new energy providers, freely opting to buy renewable energy whose production is not within regulatory requirements. Germany is still at the forefront in voluntary renewable power purchasing. Domestic customers increased from 0.8 million in 2006 to 4.9 million in 2012. This stands for 12.5% of the country's private households. In 2011, they bought 15 terawatt-hours (TWh) of green power, and in this year a further 10.3 TWh was purchased by commercial customers. Austria, Belgium (Flanders), Finland, Hungary, the Netherlands, Sweden, Switzerland, and the United Kingdom are among other major European green power markets, although the market share in these countries remains below the level of Germany. There are also green power markets in Australia, Canada, Japan, South Africa, and the United States [13].

Major industrial and commercial customers in Europe, India, Mexico, and the United States continued to turn to renewables to reduce their energy costs at the same time as increasing the reliability of their energy supply. Many set ambitious renewable energy targets in 2013, installed and operated their own renewable power systems, or signed purchase agreements to purchase directly from renewable energy project operators, by passing utilities.

In Australia, Japan, and Thailand, as well as in North America and several countries in Europe, the number of community-owned and co-operative projects also rose. Denmark has a long history of co-operatively owned projects. In 2013, close to 50% of the renewable power capacity in Germany belonged to its inhabitants and around 20 million Germans are living in regions that are considered as having 100% renewable energy [14].

Renewables have been helped by continuing advances in technologies, decreases in prices, and innovations in financing, driven largely by policy support. These developments mean that under many circumstances renewable energy is cheaper than new fossil and nuclear installations, and as a result of this, for a greater number of consumers in developed and developing countries it is more affordable. Furthermore, there is a growing awareness of renewable energy technologies and resources, and their capability to keep up with a demand for energy that is on the rise. At the same time, these renewables are creating jobs, accelerating economic development, reducing local air pollution, improving public health, and reducing carbon emissions [13].

There is also the issue of serious health problems which directly result from burning fossil fuels. The United States Environmental Protection Agency recently rated the cost of national ill health due to fossil fuel costs at between USD 362 billion and USD 887 billion per year. The European Union's Health and Environment Alliance noted that coal-fired power plant emissions cost as much as EUR 42.8 billion in health expenditure on an annual basis (IRENA) [11].

Meanwhile, pressure is on the rise for 1.3 billion people to begin accessing electricity. A great number of them live in areas which are hard to reach, thus excluding the usual large-scale power plants and transmission systems. At the same time, 2.6 billion people suffer from severe health problems due to having to use traditional biomass and cook with conventional stoves [11].

These tendencies mean that there is an obvious need for a change. Although fossil fuels drove the first industrial revolution, even in the new period of shale oil and gas, there is still the matter about their being compatible with sustainable human well-being. The time is ripe for a period of prevailing, sustainable, and cost-competitive modern renewable energy. Renewable energy technologies which supply electricity, heating and cooling, and transportation is at the moment used all over the world and trends lead us to consider that these technologies will globally continue to grow. In the recent past, renewables interested environmentally oriented people as an alternative to traditional fuels. Nowadays, it is seen that they are not only beneficial from the environmental point of view but also boost the economy, diversify the sources of revenue, and vitalize fresh advances in technology [13].

3. Renewable energy and health

3.1. Climate change and health

Climate scientists have observed that CO_2 concentrations in the atmosphere have been increasing significantly over the past century, the concentration of CO_2 in 2013 being about 40% higher than in the mid-1800s [15]. The Fifth Assessment Report from the Intergovernmental Panel on Climate Change states that human influence on the climate system is clear [16] as the rise of greenhouse gases (GHG) increase the global mean temperatures. In that sense, as stated in [17], most of the observed increase in global average temperature since the mid-twentieth century is very likely due to the observed increase in anthropogenic GHG concentrations.

The rise in temperatures has implications for life and human health. According to the World Health Organization [18], potential risks to health include deaths from thermal extremes and weather disasters, vector-borne diseases, a higher incidence of food-related and waterborne infections, photochemical air pollutants, and conflict over depleted natural resources. The valuation of these effects is complex because many effects on climate change depend on physical, ecological, and social factors. Therefore, according to the recent report by IPCC [16], all the various scientific studies that exist so far are debatable, since they are based on a number of assumptions, some of them very controversial. Also, many of these studies do not consider catastrophic changes or many other factors that can affect the economy of countries.

Despite these limitations, according to the World Health Organization [18], climate change was estimated to be already responsible for 0.2% of deaths in 2004, 85% of these being child deaths. Likewise, climate change was responsible for 3% of diarrhea, 3% of malaria, and 3.8% of dengue fever deaths worldwide. In addition, increased temperatures hastened as many as 12,000 additional deaths (not included in the above percentage because the years of life lost by these individuals were uncertain). Evenly, it will cause some 250,000 additional deaths annually between 2030 and 2050; 38,000 elderly people from exposure to heat; 48,000 from diarrhea; 60,000 from malaria; and 95,000 from child malnutrition.

Some studies have analyzed the impact of climate change on health in some countries in particular. Thus, in the study by [19] it is estimated that 166,000 deaths and about 5.5 million disability-adjusted life years (DALYs, a measure of overall disease burden) are due to cardiovascular disease, malnutrition, diarrhea, malaria, and floods.

The report titled Environmental Health Perspectives (EHP) [20] notes that the effects of climate change on health can be classified into 11 broad human health categories. Table 1 lists the main diseases associated with these categories.

Diseases	Effects of climate change on health
Asthma, respiratory allergies, and airway diseases	The global rise in asthma is indirectly related to climate change [21]. Climate change may increase the incidence and exacerbation of many respiratory allergic diseases. Some risk for respiratory disease can be clearly linked to climate change.

Diseases	Effects of climate change on health
Cancer	Climate change will result in higher ambient temperatures that may increase the transfer of volatile and semi-volatile compounds from water and wastewater into the atmosphere, changing subsequent human exposures [22]. Climate change is expected to increase heavy precipitation and flooding events, which may increase the chance of toxic contamination Depletion of stratospheric ozone, will result in increased ultraviolet (UV) radiation exposure, which increases the risk of skin cancers and cataracts [23].
Cardiovascular disease and stroke	There is evidence of climate sensitivity for several cardiovascular diseases, with both extreme cold and extreme heat directly affecting the incidence of hospital admissions for chest pain, acute coronary syndrome, stroke, and variations in cardiac dysrhythmias, though the reported magnitude of the exposure-outcome associations is variable [24-28]. There is also evidence that heat amplifies the adverse impacts of ozone and particulates on cardiovascular disease [29, 30]. Increased burden of $PM_{2.5}$ is associated with increased hospital admissions and mortality from cardiovascular disease, as well as ischemic heart disease [31, 32]. Indirectly, displacement related to disasters is frequently associated with interruptions of medical care for chronic medical conditions [33] Climate is also implicated in another indirect risk for cardiovascular disease: the incidence of certain vector borne and zoonotic diseases (VBZD) with cardiovascular manifestations [34].
Foodborne diseases and nutrition	Climate change may impact rates of foodborne illness through increased temperatures, which are associated with increased incidence of foodborne gastroenteritis Changes in ocean temperature and coastal water quality are expected to increase the geographic range of V. vulnificus and cholera [35]. Increased temperatures also affect rates of campylobacteriosis and salmonellosis. In [36] was stated a 2.5–6% relative increase in the risk of foodborne illness for every degree centigrade rise in temperature..
Heat-Related morbidity and mortality	The health outcomes of prolonged heat exposure include heat exhaustion, heat cramps, heat stroke, and death [37]. Prolonged exposure to heat may exacerbate pre-existing chronic conditions such as various respiratory, cerebral, and cardiovascular diseases [38]. The levels of pollen and other allergens are also higher in case of extreme heat.
Human developmental effects	Climate change will lead to changes in agricultural practices that might increase pesticide use. (e.g. DDT). Pesticides have been linked to decreases in fertility [39, 40] and to effects on fetal loss, child growth, and male reproductive development [41-43].

Diseases	Effects of climate change on health
Mental health and stress-related disorders	A variety of psychological impacts can be associated with extreme weather and other climate related events. Extreme weather events such as hurricanes, wildfires, and flooding, can lead to mental health disorders [44]. Prolonged heat and cold events can create chronic stress situations that may initiate or exacerbate mental disease and stress-related disorders
Neurological diseases and disorders	Effects of climate change on ocean health result in increased risks to neurological health from ingestion of or exposure to neurotoxins in seafood and fresh and marine waters, even a single low-level exposure to algal toxins [45]. Exposure to a number of agents whose environmental presence may increase with climate change may have effects on neurological development and functioning (pesticides, herbicides, heavy metals, extreme weather events etc.)
Vector borne and zoonotic diseases	Climate is one of several factors that influence the distribution of vector borne and zoonotic diseases (VBZD) such as Lyme disease, Hantavirus, West Nile virus, and malaria
Waterborne diseases	Climate change is expected to produce more frequent and severe extreme precipitation events worldwide and heavy rainfall is correlated with more than half of the outbreaks of waterborne diseases [46].
Weather-related morbidity and mortality	The health impacts of the extreme weather events include direct impacts such as death and mental health effects and indirect impacts such as population displacement and waterborne disease outbreaks

Compiled from the report by EHP [20]

Table 1. Effects of climate change on health

Likewise, climate change will have the greatest effect on health in societies with scarce resources, little technology, and frail infrastructure. After reviewing what happened in the past, it has been shown in [47] throughout history, higher temperatures have reduced economic growth mainly in poor countries, not only reducing the level of production but also growth, and with a greater effect in agriculture, industrial, and political stability.

3.2. Renewable energy, emissions, and health

To avoid the adverse impacts of climate change, the Cancun Agreements of the Conference of Parties (COP) [48] called for limiting global average temperature rise to no more than 2°C above pre-industrial values, which means decreasing CO_2 by 50–85% below 2000 levels by 2050 [17]. Among the many human activities that produce greenhouse gases, the use of energy represents by far the largest source of emissions, with CO_2 resulting from the oxidation of carbon in fuels during combustion dominating the total GHG emissions [15].

It is necessary to evaluate every step of the entire system of the fuel cycle, from the extraction of raw materials to final consumption of energy, in order to make a quantitative estimate of the risks to health from the various energy systems. According to [49], energy technologies can be classified into three groups, depending on the risks to health.

First, the fuels group: These technologies are characterized by the use of fossil fuels or biomass: coal, oil, natural gas, wood, and so on, the burning of which produces large amounts of air pollution and solid waste.

Second, the renewable group: This group is characterized by use of diffuse renewable resources with low energy density: sun, wind, water, the capturing of which requires large areas and construction of expensive facilities. Public risks are low, due to their low emissions, mostly confined to low-probability accidents, such as dam failures, equipment failures, and fires.

Third, the nuclear group: This includes nuclear fission technologies, characterized by high energy densities in the processed fuel, with corresponding low quantities of fuel to process and generating little waste to transform. Public risks are located in routine operations of reactors.

Thus, emissions given off by each technology are different and, therefore, their effect on climate change will also be distinct. In addition, technologies also differ in the way in which these emissions occur, as in some cases they are concentrated in the areas of production and in other cases in the territories of use.

Given the distinction between energy technologies, the Intergovernmental Panel on Climate Change [50] collects the lifecycle GHG emissions of each technology from estimates obtained in previous studies. According to the Intergovernmental Panel on Climate Change [50], without taking into account emissions related to land use change (LUC), the lifecycle GHG emissions normalized per unit of electrical output (g CO_2eq/kWh) from technologies powered by renewable resources are generally found to be considerably less than from those powered by fossil fuel-based resources. In that sense, renewable energy sources play a role in providing energy services in a sustainable manner and, in particular, in mitigating climate change.

The estimates of lifecycle GHG emissions for broad categories of electricity generation technologies calculated [50] show that all electricity generation technologies powered by non-renewable resources are much higher than those powered by renewable resources, except for nuclear (Figure 4). The median values for all renewable energy range from 4 to 46 g CO_2 eq/kWh, while these values range from 480 to 1,000 g CO_2 eq/kWh for non-RE (except nuclear). Among RE, photovoltaic and biopower technologies are those which emit more, up to 2–3 times above the maximum for other renewable energy technologies. Thus, the maximum value for lifecycle GHG emissions for petroleum gasoline and diesel is around 110 g CO_2 eq/MJ of fuel while the maximum value for biodiesel is 78 g CO_2 eq/MJ of fuel and for ethanol 70 g CO_2 eq/MJ of fuel. Nevertheless, not all biofuel systems are equally efficient in reducing GHG emissions compared to their petroleum counterparts. For example, ethanol from Brazilian sugarcane (45 g CO_2 eq/MJ of fuel) has lower GHG emissions than that produced from wheat and corn (70 g CO_2 eq/MJ of fuel).

Given the differences in emissions of different technologies, the potential impacts on human health due to emissions from the complete life cycle of future power generation, by using the

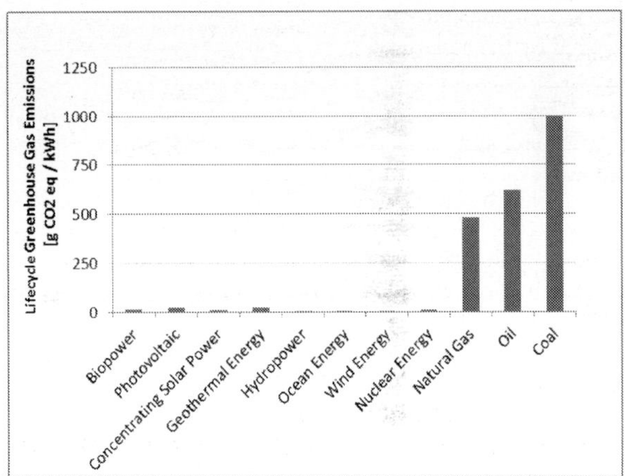

Figure 4. Estimates of lifecycle GHG emissions for electricity generation technologies (medium values)

Life Cycle Assessment (LCA) method, has been calculated [51]. LCA is a standardized method used to quantify environmental burdens and the potential impacts on human health and the environment, due to the production and consumption of goods and services. It allows for an evaluation of the environmental performance of fossil, nuclear, and renewable power gener‐ ation technologies. Health effects related to climate change, and considered in the study, include malnutrition, diarrhea, cardiovascular diseases, coastal and inland flooding, and malaria.

The results of the LCA show that total human health impacts have a similar pattern as GHG emissions according to the hierarchic and individualist perspectives: fossil electricity genera‐ tion has the highest levels of GHG emissions-human health impacts, while nuclear and renewables technologies have impacts of a scope somewhat lower, and fossil fuels with carbon capture and storage show intermediate levels. Nevertheless, when considering the egalitarian perspective, which regards the broadest range of potential health damages, and contemplates the longest time horizon, health impacts from natural gas power generation are in the same range as nuclear and some renewables [51].

3.3. Renewable energy, air pollutants, and health

In addition to reducing GHG emissions, renewable energy technologies can also offer benefits with respect to air pollution and health, compared to fossil fuels. According to the World Health Organization [52], the air pollutants that are emitted by energy technologies which have the most important impacts on human health are particulate matter (PM), nitrous oxides (NOx), sulfur dioxide (SO_2), and non-methane volatile organic compounds (NMVOC).

The IPCC highlights that non-combustion renewable energy technologies and nuclear power cause comparatively minor emissions of air pollutants [50]. The higher cumulative lifecycle emissions of NOx, SO_2, and $PM_{2.5}$ per unit of electricity generated are provoked by the combustion of hard coal, lignite, oil, and biomass in steam turbines. The higher emissions of NMVOC are provoked by the combustion of natural gas and oil in steam turbines. The SO_2 emissions from the combustion of lignite in steam turbines is around 27 g/kWh, while from the steam turbine combustion of oil and hard coal it is 14 g/kWh and 8 g/kWh, respectively. Meanwhile the highest emissions of SO_2 from renewable energies are for biomass, at around 3 g/kWh. This pattern is quite similar for $PM_{2.5}$. The NOx emissions from wood steam are around 6 g/kWh, from oil steam turbines 4.5 g/kWh, and from hard coal steam turbines 4 g/kWh. Alternatively, the maximum NMVOC emissions are registered for the natural gas steam turbines (0.7 g/kWh) and oil-fired steam turbines (0.55 g/kWh). The biogas cogeneration and roof-top PV are the renewable technologies which emit more NMVOC, with 0.25 and 0.2 g/kWh, respectively.

There are also observed differences when considering the use of different transport fuels. The use of gaseous fuels tends to reduce air pollution compared to liquid fuels [53]. The use of ethanol and biodiesel blends tends to reduce carbon monoxide (CO) and hydrocarbon emissions, compared to gasoline and diesel. Nevertheless, NOx emissions seem to be higher. Alternatively, future electric or fuel cell vehicles offer a substantial potential for reductions in air pollution if electricity or hydrogen from renewable energy sources is used as the energy carrier [54, 55].

The health effects of ambient air pollution result from a complex mixture of combustion products. Nevertheless, negative effects have been most closely correlated with three: fine PM, SO_2, and tropospheric ozone [56, 57]. Additionally, the main health impacts of these pollutants are associated with emissions by fossil fuel and biomass combustion [56, 58], the first related to transport activities. Table 2 shows the health outcomes associated with transport-related air pollutants according to the WHO report Promoting Health While Mitigating Climate Change [59].

Associated transport-related pollutants	Health outcome
Black smoke, ozone, $PM_{2.5}$	Mortality
Black smoke, ozone, nitrogen dioxide, VOCs, CAPs, diesel exhaust	Respiratory disease (non-allergic)
Ozone, nitrogen dioxide, PM, VOCs, CAPs, diesel exhaust	Respiratory disease (allergic)
Black smoke, CAPs	Cardiovascular diseases
Nitrogen dioxide, diesel exhaust	Cancer
Diesel exhaust; also equivocal evidence for nitrogen dioxide, carbon monoxide, sulfur dioxide, total suspended particles	Adverse reproductive outcomes

Source: [59]

Table 2. Health Outcomes Associated with Transport-Related Air Pollutants

It is also important to consider patterns of household fuel use, especially for developing countries. The Global Energy Assessment estimates that in 2005 about 2.8 billion people, mostly in the poorest countries, relied on solid fuels such as biomass, charcoal, and coal for cooking and other household energy needs [6]. Anyway, although both household poverty and rural location predict the use of solid fuels [60], these are still used wherever available, even in many high-income countries, as a heating fuel [61].

Poor households often used solid fuels in inefficient, poorly vented combustion devices, which results in significant waste of fuel energy and emission of toxic products of incomplete combustion. Besides, this unfit use leads to exposures that are significantly detrimental to the health of family members, particularly to women and children, who spend the most time in or near the kitchen. Very young children are especially at risk because they are highly exposed during vulnerable developmental periods [6].

The amounts and relative proportions of the various pollutants generated by solid fuel combustion depend on a number of factors, including fuel type and moisture content, stove technology, and operator behavior. Notwithstanding carbon monoxide and particles being the most commonly measured pollutants, a range of other products of incomplete combustion is found in solid fuel smoke, including phenols, oxides of nitrogen, quinones/semiquinones, chlorinated acids such as methylene chloride, and dioxins [62, 63]. Table 3 shows some of these pollutants from the combustion of biomass and fossil fuels, and their known toxicological characteristics.

Pollutant	Known toxicological characteristics
Particulates (PM_{10}, $PM_{2.5}$)	Bronchial irritation, increased inflammation reactivity, reduced mucociliary clearance, reduced macrophage response
Carbon monoxide	Reduced oxygen delivery to tissues due to formation of carboxyhemoglobin
Nitrogen dioxide (relatively small amounts from low temperature combustion)	Bronchial reactivity, increased susceptibility to bacterial and viral lung infections
Sulfur dioxide (relatively small amount from most biomass)	Bronchial reactivity (other toxic end points common to particulate fractions)
Organic air pollutants: Formaldehyde 1,3 butadiene	
Benzene	Carcinogenicity
Acetaldehyde	Co-carcinogenicity
Phenols	Mucus coagulation, cilia toxicity
Pyrene, Benzopyrene	Increased allergic sensitization
Benzo(a)pyrene	Increased airway reactivity
Dibenzopyrenes Dibenzocarbazoles Cresols	

Source: [64, 6, 63]

Table 3. Toxic pollutants from the combustion of biomass and fossil fuels.

The deployment of renewable energy should yield increased health benefits, and opportunities for policy measures combining climate change and (urban) air pollution mitigation are increasingly recognized.

Studies of the environmental benefits of clean energy initiatives tend to either focus on specific emission reduction objectives or on analyzing the overall emission reductions of multiple pollutants, including GHG and criteria pollutants [65].

A Texas Emissions Reduction Plan analysis in 2004 assessed the potential for clean energy to help meet NOx air quality requirements as part of a State Implementation Plan. Thus, the Texas Commission on Environmental Quality evaluated this plan and calculated that it achieved an annual reduction of NOx emissions of 346 tons in 2004 through energy efficiency and renewable energy. NOx reductions over the period 2007 to 2012 were projected to range from 824 tons per year in 2007 to 1,416 tons per year in 2012, which represent 0.5% and 1% of Texas NOx emissions in 2005, respectively [66, 67].

In the same vein, a Wisconsin study in 2007 measured CO_2, SO_2, and NOx emission reductions from the state's Focus on Energy Program, an energy efficiency and renewable energy project. This study found annual emission displacements of 1,365,755 tons of CO_2, 2,350 tons of SO_2, and 1,436 tons of NOx from 2001 through 2007, which respectively represent about 2%, 1%, and 2.5% of Wisconsin emissions in 2005 [67, 68].

Alternately, some studies have reassessed the economic cost of overcoming the recommended levels of air pollutants. The study performed in [69] quantifies the health cost of road-traffic related air pollution in Austria, France, and Switzerland. This study is based on the average yearly population exposure to particulate matter with an aerodynamic diameter of less than 10 μm (PM_{10}). Across the three countries (74 million inhabitants), the health costs due to traffic-related air pollution for the year 1996 amount to some 27 billion €, which represents approximately 1.7% of GDP. More recently, the study in reference [70] analyzes the economic costs associated with the hospitalization of children in the city of Seville, due to respiratory diseases caused by exposure to PM2.5 levels above the recommended limits. Moreover, a review of studies that have been conducted about this issue can be found in [71]. These authors provide a critical and systematic review of the societal costs of air pollution-related ill health studies. They find a total of 17 studies, mostly related to PM_{10} particles.

So, it can be understood that the decrease in emissions, that can be achieved through the use of less pollutant renewables, is going to be associated with less diseases and therefore with a lower economic cost. However, fewer studies have quantified the health benefits of clean energy initiatives. Methods to translate emission reductions into changes in air quality and associated health benefits can be complicated. Recently, some studies have been carried out to evaluate how renewable energy can reduce the negative effects of air pollutants [65].

The study in [72] compares the health impact of populations exposed to fine particles ($PM_{2.5}$) during their whole lifetime with two energy scenarios: the baseline (BL) scenario and the low carbon, maximum renewable power (LC-MRP) scenario. The results show that, compared to the BL scenario in which the emission factors are frozen at the 2005 level, the LC-MRP scenario, with the implementation of the current legislation on air pollution, results in a 58% reduction of $PM_{2.5}$ concentrations in Europe, reducing the loss of life expectancy by 21%. Likewise, the study concludes that a reduction of $PM_{2.5}$ concentrations by 85% and a reduction of health impact by 34% could be achieved if all technically feasible emission reduction measures were applied.

In [73], it analyzes the cost savings by the year 2050 that can be achieved when a 100% renewable energy systems is designed compared to the current systems in the case of Denmark, in which the primary energy supply has been kept constant for 35 years. The health costs are estimated on the basis of six different emissions: SO_2, NOx, CO_2, particulates ($PM_{2.5}$), mercury, and lead and are related to enumerated lost working days, hospital admissions, health damage, and deaths. The results show that while the combined health costs for the reference energy systems for 2015, 2030, and 2050 are approximately 14–15 billion DKK/year, the savings in the health costs are approximately 2 billion DKK in 2015, approximately 7 billion DKK in 2030, and approximately 10 billion DKK in 2050 when the energy system is 100% renewable.

Likewise, in [74], the authors use a simple damage function methodology to quantify some of the health co-benefits of replacing coal-fired generation with wind or small hydro in China. The authors estimate the health co-benefits of an average wind or hydro project in terms of number of hospital stays for respiratory or cardio-pulmonary conditions per year avoided due to the reductions in emissions of SO_2, NOx, and particulates.

4. Discussion

The relatively recent deployment of RE all over the world is justified in the need for substitution of fossil fuels. The first aim is to delay and reduce the future consequences of climate change provoked by GHG emissions.

Although claims for the effects of climate change have been criticized and even disputed, most countries have opted for a RE strategy path. The International Energy Agency (IEA) has devoted the Renewables Information report to highlighting the importance of RE sources, providing a comprehensive review of historical and current market trends in OECD countries, including preliminary data for 2013. As the IEA stated, the role of RE is expected to increase significantly over time in all scenarios considered with greater contributions to the power generation, heating and cooling, and transport sectors [12].

The advantages of using RE sources are still being quantified, but overall they allow countries to reduce the fossil fuels dependency, because it is considered that all countries enjoy at least one renewable resource and most of them have more than one available. Also, the benefits of using RE sources include the deployment of new technologies linked to RE sources, opening new job opportunities, and seem to be more feasible for coping with the energy demand. These actions are conducted by the so-called Green Economy.

Additionally, the RE sources contribute to reducing the GHG emissions and other air pollutants as soon as they substitute the use of fossil fuels. All these benefits of RE sources have been used to justify the Governments' promotion of RE sources. As soon as RE sources deploy positive externalities, they should be promoted.

In fact, the main difficulty is to know to what extend they should be promoted by Governments. To answer this question, many efforts have been made in order to quantify these positive effects. Additionally, they have been transformed into monetary terms with the aim of providing further information. Then, the answer will be, the RE sources should be promoted as much as the benefits that they generate.

The feed-in system has been the most usual energy policy for RE sources promotion. An attempt to evaluate the positive externalities of using RE sources in the Spanish electricity market was done by [75]. The final aim of this research was the following. Firstly, in the paper is evaluated what would happen in Spanish electricity market if pollutant energy sources were substituted by promoted RES, in terms of CO_2 avoided. Secondly, the avoided CO_2 emissions generated by RES are compared with the funds they received from the Spanish feed-in system. Third, authors calculate the economic balance of promoted RES. In 2011, approximately 10% of premiums paid to promote RES for electricity could be explained by the monetary value of CO_2 emissions avoided by not using alternative pollutant energy sources, such as coal and combined cycle.

The feed-in system has been the most usual energy policy for RE sources promotion. An attempt to evaluate the positive externalities of using RE sources in the Spanish electricity market was done by [75]. The final aim of this research was the following. Firstly, the supported RES for electricity is evaluated in terms of CO_2 emissions avoided when they are introduced in the Spanish electricity market instead of other potential polluting energy sources. Secondly, these positive environmental externalities of supported RES for electricity were compared with the funds they received from the Spanish feed-in system, in order to estimate the economic balance of this support system. The results for 2011 show that approximately 10% of premiums paid to promote RES for electricity could be explained, from an economic point of view, by the monetary value of CO_2 emissions avoided by not using alternative energy sources, such as coal and combined cycle.

However, the most recent research about the deployment of RE sources is focused not only on the benefits from the climate change point of view, but also on the positive effects on human health. The reason is that it is not only GHG emissions that have effects on human health, but also other air pollutants such as ozone precursors and particulate matter are quite dangerous.

Great efforts have been made by some countries in order to meet not only the GHG international commitments but also to follow WHO recommendations in the air quality guide [52]. However, there are still important differences among countries about the percentage of population that are affected by upper limits of air pollutants as recommended by WHO [52].

While many efforts have been made in relation to the GHG emissions and the number of countries willing to reach an agreement is becoming more numerous, there is an important lack of international commitments when considering other pollutants.

There are areas such as the European Union, where, in addition to the climate change policy, there is an air pollution policy that covers the limits of pollutants other than GHG emissions.

This air pollution policy has been demonstrated as effective since many EU countries have reduced their air emissions in the last decade.

But there are other areas or important countries, such as China or the USA, where the air pollutant emissions other than GHG are still really high.

These differences among countries in terms of air pollutant emissions and air policy have negative effects on international competitiveness. When the air emissions embodied in international trade are considered, it so happens that in those countries where no air pollutant policies have been put in place, the emissions embodied in exports are higher than those embodied in imports [76]. Therefore, areas such as the European Union are provoking more air pollution through its consumption pattern than from its production. Conversely, countries such as China and India are provoking more air pollution through their production than from their consumption pattern.

This different behavior of countries in relation to air pollution policy makes it difficult to reach an agreement in this field. Those countries with more pollution because of their consumption pattern ask for an agreement based on the production perspective, and those countries with more pollution because of their production ask for an agreement from the consumption perspective. Therefore, an international or global agreement is required if we are seeking a real solution.

As an alternative to the global agreement, the reduction in air pollution, besides the diminishing of GHG emissions, could be reached thanks to the deployment of RE sources. This is why most of the countries' efforts are not only aimed at reaching agreement on emissions, but also most of their public policies are aimed at promoting the use of renewable energy and improving energy efficiency.

The European Directive 20-20-20 goes in this direction. The countries' efforts up to 2020 should be directed to reducing GHG emissions, increasing the RE sources percentage in energy consumption, and improving energy efficiency. This strategy has been confirmed in the Roadmap to 2050.

One final idea comes into our discussion. Scientific research in different fields should help to transform the negative effects that air pollution and GHG emissions have on human health in terms of a homogenous unit of human health (or damage). This information will contribute to policy makers being able to decide the best combination of energy sources among those available, in terms of their effects on human health.

5. Conclusions

The important deployment of RE sources from an international perspective is a fact. The last reports of [11, 12] show the importance that the RE source are achieving in most countries. Although some differences are found, most countries have developed these energy sources with the aim of contributing to reducing the effects of climate change, and therefore, to reaching international commitments. Nevertheless, there are other reasons that arise which explain the deployment of RE sources.

The advantages of using RE sources are of a different nature. Firstly, from an economic perspective, the RE sources allow countries to reduce the energy dependency on fossil fuels. As at least one renewable source is available in all countries, they are a real alternative to fossil fuels. However, the development of RE sources also requires an important investment that creates job opportunities and improves new technologies. The research papers that link productive sectors and the environment are becoming more frequent. The results provide interesting information about which are the main pollutant sectors. The life cycle assessment (LCA) techniques and the input-output models might provide further information on this issue. Secondly, from the epidemiological point of view, research papers show evidence about the effects that air pollutants have on human health. Many efforts have recently been made by scientific researchers from different fields in order to display the negative effects that the different pollutants have on human health, provoking cardiovascular and respiratory diseases. The conclusions are different depending on the country and city considered, or depending on the pollutant and the other variables included in the analysis. But all of them have found that air pollutants have negative effects on human health, and have attempted to quantify them. The final step, that not all researchers follow, is to transform these negative effects into monetary units in order to be suitable for policy makers.

Therefore, the main results of those research papers with different techniques show that the most pollutant sectors from the climate change and epidemiological point of views are quite similar. They are mainly the transport and power generation sectors.

Research papers about the avoided CO_2 and air pollutant emissions that will happen due to the substitution of fossil fuels by RE sources (such as was done in [75]) will contribute to estimate the benefits of RE in terms of emissions. Additionally, a second step is needed. More studies should be carried out in order to quantify the effects on health that provoke this reduction on CO2 and air pollutant emissions due to RE sources (here the epidemiological studies have an important role).

Policy makers have to take into consideration all these positive externalities of RE sources, when evaluating the possibility of their promotion. However, this evaluation should also take into consideration that not all RE sources have equivalent positive effects. Therefore, our recommendation is that governments should be supported by recent research when deciding the most appropriate energy mix for a country.

Acknowledgements

The authors are grateful for the funding provided by the ECO2014-56399-R Project of Spanish Ministry of Economy and Competitiveness, the SEJ-132 project and the support of the Roger Torné Foundation through the Energy and Environmental Economics Chair at the University of Seville.

Author details

María del P. Pablo-Romero*, Rocío Román, Antonio Sánchez-Braza and Rocío Yñiguez

*Address all correspondence to: mpablorom@us.es

University of Seville, Seville, Spain

References

[1] Stern, N. (Ed.). (2007). The economics of climate change: The Stern review. Cambridge University press, New York.

[2] Dell, M., Jones, B. F., Olken, B. A. (2012). Temperature shocks and economic growth: Evidence from the last half century. *American Economic Journal: Macroeconomics*, 4(3), 66–95.

[3] Akachi, Y., Goodman, D., Parker, D. (2009). Global climate change and child health: A review of pathways, impacts and measures to improve the evidence base. UNICEF Innocenti Research Centre, Florence.

[4] Costello, A., Abbas, M., Allen, A., Ball, S., Bell, S., Bellamy, R., Friel, S., Groce, N., et al. (2009). Managing the health effects of climate change. *The Lancet*, 373(9676), 1693–1733.

[5] McMichael, A. J., Bertollini, R. (2011). Risks to human health, present and future. In *Climate Change: Global Risks, Challenges and Decisions*, K. Richardson et al. (eds.), pp. 114–116. Cambridge University Press, New York.

[6] GEA (2012). Global energy assessment – toward a sustainable future. Cambridge University Press, Cambridge, UK, and New York, NY, USA and the International Institute for Applied Systems Analysis, Laxenburg, Austria.

[7] Hansen, J. E. (2007). Scientific reticence and sea level rise. *Environmental Research Letters*, 2(2), 024002 (6 pp).

[8] Allison, I., Bindoff, N. L., Bindschadler, R. A., Cox, P. A., de Noblet, N., England, M. H., Francis, J. E., Gruber, N., et al. (2009). The copenhagen diagnosis. Updating the world on the latest climate science. The University of New South Wales Climate Change Research Centre (CCRC), Sydney.

[9] Lund, H., Mathiesen, B. V. (2009). Energy system analysis of 100% renewable energy systems—The case of Denmark in years 2030 and 2050. *Energy*, 34(5), 524–531.

[10] Gschwind, B., Lefevre, M., Blanc, I., Ranchin, T., Wyrwa, A., Drebszok, K., Cofala, J., Fuss, S. (2014). Including the temporal change in $PM_{2.5}$ concentration in the assess-

ment of human health impact: Illustration with renewable energy scenarios to 2050. *Environmental Impact Assessment Review*, in press.

[11] IRENA (2014). Remap2030 – A renewable energy roadmap. International Renewable Energy Agency, IRENA Innovation Technology Centre, Bonn.

[12] IEA (2014a). Renewable energy medium-term market report. International Energy Agency (IEA), OECD, Paris.

[13] REN21 (2014). Renewable 2014 global status report. Renewable energy policy network for the 21st Century. REN21 Steering Committee, Paris.

[14] IEA (2014b). World energy outlook. International Energy Agency (IEA), OECD, Paris.

[15] IEA (2014c). CO2 emissions from fuel combustion. International Energy Agency (IEA), OECD, Paris.

[16] IPCC (2013). Fifth assessment report (AR%). Climate change 2013 – The physical science basis. Cambridge University Press, New York.

[17] IPCC (2007). Climate change 2007: Impacts, adaptation and vulnerability. Cambridge University Press, New York.

[18] WHO (2009). Global health risks: Mortality and burden of disease attributable to selected major risks. World Health Organization (WHO), Regional Office for Europe, Copenhagen.

[19] Campbell-Lendrum, D., Woodruff, R. (2007). Climate change: Quantifying the health impact at national and local levels. WHO environmental burden of disease series No. 14, World Health Organization, Geneva.

[20] EHP (2010). A human health perspective on climate change: A report outlining the research needs on the human health effects of climate change. Environmental Health Perspectives (EHP) and National Institute of Environmental Health Sciences, Research Triangle Park, NC.

[21] D'Amato G., Cecchi, L. (2008). Effects of climate change on environmental factors in respiratory allergic diseases. *Clinical and Experimental Allergy*, 38(8), 1264–1274.

[22] Macdonald, R. W., Mackay, D., Li, Y. -F., Hickie, B. (2003). How will global climate change affect risks from long-range transport of persistent organic pollutants?. *Human and Ecological Risk Assessment*, 9(3), 643–660.

[23] Tucker, M. A. (2009). Melanoma epidemiology. *Hematology/Oncology Clinics of North America*, 23(3), 383–395.

[24] Ebi, K. L., Exuzides, K. A., Lau, E., Kelsh, M., Barnston, A. (2004). Weather changes associated with hospitalizations for cardiovascular diseases and stroke in California, 1983–1998. *International Journal of Biometeorology*, 49(1), 48–58.

[25] Schwartz, J., Samet, J. M., Patz, J. A. (2004). Hospital admissions for heart disease: The effects of temperature and humidity. *Epidemiology*, 15(6), 755–761.

[26] Morabito, M., Modesti, P. A., Cecchi, L., Crisci, A., Orlandini, S., Maracchi, G., Gensini, G. F. (2005). Relationships between weather and myocardial infarction: A biometeorological approach. *International Journal of Cardiology*, 105(3), 288–293.

[27] Barnett, A. G. (2007). Temperature and cardiovascular deaths in the US elderly: Changes over time. *Epidemiology*, 18(3), 369–372.

[28] Kysely, J., Pokorna, L., Kyncl, J., Kriz, B. (2009). Excess cardiovascular mortality associated with cold spells in the Czech Republic. *BMC Public Health*, 9, 1–19.

[29] Gong, H. Jr., Wong, R., Sarma, R. J., Linn, W. S., Sullivan, E. D., Shamoo D. A., Anderson K. R., Prasad, S. B. (1998). Cardiovascular effects of ozone exposure in human volunteers. *American Journal of Respiratory and Critical Care Medicine*, 158(2), 538–546.

[30] Ren, C., Williams, G. M., Morawska, L., Mengersen, K., Tong, S. (2008). Ozone modifies associations between temperature and cardiovascular mortality: Analysis of the NMMAPS data. *Occupational and Environmental Medicine*, 65(4), 255–260.

[31] Pope, C. A., Dockery, D. W. (2006). Health effects of fine particulate air pollution: Lines that connect. *Journal of the Air & Waste Management Association*, 56(6), 709–742.

[32] Jerrett, M., Burnett, R. T., Pope, C. A., Ito, K., Thurston, G., Krewski, D., Shi, Y., Calle, E., Thun, M. (2009). Long-term ozone exposure and mortality. *The New England Journal of Medicine*, 360(11), 1085–1095.

[33] Krousel-Wood, M. A., Islam, T., Muntner, P., Stanley, E., Phillips, A., Webber, L. S., Frohlich, E. D., Re, R. N. (2008). Medication adherence in older clinic patients with hypertension after Hurricane Katrina: Implications for clinical practice and disaster management. *The American Journal of the Medical Sciences*, 336(2), 99-104.

[34] Carod-Artal, F. J. (2007). Strokes caused by infection in the tropics (Ictus de causa infecciosa en el trópico). *Revista de Neurología*, 44(12), 755-763.

[35] US Food and Drug Administration (2005). Quantitative risk assessment on the public health impact of pathogenic vibrio parahaemolyticus in raw oysters. Center for Food Safety and Applied Nutrition (CFSAN), Food and Drug Administration, College Park, MD.

[36] Lake, I. R., Gillespie, I. A., Bentham, G., Nichols, G. L., Lane, C., Adak, G. K., Threlfall, E. J. (2009). A re-evaluation of the impact of temperature and climate change on foodborne illness. *Epidemiology and Infection*, 137(11), 1538–1547.

[37] Kovats, R. S., Hajat, S. (2008). Heat stress and public health: A critical review. *Annual Review of Public Health*, 29, 41-55.

[38] Luber, G., McGeehin, M. (2008). Climate change and extreme heat events. *American Journal of Preventive Medicine*, 35(5), 429-435.

[39] Cohn, B. A., Cirillo, P. M., Wolff, M. S., Schwingl, P. J., Cohen, R. D., Sholtz, R. I., Ferrara, A., Christianson, R. E., van den Berg, B. J., Siiteri, P. K. (2003). DDT and DDE exposure in mothers and time to pregnancy in daughters. *The Lancet*, 361(9376), 2205-2206.

[40] Roeleveld, N., Bretveld, R. (2008). The impact of pesticides on male fertility. *Current Opinion in Obstetrics and Gynecology*, 20(3), 229-233.

[41] Law, D. C. G., Klebanoff, M. A., Brock, J. W., Dunsuno, D. B., Longnecker, M. P. (2005). Maternal serum levels of polychlorinated biphenyls and 1,1-dichloro-2,2-bis(p-chlorophenyl)ethylene (DDE) and time to pregnancy. *American Journal of Epidemiology*, 162(6), 523-532.

[42] Longnecker, M. P., Klebanoff, M. A., Dunson, D. B., Guo, X., Chen, Z., Zhou, H., Brock, J. W. (2005). Maternal serum level of the DDT metabolite DDE in relation to fetal loss in previous pregnancies. *Environmental Research*, 97(2), 127-133.

[43] Ribas-Fitó, N., Gladen, B. C., Brock, J. W., Klebanoff, M. A., Longnecker, M. P. (2006). Prenatal exposure to 1,1-dichloro-2,2-bis (p-chlorophenyl)ethylene (*p,p'*-DDE) in relation to child growth. *International Journal of Epidemiology*, 35(4), 853-858.

[44] Fritze, J. G., Blashki, G. A., Burke, S., Wiseman, J. (2008). Hope, despair and transformation: Climate change and the promotion of mental health and wellbeing. *International Journal of Mental Health Systems*, 2, 1-13.

[45] Lefebvre, K. A., Tilton, S. C., Bammler, T. K., Beyer, R. P., Srinouanprachan, S., Stapleton, P. L., Farin, F. M., Gallagher, E. P. (2009). Gene expression profiles in zebrafish brain after acute exposure to domoic acid at symptomatic and asymptomatic doses. *Toxicology Science*, 107(1), 65-77.

[46] Curriero, F. C., Patz, J. A., Rose, J. B., Lele, S. (2001). The association between extreme precipitation and waterborne disease outbreaks in the United States, 1948-1994. *American Journal of Public Health*, 91(8), 1194-1199.

[47] Dell, M., Jones, B. F., Olken, B. A. (2012). Temperature shocks and economic growth: Evidence from the last half century. *American Economic Journal: Macroeconomics*, 4(3), 66-95.

[48] Cancun Agreements (2010). Cancun agreements. United Nations Framework Convention on Climate Change (UNFCCC), Cancun.

[49] Hamilton, L. D. (2011) Energy and health. In 53. *Environmental Health Hazards, Encyclopedia of Occupational Health and Safety*, T. Kjellström and A. Yassi (Chapter Editors), J. M. Stellman (Editor-in-Chief). International Labor Organization, Geneva.

[50] IPCC (2012). Renewable energy sources and climate change mitigation. Cambridge University Press, New York.

[51] Treyer, K., Bauer, C., Simons, A. (2014). Human health impacts in the life cycle of future European electricity generation. *Energy Policy*, Vol. 74, Supplement 1, pp. S31-S44.

[52] WHO (2006). Air quality guidelines. Global update 2005. Particulate Matter, Ozone, Nitrogen Dioxide and Sulfur Dioxide. World Health Organization (WHO), Regional Office for Europe, Copenhagen.

[53] Zah, R., Böni, H., Gauch, M., Hischier, R., Lehmann, M., Wäger, P. (2007). Ökobilanz von Energieprodukten: Ökologische Bewertung von Biotreibstoffen. Bundesamtes für Energie, Bundesamt für Umwelt, Bundesamt für andwirtschaft, Bern, Switzerland.

[54] Notter, D. A., Gauch, M., Widmer, R., Wager, P., Stamp, A., Zah, R., Althaus, H. J. (2010). Contribution of li-ion batteries to the environmental impact of electric vehicles. *Environmental Science & Technology*, 44(17), 6550-6556.

[55] Zackrisson, M., Avellan, L., Orlenius, J. (2010). Life cycle assessment of lithium-ion batteries for plug-in hybrid electric vehicles – Critical issues. *Journal of Cleaner Production*, 18(15), 1519-1529.

[56] Ezzati, M., Bailis, R., Kammen, D. M., Holloway, T., Price, L., Cifuentes, L. A., Barnes, B., Chaurey, A., Dhanapala, K. N. (2004). Energy management and global health. *Annual Review of Environment and Resources*, 29(1), 383-419.

[57] Curtis, L., Rea, W., Smith-Willis, P., Fenyves, E., Pan, Y. (2006). Adverse health effects of outdoor air pollutants. *Environment International*, 32(6), 815-830.

[58] Paul, W., Kirk, R. S., Michael, J., Andrew H. (2007). A global perspective on energy: Health effects and injustices. *The Lancet*, 370(9591), 965-978.

[59] WHO (2014). Promoting health while mitigating climate change. World Health Organization (WHO), Regional Office for Europe, Copenhagen.

[60] Masera, O. R., Saatkamp, B. D., Kammen, D. M. (2000). From linear fuel switching to multiple cooking strategies: A critique and alternative to the energy ladder model. *World Development*, 28(12), 2083-2103.

[61] Smith, K. R., Pillarisetti, A. (2012). A short history of woodsmoke and implications for Chile (Breve historia del humo de leña y sus implicaciones para Chile). *Estudios Públicos*, 126, 163-179.

[62] Jetter, J., Zhao, Y., Smith, K. R., Khan, B., Yelverton, T., DeCarlo, P., Hays, M. D. (2012). Pollutant emissions and energy efficiency under controlled conditions for household biomass cookstoves and implications for metrics useful in setting international test standards. *Environmental Science & Technology*, 46(19), 10827-10834.

[63] Smith, K. R., Frumkin, H., Balakrishnan, K., Butler, C. D., Chafe, Z. A., Fairlie, I., Kinney, P., Kjellstrom, T., Mauzerall, D. L., McKone, T. E., McMichael, A. J., Schneider, M. (2013). Annual Review of Public Health, 34, 159-188.

[64] Naeher, L. P., Brauer, M., Lipsett, M., Zelikoff, J. T., Simpson, C. D., Koenig, J. Q., Smith, K. R. (2007). Woodsmoke health effects: A review. *Inhalation Toxicology*, 19(1), 67-106.

[65] USEPA (2011). Assessing the multiple benefits of clean energy. A resource for states. State and local climate and energy program, EPA-430-R-11-014. United States Environmental Protection Agency, Washington, D. C.

[66] Haberl, J., Culp, C., Yazdani, B., Gilman, D., Fitzpatrick, T., Muns, S., Verdict, M., Ahmed, M., Liu, Z., Baltazer-Cervantes, J., Bryant, J., Degelman, L., Turner, D. (2004). Energy efficiency/renewable energy impact in the Texas Emissions Reduction Plan (TERP). ESL-TR-04/12-01. The Energy Systems Laboratory, Texas Engineering Experiment Station, Austin, TX.

[67] USEPA (2007). Technology transfer network clearinghouse for inventories & emissions factors. Emission Inventory Improvement Program. United States Environmental Protection Agency, Washington, D. C.

[68] Wisconsin Public Service Commission (2007). Public service commission of wisconsin focus on energy evaluation semiannual report (FY07, Year-end). Wisconsin Public Service Commission, Madison, Wisconsin.

[69] Sommer, H., Künzli, N., Seethaler, R., Chanel, O., Herry, M., Masson, S., Vergnaud, J. –C., Filliger, P., Horak, Jr. F., Kaiser, R., Medina, S., Puybonnieux-Texier, V., Quénel, P., Schneider, J., Studnicka, M. (2000). Economic evaluation of health impacts due to road traffic-related air pollution – An impact assessment project of Austria, France, and Switzerland. In IPCC Workshop on Assessing the Ancillary Benefits and Costs of Greenhouse Gas Mitigation Strategies, Washington DC.

[70] Pablo-Romero, M. D. P., Román, R., Gonzalez Limón, J. M., Praena-Crespo, M. (2014). Effects of fine Particles 2.5 on children's hospital admissions for respiratory health in Seville (Spain). *Journal of the Air & Waste Management Association*, in press.

[71] Pervin, T., Gerdtham, U. G., Lyttkens, C. H. (2008). Societal costs of air pollution-related health hazards: A review of methods and results. *Cost Effectiveness and Resource Allocation*, 6(19), 1-22.

[72] Gschwind, B., Lefevre, M., Blanc, I., Ranchin, T., Wyrwa, A. (2015). Including the temporal change in $PM_{2.5}$ concentration in the assessment of human health impact: Illustration with renewable energy scenarios to 2050. Environmental Impact Assessment Review, 52, 62–68.

[73] Mathiesen, B.V., Lund, H., Karlsson, K. (2011). 100% Renewable energy systems, climate mitigation and economic growth. *Applied Energy*, 88(2), 488-501.

[74] Partridge, I., Gamkhar, S. (2012). A methodology for estimating health benefits of electricity generation using renewable technologies. *Environment International*, 39(1), 103-110.

[75] García, A., Román, R. (2013). An economic valuation of renewable electricity promoted by feed-in system in Spain. Renewable Energy, 68, 51-57.

[76] Román, R., Cansino, JM., Rueda-Cantuche, JM. (2013). An air emissions analysis from the consumption perspective in EU-27. A Multi-Regional Input-Output model. V Spanish conference in Input-Output Analysis. Seville, 18-20 September.

Chapter 9

Distributed Renewable Power Sources in Weak Grids — Analysis and Control

Everton Luiz de Aguiar, Rafael Cardoso, Carlos Marcelo de Oliveira Stein, Jean Patric da Costa and Emerson Giovani Carati

Additional information is available at the end of the chapter

http://dx.doi.org/10.5772/61613

Abstract

This chapter describes the main aspects about distributed generation (DG) systems and investigates the operation of DG systems based on static power converters connected to weak grids. Initially, the concept of DG is discussed, and the main topologies for the connection of DG systems to the grid are covered. Converters used in such applications are also introduced. When connected to weak grids, DG systems based on static power converters suffer with problems related to the total harmonic distortion (THD) at the connection point. To address this issue, initially, a definition of weak grid is presented. Then, the dynamic behaviour of the most common small DG system when connected to a weak grid and the relation between the voltage harmonic distortion and the weak grid impedance are analyzed. Aiming to comply with the THD requirements, the main topologies of passive filter used in the connection of inverter-based DG units with weak grids are also discussed. Finally, a controller design that considers the grid side impedance in its formulation is developed. Experimental results are provided to support the theoretical analysis and to illustrate the performance of the grid-connected DG in a weak grid case operation scenario.

Keywords: DG connection, DG models, Harmonic distortion, Weak grids

1. Introduction

A large amount of research effort has been made in order to diversify the primary energy sources and to accommodate the growth in consumption. This consumption growth, allied to the limited power generation capacity of traditional power plants, is encouraging the development of distributed generation (DG) systems. Figure 1 illustrates a simplified diagram of a grid-connected DG concept.

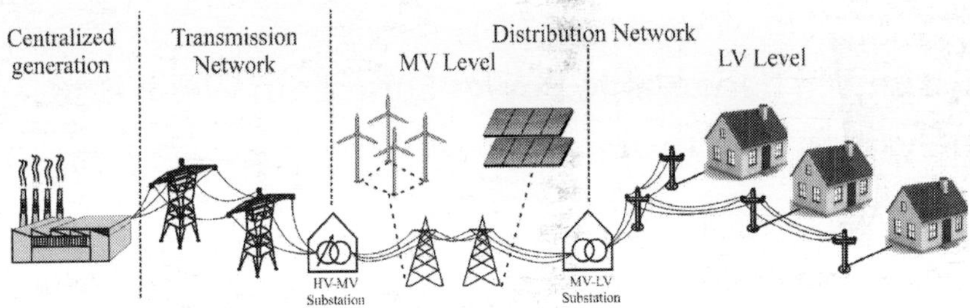

Figure 1. Simplified diagram with the most frequently deployed grid-connected DG concept.

The use of generation near to the consumption centres instead of the traditional large centralized power plants provides many advantages such as reduction of the power losses in transmission and distribution lines, improvement in power quality, reduction of the power usage of the high voltage (HV)/medium voltage (MV) transformers and reliability improvement of the power supply. It also provides less environmental impact. However, the DG systems are frequently connected in rural areas where weak grids are prominent. MV radial distribution feeders serving relatively wide areas are common in rural areas of developing countries. Usually, these grids are connected to radial subtransmission systems (up to 110 kV) that can also cover long distances. As a result, high short-circuit impedances characterize these distribution networks. Consequently, noticeable voltage differences between different locations are to be expected and certain power quality issues can arise.

The main forms of primary sources used in DG are wind generation, photovoltaic generation and hydro generation. Other primary sources such as fuel cells, geothermal generation and cogeneration are less used. Particularly, wind and photovoltaic generations are becoming more cost-effective because of the evolution of material technology and the advances in power electronics solutions.

DG systems based on photovoltaic or wind depend on the use of power converters to adequately process the power to allow the connection to the grid. The use of such converters allows the use of same DG system in grid-tied operation or, in some cases, in stand-alone operation. In grid-tied mode, the power is injected in the grid, while in the stand-alone operation, the DG is the electrical supply for a local load that is not connected to the grid. Figure 2 illustrates this concept. In this figure, a hybrid DG system composed of a wind generator and a solar panel is connected to the grid. In addition, a stand-alone DG system supplies a local load and possesses the capability to store energy that can be used when the primary energy source is unavailable.

In the electrical system depicted in Figure 2, the synchronous generator G_2 represents a large hydroelectric power plant and it is connected to the step-up transformer T_4 through the impedance Z_{g2}. The transformer T_4 raises the voltage from $\mathbf{B}_6$ to $\mathbf{B}_5$ $(V_{B_5}>230\text{kV})$. From the bus $\mathbf{B}_2$, the energy generated by G_2 reaches the step-down transformer T_2 passing through the line impedance Z_{LT2}. The transformer T_2 takes the transmission voltage level and adapts itself to transmission-level voltage to the primary distribution voltage (13.8 kV).

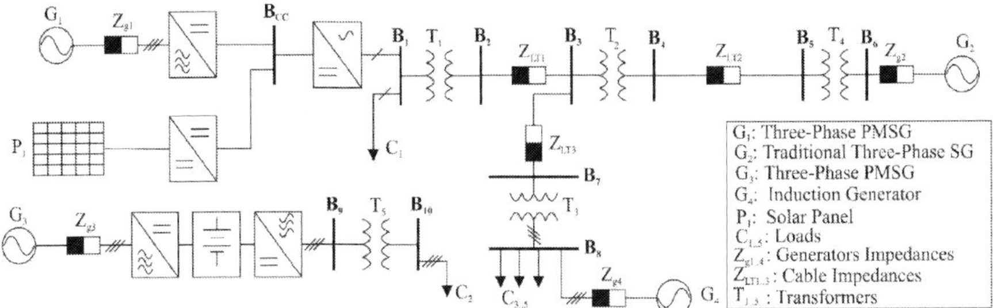

Figure 2. Different primary energy sources grid connections.

Distribution lines whose impedances are Z_{LT1} and Z_{LT3} and the power transformers T_1 and T_2 are shown in the central region of Figure 1. The central region between the buses $\mathbf{B}_2$, $\mathbf{B}_3$ and $\mathbf{B}_7$ marks the interconnection between DG units and the centralized generation G_2.

The $\mathbf{B}_1$ bus from Figure 2 represents the connection point of a single-phase hybrid wind/ photovoltaic distribution generation unit with the grid. Note that the coupling of the two forms of primary energy (G_1 – wind and P_1 – photovoltaic) is done through a shared DC bus represented by $\mathbf{B}_{CC}$. The inverter of the hybrid DG unit located between the buses $\mathbf{B}_{CC}$ and $\mathbf{B}_1$ is connected to the local load C_1 and the distribution transformer T_1. The inverter of the DG unit works at secondary distribution voltage level (e.g. 220 V line-to-line root mean square, RMS).

The bus $\mathbf{B}_8$ in Figure 2 represents the connection point of a wind generation unit to the three-phase distribution grid. This unit is formed by an induction generator with a gearbox. The induction generator allows generation of constant frequency independent of the shaft speed. The equivalent series impedance of the induction generator G_4 is represented by Z_{g4}. The nominal voltage of the G_4 connection point with the distribution grid is the secondary distribution voltage. The generator G_4 shares the connection point with the loads C_3–C_5, represented by $C_{3..5}$, which are typical residential or commercial consumption loads at the secondary distribution voltage level.

Figure 2 also depicts the stand-alone use of DG. This situation can arise from an intentional or unintentional disconnection of some subsystem from the main grid where the load is critical and cannot tolerate power outages. The bus $\mathbf{B}_{10}$ represents the connection point of the stand-alone distribution generation with the critical load C_2. In this example, the DG is connected to the load through a three-phase inverter and a transformer T_5. Since the load C_2 is critical, a battery bank is used to store energy to be used in the cases that the prime source is not available.

As previously mentioned, several DG systems use static power converters to connect the primary energy source to the grid. These converters use passive filters, based on inductive and capacitive elements, to mitigate the harmonic content of the pulse width modulation (PWM). When the DG connection is made at the secondary distribution voltage level, in several cases, the length of the distribution line is large so that the equivalent series impedance between the

DG system and the substation transformer is not negligible. In those cases, the interaction of the coupling filter of the inverter with the grid impedance can lead to undesirable resonances that must be considered in the design stage of the control system. In addition, the dynamic interaction of the DG with a weak grid can degrade the power quality even if the resonance is not an issue.

This chapter considers a wind single-phase DG system and explores the problem of the power quality degradation when the system is connected to weak grids. In addition, the main concepts of DG are covered, such as main topologies for DG connected to the grid, power converters for DG, definition of weak grids and the effects of the use of inductive (L) or inductive–capacitive–inductive (LCL) filters in weak grids connection. It also presented a complete control system design model, some simulations and experimental results to illustrate the analysis.

2. Renewable energy sources in DG

Figure 2 showed a typical distribution grid in which different possible connection topologies for DG were illustrated. In this section, further attention is given to describe possible connection topologies for DG. The main static power converters used in such applications are also addressed. Then, a mathematical model of a commonly used DG system, adequate for the purposes of dynamic analysis and control design, is developed.

2.1. Topologies to connect DG into the grid

Two main approaches are usually used to connect DG systems to the grid. In the first one, an AC bus is shared among the DG units. Each DG unit has its own DC bus and output inverter. In the second approach, a DC bus is shared among the different generator units. Then a single inverter is responsible for the connection of the DG to the grid.

Figure 3(a) illustrates the first case and represents two different DG systems based on power inverters sharing the same AC bus, denoted by $\mathbf{B}_{AC}$. The outputs of the inverters are synchronized with the voltages at the connection point $\mathbf{B}_{AC}$. A transformer T_1 connects the system and the load C_1 to the main grid.

Figure 3(b) depicts the second case where two DG units share the same DC bus, for example, photovoltaic arrangement and a permanent magnet synchronous generator (PMSG) with controlled rectifier. Note that, in the case of DC coupling, there are no individual inverters. However, DC/DC (for photovoltaic unit) and AC/DC converters (for wind generation) are necessary. It is worth pointing out in Figure 3(b) that the synchronization with the AC bus is done by only one inverter, which connects the generation unit to the main grid through a transformer T_1. The choice of coupling the units through a unified DC bus has the advantages of allowing centralized control of parameters as power, voltage and current. With the AC coupling, a single synchronization algorithm with the utility grid can be used.

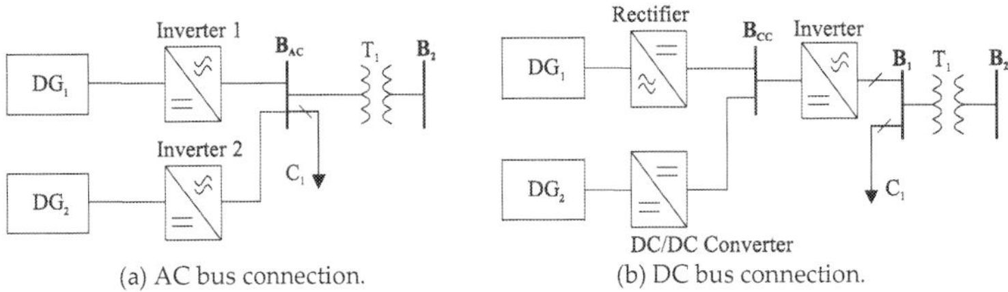

(a) AC bus connection.   (b) DC bus connection.

Figure 3. AC bus and DC bus DG grid connection topologies.

Figure 4 shows a hybrid bus connection where some DG units are grouped in a DC bus and other units are grouped in an AC bus. It can be noted that DG_3 and DG_4 units are grouped into the AC bus via separate converters, as well as in Figure 3(a). The difference is that in Figure 4, there is also a group of DG units sharing the same single DC bus. As can be noted, DG_1 and DG_2 units are engaged on the bus B_{CC} by means of suitable converters. The rectifier of DG_1 controls the B_{CC} voltage. Likewise, the DC/DC converter for DG_2 regulates the DC output voltage in order to deliver power to the B_{CC} bus.

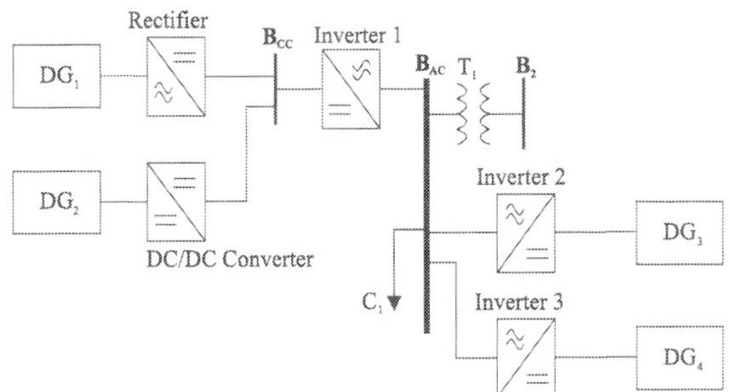

Figure 4. Hybrid bus connection.

2.2. Static power converter for DG applications

The most common converter topology in DG is the full-bridge converter in an AC/DC/AC configuration known as a back-to-back converter. In this configuration, two full-bridge converters share the same DC bus. One converter acts as a rectifier (on the generator side) and the second converter acts as the inverter (connected to the grid). When insulated gate bipolar transistors (IGBTs) are used, the converter allows bidirectional flow of energy. The schematic diagram of a full-bridge three-phase AC/DC/AC converter is shown in Figure 5.

The topologies of static converters used in DG units depend on the generator technology. In photovoltaic systems, the voltage and current of the photovoltaic panel are continuous and generally have low amplitude. The connection of PV systems to the distribution network leads

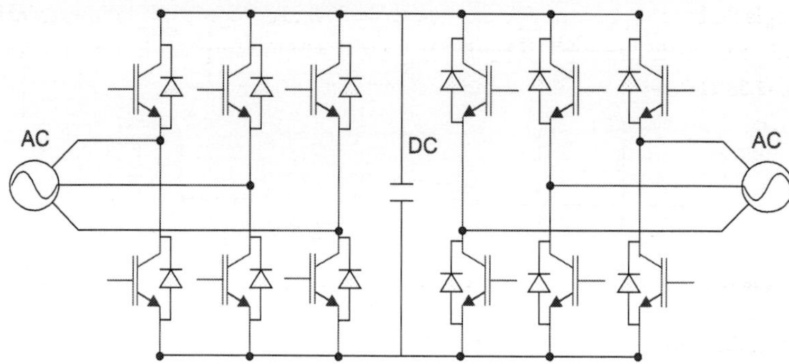

Figure 5. AC/DC/AC back-to-back converter for three-phase connection.

to an arrangement of the panels in the DC bus [1] or the uses of a DC/DC [2] boost converter. A full-bridge inverter is usually used at the output of the DG system for the grid connection.

Wind generators have alternating output voltage and current. Usually, PMSGs, doubly fed induction generators (DFIG) and traditional synchronous generators also employ power inverters for the connection with the grid. A DFIG-based system has a power conversion rating of 30% of the nominal power of the wind turbine.

In the case of the PMSG, the frequency is directly proportional to the machine rotor speed [3]. Thus, generation units with PMSG without static power converters must have a mechanical speed adjustment system. In certain applications, the mechanical speed control has a slow time response that cause variations in the frequency generated. In the cases in which mechanical adjustment of the angle of the blades is insufficient, the DG system will work with variable voltage frequency. To cope with that problem, a static power converter is mandatory.

Applications of PMSG with variable speed use an inverter to synchronize the generation with the grid (for grid-tied applications) or to ensure constant frequency and voltage at the load (for stand-alone applications). For PMSG-based systems, an AC/DC/AC converter may also be used.

Figure 6 illustrates the elements of an AC/DC/AC converter used to connect a DG unit to a single-phase grid. In this converter, a three-phase noncontrollable rectifier to lower the converter cost is used on the machine side converter (MSC). On the grid side converter (GSC), a full-bridge inverter is responsible for the connection with the grid.

The DC bus may consist of batteries, capacitors or super capacitors. Batteries have improved the stability of bus voltage level but have a high charge and discharge time. Capacitors have a low charge time constant, but in return, the constancy of the voltage on the bus is impaired compared with the battery systems. Finally, circuits with super capacitors promise quick charge maintaining a constant voltage at DC bus, but at a higher cost than the other two technologies.

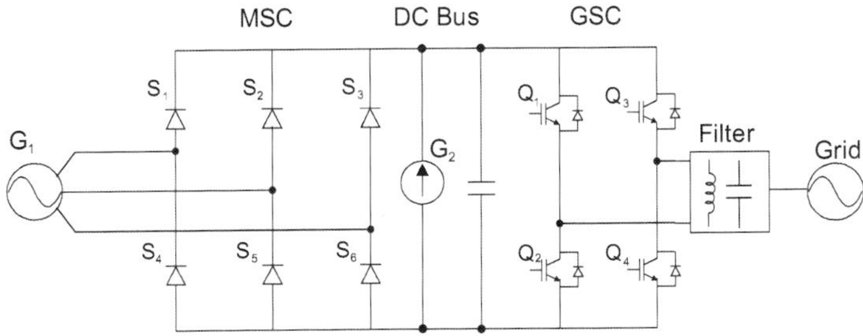

Figure 6. AC/DC/AC converter for single-phase connection.

Another important feature of the AC/DC/AC circuit in Figure 6 is the possibility to regulate the DC bus voltage using decoupled controllers. The DC bus voltage control gives the possibility of hybrid generation systems where the different generations are connected at the DC bus. This is illustrated in Figure 6, for example, where a photovoltaic unit G_2 (with regulated DC output voltage) and a wind power unit G_1 are connected to the grid by the DC bus.

The connection of wind and photovoltaic units to the distribution network is extensively discussed in the literature. With the advances in power electronics, full-power switched converters or full-bridge AC/DC/AC converters are widely used to control the current and voltage at the connection point. The strategy of PWM is used in most applications, which use the full-power converters. The PWM is used to synthesize the control actions needed to ensure that the reference tracking of the output quantities (current, voltages, frequency or power) are achieved. Several PWM strategies are found in the literature. The more common techniques are the following: three-level sinusoidal modulation, two-level sinusoidal modulation, space vector modulation and multilevel modulation. Each modulation strategy is associated with a different frequency spectrum obtained by the Fourier series of the output voltage of the inverter.

The voltage that arises from the switching of the PWM logic is a discontinuous function and has high total harmonic distortion (THD). To ensure minimum levels of power quality, low-pass filters are used at the output of the inverters used in wind and photovoltaic generations.

2.3. Mathematical models for DG primary sources

About the nature of primary energy, the DG units can be classified as wind, photovoltaic, hydro, biomass and cogeneration. Each form of conversion of primary energy into electrical energy has its own dynamic characteristic. This dynamic feature determines the design and operation and the connection of the device to the load. The most promising primary energy sources for DG, which has become widespread in recent years, are wind, solar and hydro power plants.

The main DG power primary sources are as follows:

- Wind generation

- Photovoltaic generation

- Hydro generation

- Cogeneration

- Biomass generation

Cogeneration plants use heat energy remaining in industrial processes, sometimes called process heat, to generate steam for moving pressure turbine electric generators. The heating and water vapour generation can also be performed through combustion of biomass. Biomass is composed of plant remains, such as sugarcane bagasse.

The units of wind and photovoltaic generation, among the primary sources cited, are the most applicable forms for micro- and mini-generation residential units. However, the generation of hydroelectric units have high power density, high efficiency and dominant technology resulting in a reduced cost per watt installed.

As this chapter deals with the connection of small-scale DG to the single-phase mains, we chose to implement the most appropriate generation technologies for residential applications.

The frequency response of a single-phase generation system from the common connection point aspect is dependent on the following two excitation functions: the inverter output voltage and the voltage of the single-phase grid. The connection system analysis can be made considering the grid voltage as a disturbance in the control diagram. The frequency response of a quantity in a system with two excitation sources is given by superposition of responses from each excitation source individually, considering the other null. Consider that a source of zero voltage is the same as replacing the source to be annulled by a short circuit.

Consider a generic system with five complex impedances represented by Z_n, depending on the complex frequency s. Figure 7 shows the general diagram of impedance considering two ideally sinusoidal sources, V_1 and V_5. The generic model presented can be applied to model the connection point of a single-phase inverter to the power grid, considering the inverter output voltage $V_{inv}(s)=V_1(s)$, $V_4(s)=V_{PCC}(s)$ and $V_{grid}(s)=V_5(s)$. The other definitions taken to the generic diagram are the following: $I_1=I_{inv}$, $I_2=I_C$, $I_4=I_{local}$, $I_5=I_{grid}$.

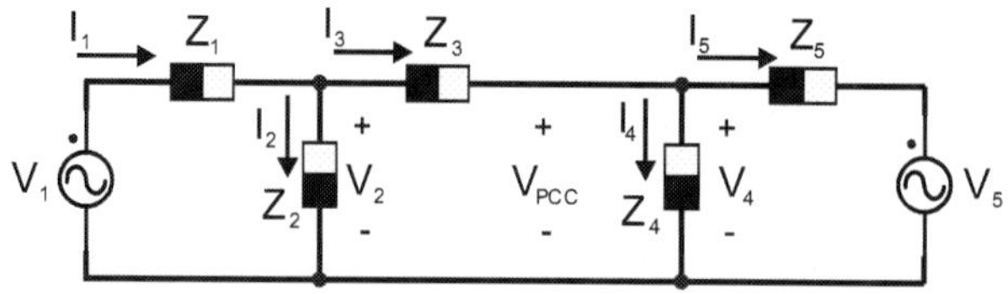

Figure 7. Generic circuit with complex impedances at the point of common connection.

The complex impedances of Figure 7 are defined in function of the filter parameters and the grid parameter. According to Figure 7, $Z_1 = sL_1$, $Z_2 = \dfrac{1}{sC}$, $Z_3 = sL_3$, $Z_4 = R_{local}$, $Z_5 = sL_{grid}$.

The single-line diagram that represents the connection of a distributed generating unit to the grid and the local load through a full bridge converter is shown in Figure 8. The diagram in Figure 8 is based on the generic circuit of Figure 7.

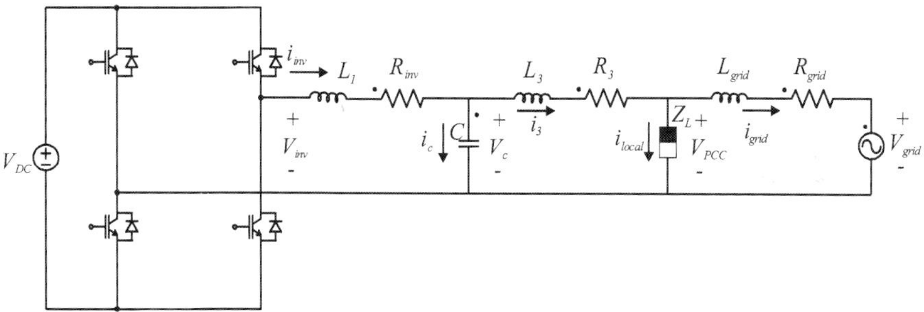

Figure 8. Full-bridge converter connected to the single-phase grid and a local load through an LCL filter.

The Laplace transform of the filter output current and of the point of common coupling (PCC) voltage can be written, respectively, by the expressions:

$$I_3(s) = F3_{inv}(s)V_{inv} + F3_{grid}(s)V_{grid} \tag{1}$$

and

$$V_{PCC}(s) = F4_{inv}(s)V_{inv} + F4_{grid}(s)V_{grid}, \tag{2}$$

wherein the transfer functions $F3_{inv}(s)$, $F4_{inv}(s)$, $F3_{grid}(s)$, $F4_{grid}(s)$ are defined considering the superposition principle applied to the sources $V_1(s)$ e $V_5(s)$ from Figure 7. The transfer functions, this way, are obtained as follows:

$$F3_{inv} = \left.\frac{I_3(s)}{V_{inv}(s)}\right|_{V_{grid}=0} = \frac{s\vartheta_1 + \vartheta_0}{s^4 + s^3\alpha_3 + s^2\alpha_2 + s\alpha_1}, \tag{3}$$

$$F3_{grid} = \left.\frac{I_3(s)}{V_{grid}(s)}\right|_{V_{inv}=0} = \left.\frac{I_3(s)}{V_5(s)}\right|_{V_1=0} = \frac{s^2\vartheta_2 + \vartheta_1}{s^4 + s^3\alpha_3 + s^2\alpha_2 + s\alpha_1}, \tag{4}$$

$$F4_{inv} = \frac{V_{PCC}(s)}{V_{inv}(s)}\bigg|_{V_{grid}=0} = \frac{sR_{local}\varpi_1}{s^4 + s^3\alpha_3 + s^2\alpha_2 + s\alpha_1}, \tag{5}$$

$$F4_{grid} = \frac{V_{PCC}(s)}{V_{grid}(s)}\bigg|_{V_{inv}=0} = \frac{V_4(s)}{V_5(s)}\bigg|_{V_1=0} = \frac{s^3\delta_3 + s\delta_1}{s^4 + s^3\alpha_3 + s^2\alpha_2 + s\alpha_1}. \tag{6}$$

In the expressions (3)–(6), the constants that multiplies the frequency 's' are as follows:

$$\alpha_0 = \frac{R_{local}}{CL_1L_3L_{grid}}, \alpha_1 = \frac{R_{local}}{CL_3L_{grid}} + \frac{R_{local}}{CL_1L_{grid}} + \frac{R_{local}}{CL_1L_3}, \alpha_2 = \frac{1}{C}\left(\frac{1}{L_3} + \frac{1}{L_1}\right), \alpha_3 = R_{local}\left(\frac{1}{L_3} + \frac{1}{L_{grid}}\right),$$

$$\delta_1 = \frac{R_{local}}{CL_3L_{grid}} + \frac{R_{local}}{CL_1L_{grid}}, \delta_3 = \frac{R_{local}}{L_{grid}} \text{ and } \varpi_1 = \frac{1}{CL_1L_3}.$$

3. Weak grids

The grid can be represented by an alternating current voltage source having an internal impedance Z_{grid}. The impedance seen from the DG system connection point may vary depending on local characteristics and the voltage level (distribution or transmission). For a system connected to the distribution network, for example, the impedance seen from the DG connection point has contributions of the internal impedance of the input, the line impedances of the transformers from the distribution bus to the transmission, the impedances of the voltage regulators in the passage and the loads connected across the line. The block diagram in Figure 9 shows the power flow from the connection point to the distribution substation bus to the DG system connected to the distribution grid.

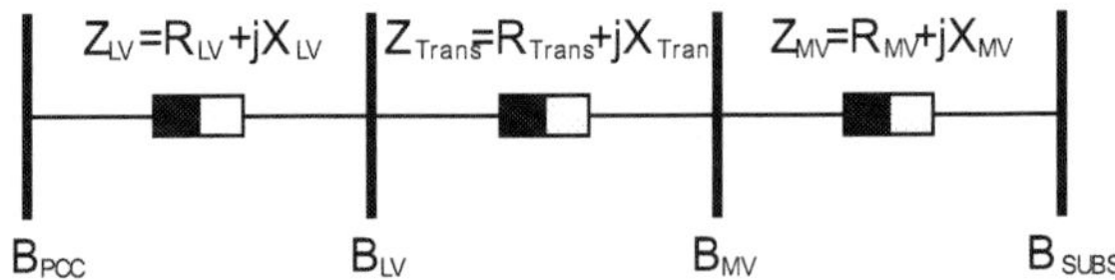

Figure 9. Power flow from the connection point to the substation bus.

Figure 9 shows the equivalent series impedances for each element from the connection point to the distribution substation bus. The connection point is represented by B_{PCC}. The electrical low-voltage network from the connection point to the secondary voltage terminals of the distribution transformer has a series of equivalent impedance represented in Figure 9 by $Z_{LV}=R_{LV} + jX_{LV}$. The secondary voltage terminals of the distribution transformer are represented by B_{LV}. The equivalent series impedance of the distribution transformer, referred to the low-voltage side, is represented in Figure 9 by $Z_{Trans}=R_{Trans} + jX_{Trans}$. The primary distribution voltage terminals, HV transformer, are represented in Figure 9 by B_{HV}. The distribution line primary impedance, from the HV terminals of the transformer to the distribution substation, referred to the low-voltage side is represented by $Z_{HV} = R_{HV} + jX_{HV}$. Finally, the substation is represented by the bus B_{SUBS}. The total impedance Z_{total} is given by the expression:

$$Z_{total} = \sqrt{R_{total}^2 + X_{total}^2},$$

(7)

and, as can be seen in Figure 9, $R_{total} = R_{LV} + R_{Trans} + R_{HV}$ e $X_{total} = X_{LV} + X_{Trans} + X_{HV}$.

Define the fundamental component of the PCC voltage by $\hat{V}_{PCC} = |V_{PCC}| e^{j\delta}$ and the infinite bus voltage fundamental component by $\hat{V}_{GRID} = |V_{GRID}| e^{j0}$. The active power transferred from the PCC to the grid is given by the expression:

$$P_{12} = \frac{V_{PCC} V_{GRID}}{|Z_{total}|} \sin(\delta - \alpha_z) + \frac{V_{PCC}^2 R_{total}}{|Z_{total}|^2},$$

(8)

where $\alpha_z = \operatorname{arctg}\left(\dfrac{R_{total}}{X_{total}}\right) + \dfrac{V_{PCC}^2 R_{total}}{|Z_{total}|^2}$ and δ is the load angle [3].

It can be noted in expression (8) that the transferred power is inversely proportional with the amplitude of the impedance between the PCC and the grid, if the total resistance is considered zero. Moreover, it is noted in the first term of expression (8) that the power P_{12} varies with the sine of the load angle minus the angle α_z. The second term of the expression (8) represents the active power dissipated in the equivalent resistance between the PCC and the grid. As greater is the power transfer capability between the PCC and the grid, the stiffer is the grid. There are two main indicators that measure the strength of a network: The short circuit ratio at the connection point, defined by the SCR or by inductive–resistive ratio IRR. As shown in [4] and [5], SCR is defined by the expression:

$$SCR = \frac{V_{PCCnom}^2}{Z_{total} S_{nominal}}.$$

(9)

where $S_{nominal}$ is the rated apparent power of the DG unit, Z_{total} is the equivalent Thevenin impedance, seen from the connection point, and V_{PCCnom} is the generated rated voltage for the DG connection point. Weak grids are characterized for SCR < 10.

Another way of measuring the grid strength is by the IRR ratio, defined by the ratio between the equivalent reactance view of the connection point and the equivalent resistance R seen from PCC, namely

$$IRR = \frac{X_{total}}{R_{total}}.$$

(10)

The grid is considered weak if IRR < 0.5 [4].

The grid is considered weak if the SCR indicator or the indicator IRR reaches values below 10 and below 0.5, respectively. In this chapter, simulation analysis and experimental results are given disregarding the resistive portion of the grid impedance, that is, considering $R_{total}=0$. This choice leads to an IRR indicator that tends to infinity, and thus the network is classified as poor only by the SCR indicator. Another direct consequence from the equation (8) is that if $R_{total}=0$, then $\alpha_z=0$, and the second term of equation (8) is zero too. If this assumption is made, the power transferred by the DG power unit is inversely proportional to the total reactance and equation (8) becomes

$$P_{12} = \frac{V_{PCC}V_{GRID}}{X_{total}}\sin(\delta).$$

(11)

The choice for ignoring the resistive portion is made to simplify the numerical and experimental analyses. Since R_{total} is equal to zero, the impedance of the grid is simple X_{total}. Note that if $X_{total}=L_{total}\omega$, with L_{total} being the total inductance from the PCC to the HV bus and ω is the angular frequency of the distribution network, then it is possible to derive the total inductance threshold between a strong network and a weak network. From equation (9) results,

$$L_{total} = \frac{V_{PCCnom}^2}{\omega\, S_{nominal}SCR}.$$

(12)

Knowing the rated characteristics of the generation system, from the equation (12), by replacing SCR = 10, we can determine the inductance threshold between a weak network and a strong network. The impedance grown from PCC to the HV bus limits the active power supplied by the DG unit. Beyond this problem, the switched AC/DC/AC converters face another problem when connected to weak grids: The harmonic distortion at PCC. The DG system inverter is seen from the PCC as a harmonic generator. The low-pass filter parameterization in the inverter output is very important to the performance and stability of the DG system. The main filters used in the literature, considering the weak grid problems, are explained in the next section.

4. Analysis of L and LCL filters connected in weak grids

The most used low-pass filters in renewable DG systems (RDGS) are generally passive and can be implemented with L, LC, LCL or LLCL topologies. The L filter topology is indicated [6] for low-power generation connected to the grid and without local load. The problem of using L filters becomes apparent when the generation unit is located at a point in the network where there is a high grid series equivalent impedance. For these cases, the effect of the voltage drop across the line impedance results in a higher THD for both current and voltage at the connection point [7]. When a local load is considered, the problem becomes even more evident. The voltage and current THD at local load connected at the PCC depend on the grid parameters, the characteristics of the passive filter, the harmonic spectrum of the output current of the inverter and the characteristics of the local load current. The problem of high current THD at the PCC can be mitigated by using a capacitive filter. Therefore, LC and LCL filter topologies are most applicable to systems with local load and high-impedance network. LCL filters are preferable to the LC filter by presence of the inductive element in line with the series grid impedance.

In this section, several passive filters' mathematical models will be shown, in both time domain and complex domain. Simulation results are presented to exemplify the model analysis.Figure 10 shows grid voltage and current waveforms that were obtained using computational simulations, considering a 2.2-kW generation setup, an inductive filter with L=2 mH, and a strong grid where the equivalent line impedance is L_{grid}=2µH. The THD of the voltage waveform shown in Figure 10 is 0.069%, while the current waveform THD is 3.427%. In this case, both waveforms are considered of good quality, since they present low THD, below 5%.

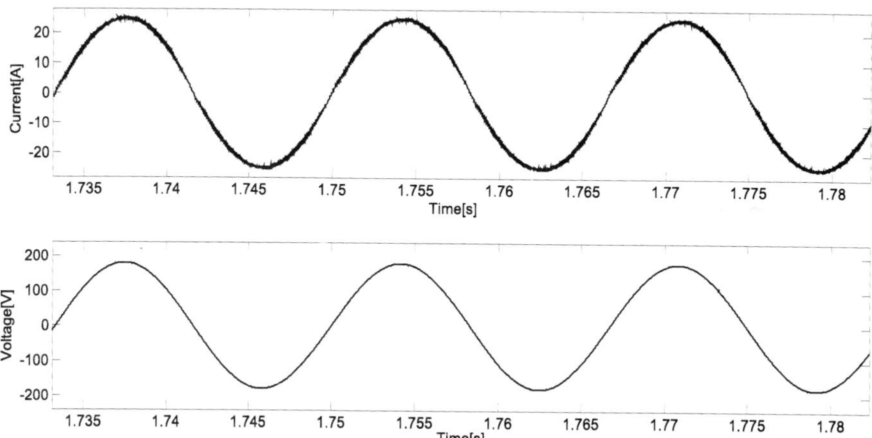

Figure 10. Current and voltage waveforms in PCC of RDGS connected by L filter to a strong single-phase grid.

In weak grids, the THD of PCC voltage grows considerably when the same inductor of previous case is used. This effect is shown in Figure 11. In this case, the line inductance is set to L_{grid}=1.5mH to evaluate the response of same RDGS connected to a weak grid.

In Figure 11, the current THD for a weak grid is 2.116%, lower than the THD for the same system connected to a strong grid. However, the voltage THD in same figure is 27.34%. It can be noted that voltage THD grows significantly in PCC for a weak grid. Moreover, when local loads are connected in PCC, HV THD is unacceptable.

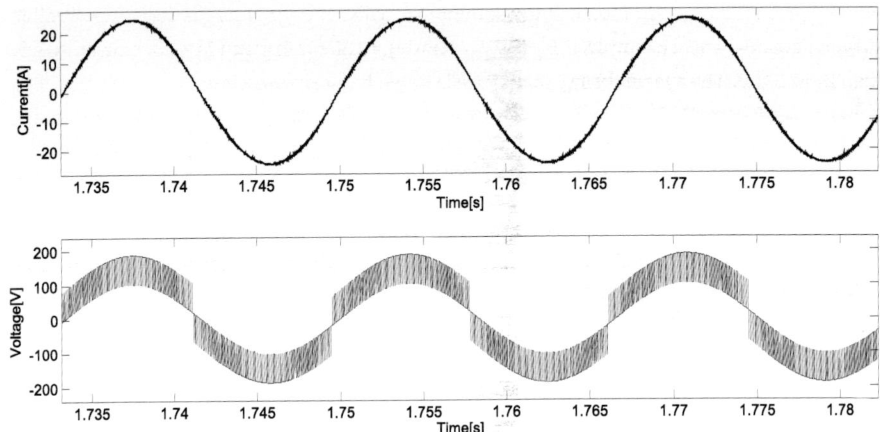

Figure 11. PCC current and voltage of single-phase RDGS connected to weak grid using L filter.

In most cases where weak grids are connected to RDGS with inductive (L) filters, a significant raise of physical size of connection inductor in order to reduce the voltage THD is necessary. In most situations, this rising makes the implementation unviable. On the other hand, from the controller viewpoint, the utilization of pure inductive filters only inserts a first-order dynamic in connection point such that the current control is generally easier than in other filter topologies.

The LCL filter is usually a good option to weak grids applications. The LCL filter inductances are generally lower than equivalent one in the L filter for the same grid impedance and the same voltage THD. Since now the LCL filter has three dynamical elements, the controller must be able to operate with an insertion of a third-order dynamic in PCC. This condition requires more analysis and careful designing of the current controller. When an LCL filter is considered the choice of filter, resonance frequency must take into account the switching frequency and fundamental frequency. A reasonable choice discussed in the literature is the resonance frequency to be ten times the fundamental frequency and half the switching frequency. This choice allows high frequencies of PWM modulation to be sufficiently attenuated, while the resonance frequency does not affect the control variables in fundamental frequency.

A design approach for LCL filter to be used in weak grids are presented and discussed in a sequence.

4.1. Design of an LCL passive filter for weak grid connection

As in [8] and [9], the system base impedance can be defined by the following expression:

$$Z_b = \frac{E_n^{\ 2}}{P_n} \tag{13}$$

where Z_b is the base impedance, E_n is the nominal RMS line voltage and P_n is the nominal active power. As a numerical example, if $E_n = 127V_{RMS}$ and $P_n = 2.2kW$ are given in a low power RGDS, then $Z_b = 7.331\Omega$. The system base capacitance is given by the expression:

$$C_b = \frac{1}{\omega_g Z_b} \tag{14}$$

where ω_g is the fundamental angular frequency of the system. Thus, considering a grid with 127 V RMS nominal voltage and 60 Hz nominal frequency, the base capacitance $C_b = 361.8\ \mu F$. Moreover, the base capacitance relates to filter capacitance (C_f) through γ_{cap} as follows:

$$C_f = \gamma_{cap} C_b. \tag{15}$$

As practical design recommendation [8], the maximum capacitance of the LCL filter (C_{fmax}) is recommended not to exceed 5% of base capacitance C_b such that $\gamma_{capmax} = 0.05$. In this way, $C_f = 10\mu F$ can be a good choice, and it will be considered in the next analysis.

The filter output equivalent inductance is denoted by L_0 and is given by the sum of grid inductance L_{grid} and filter output inductance L_3, where local load is not considered. The parameter L_0 relates to input inductance L_1 by a factor γ_i such that

$$L_0 = \gamma_i L_1. \tag{16}$$

In [9], it is shown that the voltage amplitude in the switching frequency for three-level PWM modulation, which corresponds to 200th harmonic, is 0.473 pu of nominal peak voltage. In [8], an important relationship for the inductor current amplitude in the inverter side and the inverter output voltage is given, which is valid for the switching frequency $f_{sw} = h_{sw} 60$, such that

$$\left| \frac{i_{L_1}(h_{sw}\omega_g)}{u(h_{sw}\omega_g)} \right| \approx \frac{1}{\left| h_{sw}\omega_g L_1 \right|}. \tag{17}$$

From equation (17), it is possible to compute L_1 if the current amplitude in the switching frequency and voltage harmonic order are known. The procedure for determining L_1 using expression (17) starts from the specification of maximum per unit amplitude for the desired current in L_1. In [9], the maximum amplitude is set to 0.3% for harmonic currents above 35th order. The second stage to obtain L_1 is to determine the per unit amplitude of the inverter output voltage in the switching frequency. For the given values and using a spectral analysis,

u_{sw}=0.473 pu is obtained. In order to attend the minimal specifications in standards, it should be considered that i_{L_1} is lower than 0.3%. Thus, consider the design choice i_{L_1}=0.1%=0.01pu, such that the expression (17) can be rearranged to solve for L_1 that results in the equation:

$$L_1 \approx \frac{\left|u(h_{sw}\omega_g)\right|}{\left|h_{sw}\omega_g i_{L_1}(h_{sw}\omega_g)\right|} = \frac{0.473}{200\cdot 2\cdot\pi\cdot 60\cdot 0.01} = 822\,\mu H. \tag{18}$$

When considering practical applications, an approximation L_1=1mH can be used for the inductor given in (18).

The relationship between the currents of grid side inductor i_{L_1} and the inverter side inductor i_{L_2} at the switching frequency (h_{sw}harmonic) is given by

$$\frac{i_{L_1}(h_{sw})}{i_{L_2}(h_{sw})} = \frac{1}{\left|1+\gamma_1\left(1-L_1 C_b 2\pi f_{sw}\gamma_2\right)\right|}. \tag{19}$$

In (19), $\gamma_1$is the proportionality factor between L_0 and L_1. Considering design parameters, Figure 12 shows $i_{L_1}(h_{sw})/i_{L_2}(h_{sw})$ for a range of γ_1 from 0.5 to 4.5.

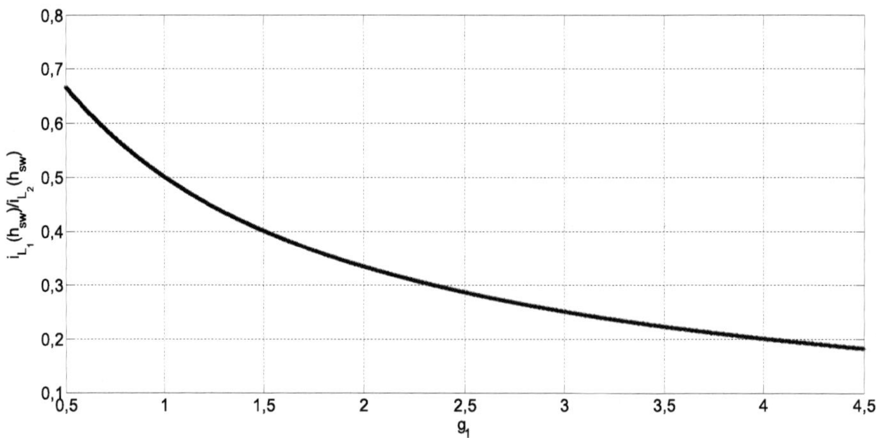

Figure 12. $i_{L_1}(h_{sw})/i_{L_2}(h_{sw})$ for a variation of γ_1.

Figure 12 shows that 0.3% attenuation in current is obtained for γ_1>2.34. Thus, from (19), the filter output equivalent inductance is determined as L_0=L_2+L_g=1.92mH. For practical implementation, the value L_0=2mH is considered.

The resonance frequency must be verified from the design parameters obtained in the previous procedure (L_1=1mH, C_f=10μF,L_0=2mH). As in [8], it is considered that the resonance frequency definition is given by

$$\omega_{res} = \sqrt{\frac{L_1 + L_0}{L_1 L_0 C_f}} = 12247\frac{\text{rad}}{s} = 1949\text{Hz}. \tag{20}$$

Taking into account the required interval for the resonance frequency $10f_s < f_{res} < 0.5f_{sw}$, where f_{sw}=12kHz, it can be seen that the designed parameters can be used for practical implementation. From these LCL-designed parameters, a numerical simulation is performed considering a weak grid and the obtained PCC voltage and current are shown in Figure 13.

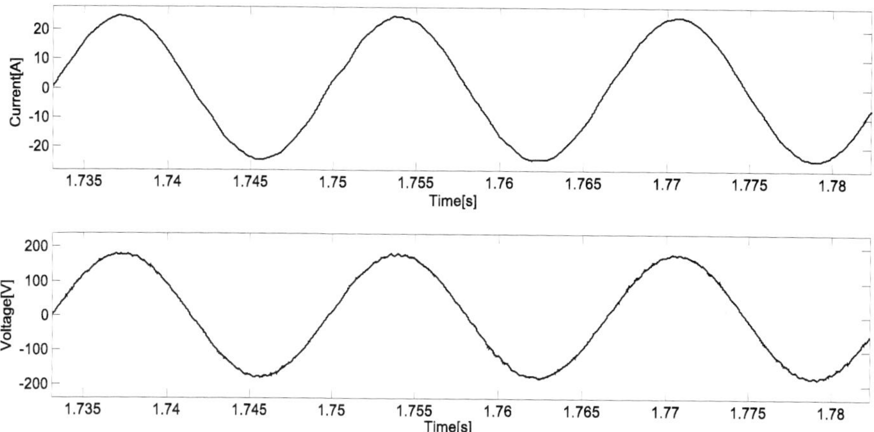

Figure 13. PCC current and voltage for a single-phase RDGS connected to a weak grid using an LCL filter.

The current THD for the waveform shown in Figure 13 is 1.31%. It can be seen that the use of LCL filter in the connection of RDGS to a weak grid resulted in a lower current THD compared to the case of an L filter being connected to a strong grid. The voltage THD obtained for the waveform of Figure 13 is 1.22%, much lower (THD^V_{LCL}= 0.04 THD^V_L) than voltage THD for a pure L filter in same conditions in Figure 11.

5. Controlled converters connected to weak grids

Many control topologies applied to DG are available in the literature, in static and synchronous reference frame (d–q axis). The proportional–integral (PI) controllers can be utilized in static reference frame, but the zero steady-state error is not achieved for this case. The static reference frame PI has the advantage of being easily parameterized. There are others topologies that achieve zero steady-state error. The proportional–resonant (PR) controller reaches zero steady-state error with good dynamic performance, considering a fixed fundamental frequency [10] [11]. Otherwise, as is presented in [12], the application of PR controllers in single-phase grids with LCL filter has limited stability margins. Other researches, like in [13] and [14], utilize discrete deadbeat controllers, which have a strong dependency on the electric parameters variations.

An interesting solution to control a voltage source inverter (VSI) connected to a single-phase grid with zero steady-state error is to utilize synchronous reference frame controllers. The alternate quantities are synchronized with the PCC frequency and become continuous because of the Park transformation [15], [16], [17].

In the synchronous reference frame, the reference voltage and current quantities are constants. This way is possible to guarantee good dynamic response and zero steady-state error, utilizing simple synchronous reference frame PI controllers. Considering the LCL filter connection, some studies show that the gain in the resonance frequency can be damped by inserting a damping resistor in series with the capacitor called the damping resistor [18]. The damping can be passive (fixed) and active (switched).

Particularly for the case of single-phase connection of DG unit, the general diagram that represents all the elements involved in the conversion and power conditioning process is shown in Figure 14.

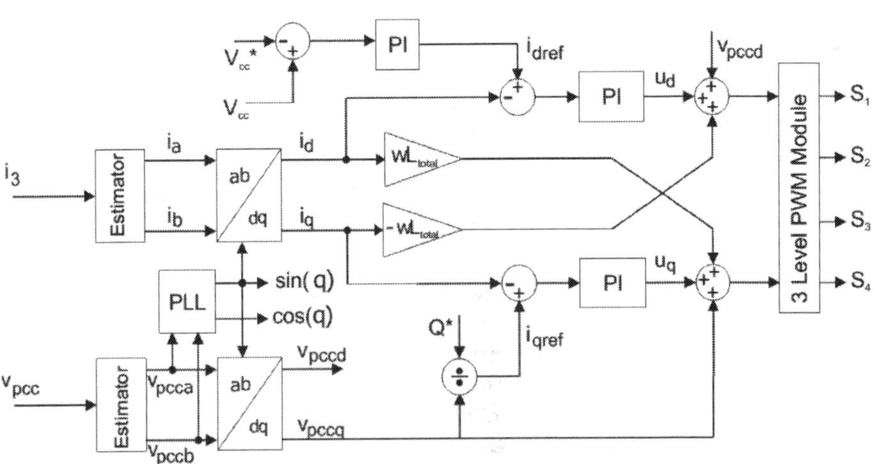

Figure 14. PCC voltage and current for single-phase weak grid connection through LCL filter.

The capacitive reactance is given by $X_c = 1/\omega C$, inversely proportional to the fundamental frequency of the power grid. Therefore, an approach to low frequencies in the vicinity of 60 Hz is to disregard from the control analysis the dynamics of the capacitive element from LCL filter calculated above. This consideration, as will be shown in sequence, not only significantly changes the characteristic low-frequency system but also simplifies the fourth-order system, shown above, resulting in a first-order system.

If the LCL filter capacitor is dismissive of the methodology of parameterization of the controller, then the equivalent circuit of inverter output voltage connection to the connection point is simplified, as shown in Figure 15.

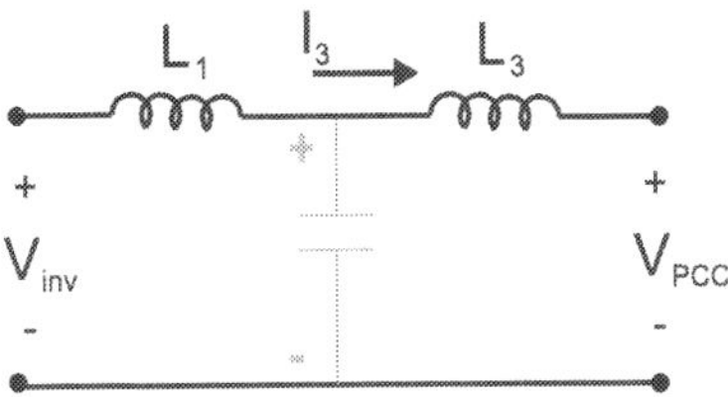

Figure 15. Simplified diagram for the connection point.

From Figure 15, disregarding the L_1 and L_3 inductors internal resistances, results in

$$V_{inv} = (L_1 + L_3)\frac{di_3}{dt} + V_{PCC}.$$ (21)

Considering an additional β axis shifted 90º from the natural axis α, the currents derivative in the filter is

$$(L_1 + L_3)\begin{bmatrix} \dot{i}_{3\alpha} \\ i_{3\beta} \end{bmatrix} = \begin{bmatrix} 1 & 0 & -1 & 0 \\ 0 & 1 & 0 & -1 \end{bmatrix}\begin{bmatrix} V_{inv\alpha} \\ V_{inv\beta} \\ V_{PCC\alpha} \\ V_{PCC\beta} \end{bmatrix}.$$ (22)

Consider the original system of the Figure 8. Disregard the capacitor branch, and then $i_{inv_{dq}} = i_{3_{dq}}$. With that, the line inductance from inverter to PCC becomes

$$L_{total} = L_1 + L_3.$$ (23)

The simplified dynamic system is represented by the ratio

$$\begin{bmatrix} \dot{i}_{3d} \\ i_{3q} \end{bmatrix} = \begin{bmatrix} i_{3d}\omega \\ i_{3q}\omega \end{bmatrix} + \frac{1}{L_{total}}\begin{bmatrix} V_{inv\,d} - V_{PCCd} \\ V_{invq} - V_{PCCq} \end{bmatrix}.$$ (24)

Note that there is a coupling term $\pm\omega i_{dq}$, a voltage component $V_{PCC_{dq}}$ and $V_{inv_{dq}}$. In order to obtain a decoupled PI controller, the control laws $u_{d_{simplified}}$ and $u_{q_{simplified}}$ are defined by

$$u_{d_{simplified}} = V_{invd} - V_{PCCd} + \omega i_{3q} L_{total}$$

(25)

and

$$u_{q_{simplified}} = V_{invq} - V_{PCCq} - \omega i_{3d} L_{total}.$$

(26)

The two decoupled expressions for de grid current in rotating frame, depending on the control actions are

$$u_{d_{simplified}} = L_{total} \frac{di_{3d}}{dt}.$$

(27)

and

$$u_{q_{simplified}} = L_{total} \frac{di_{3q}}{dt}.$$

(28)

Note that the equations (27) and (28) are similar. Applying the Laplace transform in equations (27) and (28) can be obtained:

$$G_{3d_{simplified}}(s) = \frac{i_{3d}(s)}{u_{d_{simplificado}}(s)} = \frac{1}{sL_{total}},$$

(29)

and

$$G_{3q_{simplified}}(s) = \frac{i_{3q}(s)}{u_{q_{simplificado}}(s)} = \frac{1}{sL_{total}}.$$

(30)

In this example, it is desired to control the current loop using a PI type controller. Thus, the control output for this type of controller is defined by the expressions

$$u_{d_{simplified}} = K_{P3d} e_{i_{3d}} + K_{i_{3d}} \int e_{i_{3d}} dt$$

(31)

and

$$u_{q_{\text{simplified}}} = K_{p_{3q}} e_{i_{3q}} + K_{i_{3q}} \int e_{i_{3q}} \, dt, \tag{32}$$

in which the current error $e_{i_{3d}}$ and $e_{i_{3q}}$ can be calculated through the relations

$$e_{i_{3d}} = i_{3d_{\text{ref}}} - i_{3d} \tag{33}$$

and

$$e_{i_{3q}} = i_{3q_{\text{ref}}} - i_{3q}. \tag{34}$$

Applying the Laplace transform in equations the following can be obtained:

$$G_{pi_{3d}}(s) = \frac{u_{d_{\text{simplified}}}(s)}{e_{3d}(s)} = K_{p_{3d}} + \frac{K_{i_{3d}}}{s} \tag{35}$$

and

$$G_{pi_{3q}}(s) = \frac{u_{q_{\text{simplified}}}(s)}{e_{3q}(s)} = K_{p_{3q}} + \frac{K_{i_{3q}}}{s}. \tag{36}$$

Define the transfer function $\dfrac{I_{3_{\text{simplified}}}(s)}{V_{\text{inv}}(s)}$ in the natural reference by

$$G_{\text{simplified}}(s) = \frac{I_{3_{\text{simplified}}}(s)}{V_{\text{inv}}(s)} = \frac{1}{L_{\text{total}}s}. \tag{37}$$

Figure 16 shows the Bode diagram of the complete transfer function $I_{3_{\text{simplified}}}(s)/V_{\text{inv}}(s)$ and simplified transfer function of equation (37) superimposed to the same filter parameters calculated in the previous section. Note that there is a linear region above 6284 rad/s (1 kHz), in which the two transfer functions correspond.

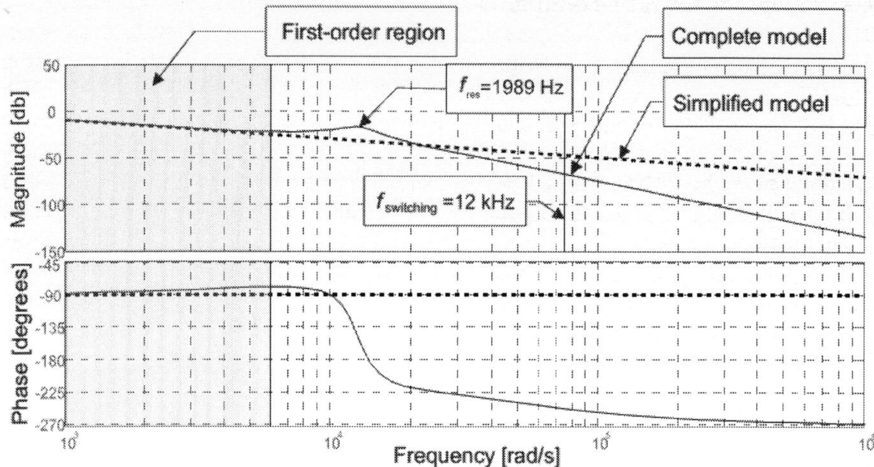

Figure 16. Bode diagram with complete and simplified models.

For the first-order region of Figure 16, the current controller design for the system of fourth-order plant can be simplified by the project with the first-order plant, modelled by the transfer function of equation (37).

Gains for a classic PI controller are given by the expressions:

$$Kp = \frac{2\xi\omega_c L_1}{\sqrt{2\xi^2 + 1 + \sqrt{\left(1 + 2\xi^2\right)^2 + 1}}} \tag{38}$$

and

$$Ki = \left(\frac{\omega_c}{\sqrt{2\xi^2 + 1 + \sqrt{\left(1 + 2\xi^2\right)^2 + 1}}}\right)^2 L_1. \tag{39}$$

The grid current control parameterization in d–q axis, considering the first-order region in Figure 16, follows the equations (38) and (39). Thereby, it was considered a damper factor $\xi = 1$, $L_{total} = 1.5$mHand a cut frequency of $\omega = 1200\ \pi$ rad/s and the gains were calculated. Therefore, $Kp_{i3d} = Kp_{i3q} = 4.55$and $Ki_{i3d} = Ki_{i3q} = 3459$ was obtained. The discrete gains calculated, considering the Euler discretization method, are $Kpd_{i3d} = Kpd_{i3q} = 4.55$ and $Kid_{i3d} = Kid_{i3q} = 0.288$.

6. Practical application example

One way to apply the models and control techniques discussed above is to implement a microcontroller system as shown in Figure 17. The implemented system has a mechanical interface powered by a 'wind turbine emulator', which regulates the torque and the speed of a three-phase PM generator. The three-phase generator used, a W-Quattro WEG 2.2 kW, has a nominal line voltage of 220 V at no load. Both the three-phase rectifier and the inverter used are composed of SKM50GB063D Semikron IGBT modules. The DC bus is composed of an equivalent capacitance 32 mF and supports up to 500 VDC. The IGBT modules drive are accomplished through Semikron SKHI-22BR drivers, which have current and voltage surge protection circuit and automatic blocking on error. The control system shown in Figure 17 applies a TMS320F28069 microcontroller, from the Piccolo family, which has 90 MHz operating frequency, co-math coprocessor, real-time debugging and additional six PWM modules for drivers. The voltage and current measurements necessary for the control algorithm are performed by means of transducers.

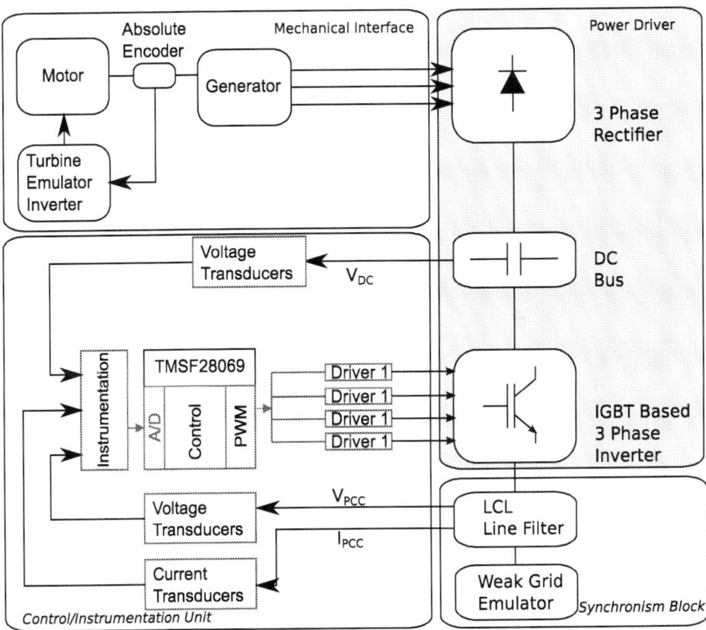

Figure 17. Block diagram of the setup rig.

The transducers used for current were the LEM-LA55P and for voltage the LEM-LV25P. An instrumentation circuit, specially developed for the application described in this section, conditions the transducer output signals. Basically, the instrumentation circuitry comprises an attenuation of the common mode noise step, a filtration step and a protection step, which is implemented with instrumentation amplifiers INA128P, AD706 and rail-to-rail modules for protection.

The assembly of the elements of Figure 17 results in the laboratory arrangement shown in Figure 18.

The experimental validation of the current controllers parameters, calculated in Section 4, was made by adjusting the filter output current. The isolated load considered in the experiments is purely resistive, with 12.5Ω of effective resistance.

The LCL filter used for this current loop test has the parameters calculated in Section 4. It was considered steps in the $i_{d\,\mathrm{ref}}$ reference, while maintaining constant $i_{q\mathrm{ref}}=0$. The current reference steps were 1A, 2A, 3A and 4A. Figure 19 shows the voltage and current responses to changes in the reference current.

Figure 20 shows the waveforms in detail considering the current transition from 3A to 2A in Figure 19.

It should be noted in Figure 20 that there is a distortion in the voltage and current zero crossing at the connection point. This distortion is caused by the dead time of the Driver SKHI 22BR, by default equal to 2.3 ms. The results presented in Figure 20 show that voltage at the connection point has low harmonic distortion when using LCL filter.

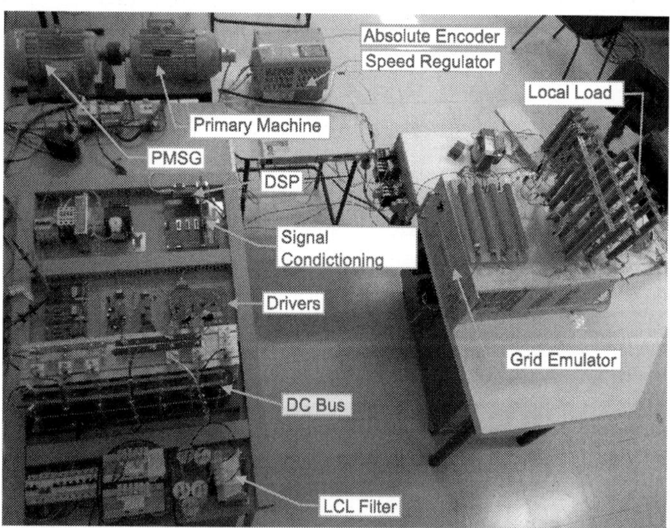

Figure 18. Test Rig.

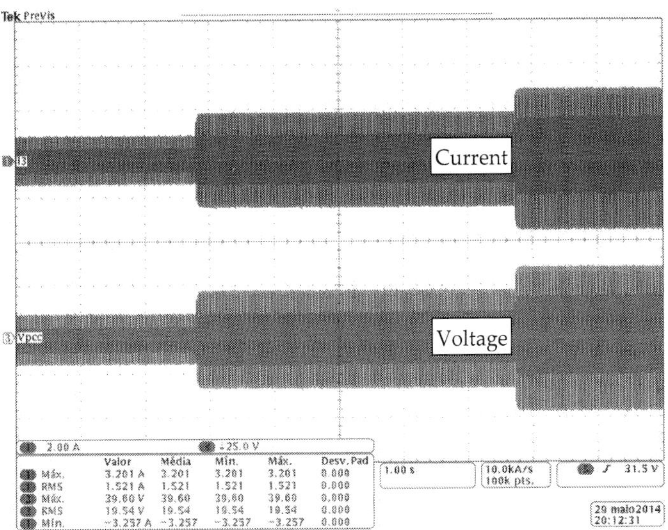

Figure 19. Current and voltage for reference changes in i_d.

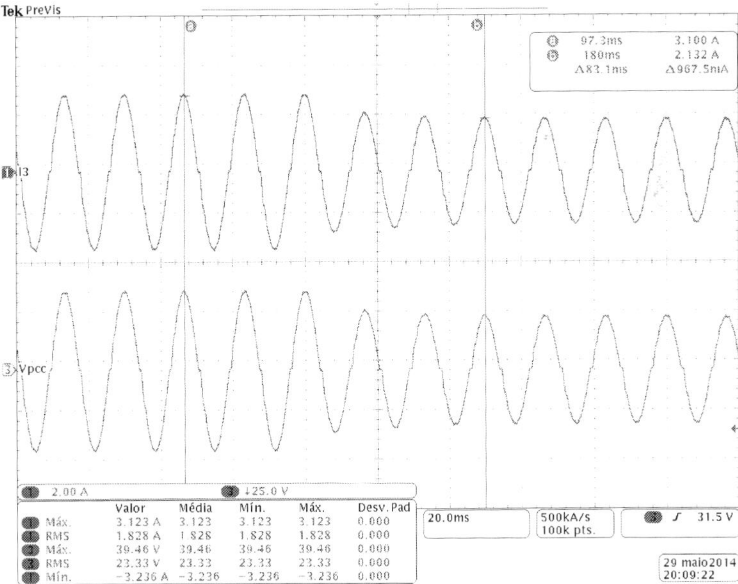

Figure 20. Detail of reference current transition from 3A to 2A.

7. Conclusions

It is clear that DG offers many advantages by including generation near to the consumption centres. The main advantages include fewer losses in power transmission, less environment impact and more reliability in power supply. However, the DG also offers new challenges, since, in several cases, it is close to low voltage loads. It is necessary to increase the voltage in connection point to achieve power flow that can lead to an overvoltage in distribution loads. In addition, this power generation needs to deal with resonances since, in most cases, high distribution line impedances are combined with the passive filters of the converters used in DG schemes. In this chapter, these effects were analyzed considering weak grids. Simulation analysis shows that it is possible to deal with these issues using well-designed control schemes and LCL converter filter. In addition, an experimental 2.2-kW prototype was used to show the feasibility of DG for low-power consumer structures.

Author details

Everton Luiz de Aguiar, Rafael Cardoso, Carlos Marcelo de Oliveira Stein, Jean Patric da Costa* and Emerson Giovani Carati

*Address all correspondence to: jpcosta@utfpr.edu.br

Department of Electrical Engineering, Campus Pato Branco, Federal University of Technology, Paraná, Brazil

References

[1] G. Grandi, D. Casadei and C. Rossi, "Direct coupling of power active filters with photovoltaic generation systems with improved MPPT capability," in *Power Tech Conference Proceedings*, Bologna, 2003.

[2] B. Ho and H.-H. Chung, "An integrated inverter with maximum power tracking for grid-connected PV systems," *IEEE Transactions on Power Electronics*, pp. 953–962, Volume 20, July 2005.

[3] A. Fitzgerald, C. K. Jr and S. D. Umans, Electric Machinery, McGraw-Hill, 6th edition, 2005.

[4] S. Grunau and F. W. Fuchs, "Effect of Wind-Energy Power Injection into Weak Grids," in *European Wind Energy Association 2012 Annuan Proceedings*, 2012.

[5] G. V. T. Prasad, *Impact of photovoltaic generators and electric vehicles on a weak low voltage distribution grid, University of British Columbia*, 2012.

[6] R. Teodorescu, M. Liserre and P. Rodriguez, Grid Converters for Photovoltaic and Wind Power Systems, New Delhi, India: John Wiley & Sons, Ltd., 2011.

[7] E. L. d. Aguiar, E. G. Carati and J. P. d. Costa, "Single Phase Wind Generation Distributed System Connected to Weak Grids: Analysis and Implementation," in *PCIM South America*, São Paulo, 2014.

[8] I. J. Gabe, *Contribution to the control of voltage source PWM inverters connected to the grid through LCL-filters.* Santa Maria: UFSM, Master Thesis, 2008.

[9] T. Wang, Y. Zhihong, S. Gautam and Y. Xiaoming, "Output filter design for a grid-interconnected three-phase inverter," in *Power Electronics Specialist Conference, 2003. PESC '03. 2003 IEEE*, 2003.

[10] R. Carnieletto, D. Ramos, M. Simoes and F. Farret, "Simulation and analysis of DQ frame and P x002B;Resonant controls," in *COBEP '09. Brazilian Power Electronics Conference, 2009*, 2009.

[11] P. C. Loh and D. Holmes, "Analysis of multiloop control strategies for LC/CL/LCL-filtered voltage-source," *IEEE Transactions on Industry Applications*, vol. 41, pp. 644–654, 2005.

[12] J. Xu, S. Xie and T. Tang, "Evaluations of current control in weak grid case for grid-connected," *Power Electronics, IET*, vol. 6, pp. 227–234, 2013.

[13] T. Komiyama, K. Aoki, E. Shimada and T. Yokoyama, "Current control method using voltage deadbeat control for single," in *Industrial Electronics Society, 2004. IECON 2004. 30th Annual Conference*, 2004.

[14] N. a. K. M. Liying Wang and Ertugrul, "Evaluation of dead beat current controllers for grid connected converters," in *Innovative Smart Grid Technologies - Asia (ISGT Asia), 2012 IEEE*, 2012.

[15] R. Zhang, M. Cardinal, P. Szczesny and M. Dame, "A grid simulator with control of single-phase power converters in," in *Power Electronics Specialists Conference, 2002. PESC '02. 2002 IEEE*, 2002.

[16] A. Roshan, R. Burgos, A. Baisden, F. Wang and Boroyevich, "A D-Q Frame Controller for a Full-Bridge Single Phase Inverter Used," in *Applied Power Electronics Conference, APEC 2007 - Twenty Second Annual*, 2007.

[17] B. Bahrani, A. Karimi, B. Rey and A. Rufer, "Decoupled dq-Current Control of Grid-Tied Voltage Source Converters," *IEEE Transactions on Industrial Electronics*, vol. 60, pp. 1356–1366, 2013.

[18] X. Wang, F. Blaabjerg and M. Liserre, "An active damper to suppress multiple resonances with unknown frequencies," in *Applied Power Electronics Conference and Exposition (APEC), 2014*, 2014.

Chapter 10

Planning Tools for the Integration of Renewable Energy Sources Into Low- and Medium-Voltage Distribution Grids

Jean-François Toubeau, Vasiliki Klonari, Jacques Lobry, Zacharie De Grève and François Vallée

Additional information is available at the end of the chapter

http://dx.doi.org/10.5772/61758

Abstract

This chapter presents two probabilistic planning tools developed for the long-term analysis of distribution networks. The first one focuses on the low-voltage (LV) level and the second one addresses the issues occurring in the medium-voltage (MV) grid. Both tools use Monte Carlo algorithms in order to simulate the distribution network, taking into account the stochastic nature of the loading parameters at its nodes. Section 1 introduces the probabilistic framework that focuses on the analysis of LV feeders with distributed photovoltaic (PV) generation using quarter-hourly smart metering data of load and generation at each node of a feeder. This probabilistic framework is evaluated by simulating a real LV feeder in Belgium considering its actual loading parameters and components. In order to demonstrate the interest of the presented framework for the distribution system operators (DSOs), the same feeder is then simulated considering future scenarios of higher PV integration as well as the application of mitigation solutions (reactive power control, P/V droop control thanks to a local management of PV inverters, etc.) to actual LV network operational issues arising from the integration of distributed PV generation. Section 2 introduces the second planning tool designed to help the DSO, making the best investment for alleviating the MV-network stressed conditions. Practically, this tool aims at finding the optimal positioning and sizing of the devices designed to improve the operation of the distribution grid. Then a centralized control of these facilities is implemented in order to assess the effectiveness of the proposed approach. The simulation is carried out under various load and generation profiles, while the evaluation criteria of the methodology are the probabilities of voltage violation, the presence of congestions and the total line losses.

Keywords: Design of experiments, Dispersed generation, Distribution systems, Monte Carlo simulation, Smart metering

1. Introduction

Following the Kyoto Protocol and in the context of a liberalized energy market, decentralized generation, often based on renewable energy sources, is emerging in medium- voltage (MV) and low-voltage (LV) distribution grids. Those networks are becoming more and more active with power flows and voltage profiles influenced by both generation and consumption. In the future, given the '20–20–20' objectives of the European Union and the more ambitious objectives in Wallonia (Belgian region), the expected penetration of decentralized generation in distribution grids could lead to a critical behaviour of the system that has not been initially designed and sized to face power injections coming from dispersed units. This major paradigm shift in distribution grids is even strengthened by the numerous financial incentives granted to the renewable generation for ensuring their profitability. Indeed, this financial support has progressively driven down the electricity prices on power markets to the point of making the conventional generation less profitable, which already led to the closing or the mothballing of more than 50 GW of gas plants in Europe during the last few years [1].

Consequently, in the MV grids, the main problems due to a massive integration of distributed generation (mainly wind parks) are due to local congestions of power lines and/or to over-voltages in the neighbourhood of the dispersed generation units. Concerning the LV grids, the major problem that can be met by the distribution system operators (DSOs) comes from overvoltage in the neighbourhood of photovoltaic (PV) installations. Such a situation could happen during periods of low consumption and high generation of those decentralized units, especially on long power lines. Furthermore, reverse power flows going up the line towards the MV/LV transformer can then degrade the network stability.

The DSO is responsible for the security of its system and power quality. It is therefore important to develop planning tools that ensure secure and reliable operating conditions within the whole grid. Moreover, it is also essential for DSOs to avoid early damages of the network equipment as well as to optimally manage the investment decisions and the maintenance plans. However, as the intermittent generation is often difficult to predict due to its stochastic behaviour, the planning studies become more complex to carry out. Indeed, it is now indispensable to use probabilistic approaches to model the different network components, since the traditional deterministic worst-case scenarios yield much more restrictive conclusions that can lead to overestimated reinforcements [11].

In that way, two probabilistic tools are described in this work. Both are based on Monte Carlo (MC) simulation whose general principle is to generate a large number of system states in order to provide consumption and generation patterns that are representative of the actual behaviour of all customers. However, it is important to mention that long-term analyses are focused on modelling realistic scenarios of the system evolution rather than making an accurate prediction. It is indeed utopian to expect a precise hourly model of the wind farm generation for a whole year. Practically, two different types of Monte Carlo simulation are implemented. The first one is a non-sequential simulation that generates a set of system states independent of each other. This method is combined with the design of experiments (DOE) methodology in order to proceed to an optimal positioning as well as to a pre-sizing of voltage compensation devices (capacitor banks, storage units, static VAR compensators (SVCs), etc.) at the MV level. The purpose is to help the DSO in optimally managing its systems with adequate investments. The first tool requires aggregated load profiles at the MV/LV interface. Those statistical profiles are typically obtained, thanks to a second tool developed for the safe integration of PV units in LV grids. The latter is implemented in a pseudo-chronological Monte Carlo environment and is based on quarter-hourly energy flows recorded by smart meters (SMs). Practically, thanks to statistical profiles established by the use of these measurements, the tool is able to compute different parameters throughout a specified period of time such as the voltage profile along the studied feeder or the distribution of the power profile at the MV/LV substation. In this chapter, in order to demonstrate its interest for DSOs and to introduce the potential of some technical solutions (reactive power control, P/V droop control thanks to a local management of PV inverters, etc.) to alleviate the influence of PV generation in LV grids, this tool will be applied to a practical case study.

This chapter is thus divided into two main parts, each one focusing on one of both the above-mentioned probabilistic tools. Section 1 concerns the description of the planning at the LV level. More particularly, the extraction of the energy flow data coming from SM installed in the LV-Belgian distribution grid as well as their use in the Monte Carlo procedure is presented. Then, the potential of several power- or voltage-control strategies is evaluated for an existing LV area subjected to abnormal operating conditions. Section 2 relates to the planning tool studying investment plans that may have to be taken in critical areas of the MV grid.

2. Probabilistic analysis of low-voltage networks

The techno-economic analysis of LV feeders is typically governed by a set of basic rules that define voltage, voltage unbalance and current limits. According to these rules, which are most often precisely determined by regional or national standards, voltage along the feeder should fall within a defined band during a minimum percentage of time. As far as the voltage unbalance is concerned, an upper limit should also be respected during a minimum percentage of time. The limits that concern the value of the current do not often apply in LV feeders apart from urban networks with high load density.

When adding distributed photovoltaic generation to an LV feeder, voltage profile is the parameter that is mostly affected. Indeed, during periods of high PV injection and low consumption, reverse power flows towards the head of the feeder can occur. These reverse power flows result in a voltage rise at the end of the feeder. According to the European EN 50160 standard [2] and its national implementations (i.e. the voltage cannot vary more than 10% around its nominal value), if the upper limit of the root mean square (rms) voltage is exceeded at a certain PV node, the unit must be temporarily cut off [2], [3]. This causes a loss of generated PV power, which means a loss of income for the PV owner. Moreover, these frequent cut-offs, combined with the randomly changing loading profile along LV feeders, affect the distributed power quality and accelerate the degradation of network components. The operational cost of the network, currently assigned to the DSO, is therefore increased due to the accelerated ageing of its components.

The development of technical solutions to such problems has, from the DSO point of view, a threefold objective, namely the security level of the network, the increase of PV-hosting capacity of the feeder and, naturally, the investment cost. For a given PV-hosting capacity, the optimal design of a technical solution could be roughly determined by a graph similar to Figure 1. On the other hand, for a selected security level, the optimal hosting capacity in a feeder could be based on a graph similar to Figure 2.

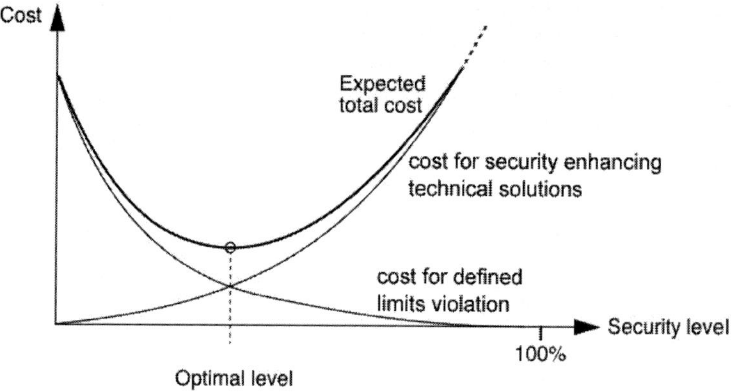

Figure 1. Schematic of cost-security.

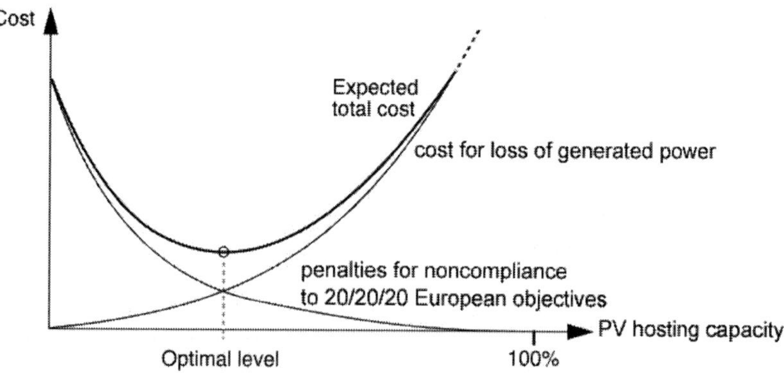

Figure 2. Schematic of cost-hosting capacity.

For designing adequate technical solutions to current problems while increasing the PV-hosting capacity, the DSO needs to locate the critical points and quantify the voltage magnitude and unbalance violation risk in every LV feeder. Currently, the DSO evaluates the LV network operation using deterministic analysis tools: the energy exchanges at the nodes of a network with distributed PV generation are determined based on worst-case scenario. For example, as far as the evaluation of overvoltage risk is concerned, the DSO makes its calculations considering the scenario of the highest expected PV injection and the lowest consumption load along each specific feeder. Although worst-case approach is safe for the protection (100% security level) of the electrical system, it focuses on rare extreme operation states of the network instead of the most possible ones. As a result, the worst-case approach often leads to oversized and costly technical solutions that do not consider the time fluctuation of PV injection and the randomness of consumption loads. Taking this time dependence into account will allow a refined optimized design of mitigation techniques in terms of efficiency and investment cost, closer to the optimal level that is roughly represented in Figure 1. For the above reasons, a doubt concerning the effectiveness of the deterministic worst-case models has recently arisen in literature [4], [5].

In order to take into account the time variation of the PV injection and the random behaviour of the loads and PV installations along an LV feeder, probabilistic analyses were introduced in late research works [6, 9]. These methodologies are based on analytical or numerical approaches, such as the Monte Carlo simulation, aiming at simulating the uncertainty of PV injection and residential loads. Nevertheless, most of these studies propose advanced probabilistic models based on solar irradiation data for the PV panels and meteorological data for the consumption loads. Such data are rarely directly accessible to the DSO and most of them do not consider the efficiency of the PV cells. Currently, DSOs in several countries are installing smart metering devices that record voltage, current as well as active and reactive powers with a high time resolution (10 or 15 min), even for domestic customers. According to European Photovoltaic Industry Association (EPIA), large amounts of such data will be available at the horizon of 2020 [10]. The processing of these data is, nevertheless, still a challenging task that will have to be addressed.

2.1. Statistical modelling of the energy flows at LV nodes

A procedure for building statistical profiles of energy flows at LV nodes and of the voltage at the MV/LV substations is presented in references [11, 13] and will be summarized here. The data are recorded by SM devices that have been installed by the DSO in the Belgian grid but the procedure can be extended to any LV grid with such metering devices. The SM configuration of every PV customer is presented in Figure 3.

The SM1 records the total quarter-hourly (15 min) injected energy towards the network (E_{inj}) and consumed energy by the customer (E_{cons}). It is therefore the metering device that records the energy flow at the interface between the customer and the LV network. The SM2 records the total 15-min energy generated by the PV unit ($E_{PV,inj}$) and consumed energy ($E_{PV,cons}$) by the power electronic devices that are necessary for its connection to the network. These four values have been recorded on a 15-min basis for all PV customers over a long period (>2.5 years).

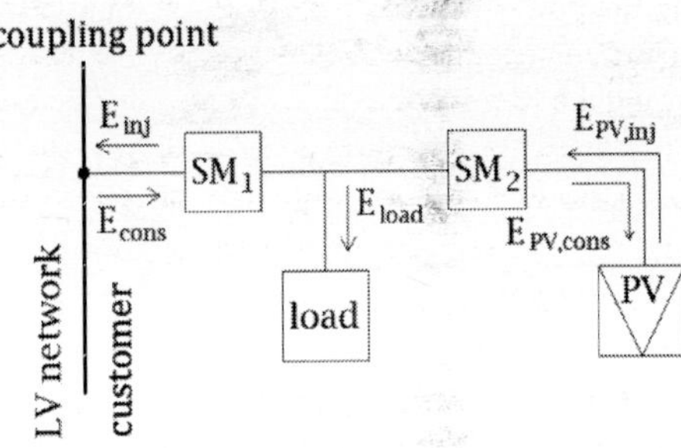

Figure 3. Smart metering configuration of every customer with PV unit, connected to the LV feeder.

For the purpose of evaluating the overvoltage risk along a feeder, the SM1 data, namely the energy flows at the coupling point of the PV customer to the network, are most preferably used. These 15-min values are therefore used for the creation of typical day profiles (TDPs) for every individual quarter of an hour in a day (96 quarters of an hour). Created independently for each node (customer), the TDPs reflect the variation of energy flow, separately for injection and consumption, for every 15-min interval in a day, during the studied period. Similar TDPs are also created to reflect the variation of the voltage at the head of the feeder, based as well on 15-min SM voltage data recorded at the LV side of the MV/LV transformer. It is worth noting that such profiles can be very easily created for various averaging time steps of SM devices. In most cases, SM devices in LV networks in Western Europe record energy flows on a 15- or 10-min time resolution.

The TDPs are then statistically transformed into 96 cumulative distribution functions (CDFs), that is, one for each quarter of an hour of the typical day, and they will be used in the framework of an MC probabilistic load flow.

2.2. Probabilistic load flow using a Monte Carlo method

The Monte Carlo simulation is a mathematical technique that allows accounting for the random behaviour of some variables within the studied system. In our case, the stochastic variables are the consumption and the generation of each customer of the LV grid during the day. The principle relies on the repeated random sampling of these variables in order to accurately simulate the system and to obtain numerical outcomes (i.e. probability of overvoltage at each node of the LV grid).

The above-mentioned CDFs are sampled under an MC framework in order to generate randomly 15-min network states. A balanced power-flow algorithm then computes the voltage profile along the feeder considering the sampled energy flow values at each node and the sampled voltage at the head of the feeder. The procedure is repeated numerous times in order to test a large amount of possible combinations. For obtaining a good convergence threshold (<0.1%) on the results, it is shown that 10,000 simulations is a good trade-off to keep acceptable computation times [11].

However, for performing the power-flow computation, the total 15-min energy-flow values must be transformed into instantaneous power values at each node. The injected or consumed power is rarely stable during the 15-min interval, which means that the total recorded energy is very often injected or consumed in a very short (<<15 min) period of time. Indeed, it has been shown in reference [12], thanks to advanced grid analysers installed by the DSO at critical network points that the power flow at the coupling point of a grid customer is highly fluctuating within the 15-min intervals. Therefore, the peak values of injected or consumed power might be much higher than the average ones that consider a stable power flow.

In the proposed probabilistic framework, for the sake of security, these instantaneous power values are not considered equal to the average values, which would be computed by dividing the total injected or consumed energy by the respective period of time. As explained in reference [12], TDPs are also created for the time repartition of the energy flow within the 15-min intervals based on 3-min recordings of advanced grid analysers at limited critical points of the network. These TDPs are also integrated in the MC algorithm, once they have been transformed into CDFs, in order to provide the sampling frame for the time repartition factor f_i for each node i. This factor practically defines the peak injection or consumption power values.

The computed peak power values are therefore the ones considered in the power-flow algorithm. The rms voltage in this way is calculated node by node for each 15-min network state, which is therefore characterized by the following values:

$$
\begin{aligned}
&E_{\text{cons},i},\ E_{\text{inj},i},\ f_i \text{ for nodes } i = 2:N \\
&V_{\text{MV/LV}} \text{ for node } i = 1 \\
&P_i = \frac{E_{\text{inj},i}}{f_i} \text{ for nodes } i = 2:N
\end{aligned}
\tag{1}
$$

These 15-min values are compared to the voltage band that is imposed by national, regional or local values. In the case of Europe, such limits are imposed by the EN 50160. In this way, the overvoltage probabilities can be determined, node by node, for the studied period.

2.3. Illustration on an LV-Belgian network

The LV feeder, which is illustrated in Figure 4, makes part of the LV-Belgian distribution network. This feeder (technically described in reference [12]) is simulated with the presented probabilistic framework for every month of a typical year, considering the 15-min SM data recorded over a period of two years. In total, 16 clients are connected to the feeder, among which three are equipped with a PV installation (5 kVA at node 4, 2.65 kVA at node 12 and 5 kVA at node 13). In order to clearly demonstrate the influence of the PV generation on the voltage variation, six more PV clients have been considered at nodes 5, 6, 8, 10, 11 and 14 for this simulation. For these additional units, the same PV injection SM data as for the PV unit at node 13 were considered.

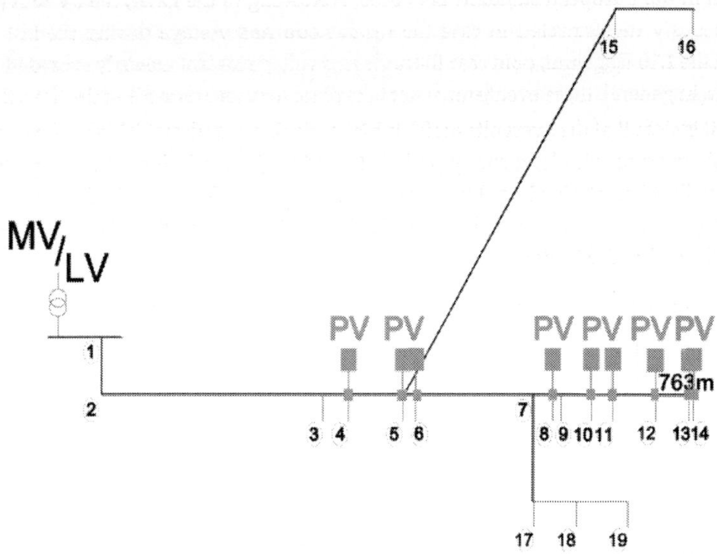

Figure 4. The simulated LV feeder.

Time variation for a given MC simulation of the voltage profile at node 14 is shown in Figure 5. Calculated with a deterministic analysis, the voltage would have been a stable worst-case value that would characterize the whole period of a day. However, based on the current analysis and the use of SM data, it is clearly shown that the voltage rises only during certain hours of the day, usually between 11:00 A.M. and 16:00 P.M.

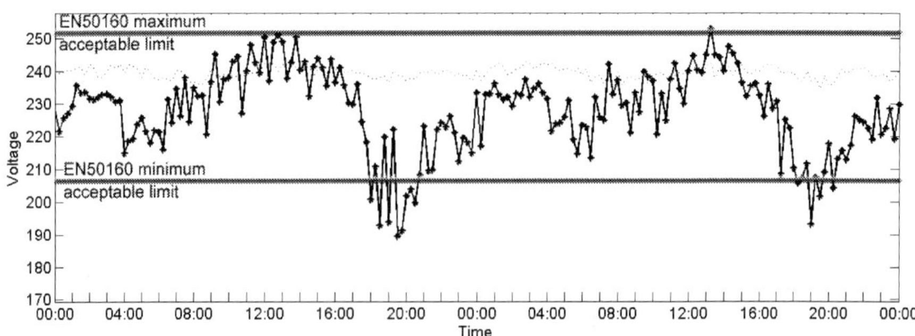

Figure 5. Voltage at PV node 14 during two typical days in April, calculated on a 15-min basis.

Thanks to the probabilistic simulation, the overvoltage risk can be defined for each node during the studied period. This risk is defined as the probability of exceeding the limit ($>1.10\ V_{nom}$)

suggested in the European standard EN 50160. According to the latter, the PV unit must be instantaneously disconnected in case the mean node rms voltage during the last 10 min exceeded the 1.10 V_{nom} limit, or in case the node rms voltage instantaneously exceeded the 1.15 V_{nom} limit. In general, these events must not take place for more than 5% of the time. Based on Figure 6, it is clear that the overvoltage risk is higher during months with high solar irradiation (June to August) or with long sunny periods (March to June). Such events are expected for hours usually characterized by low energy consumption along the feeder and high PV generation, since reverse power flows can lead to instantaneous rise of the voltage profile towards the end of the feeder.

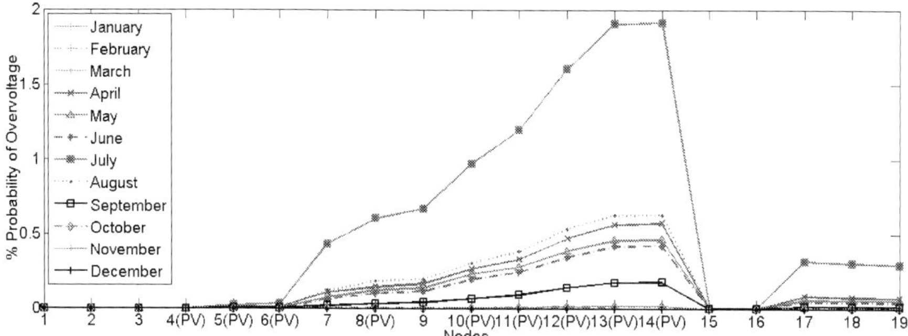

Figure 6. Overvoltage risk at the nodes of the simulated feeder for every month of a typical year.

A significant voltage drop takes place between 18:00 and 21:00 hrs, due to peak electricity demand, which increases undervoltage risk (<0.90 V_{nom}) during this period. Based on the probabilistic simulation of the same network for the period of a 'typical year', the average undervoltage risk for each node is illustrated in Figure 7. As expected, this risk rises between November and March due to higher electricity demand for heating.

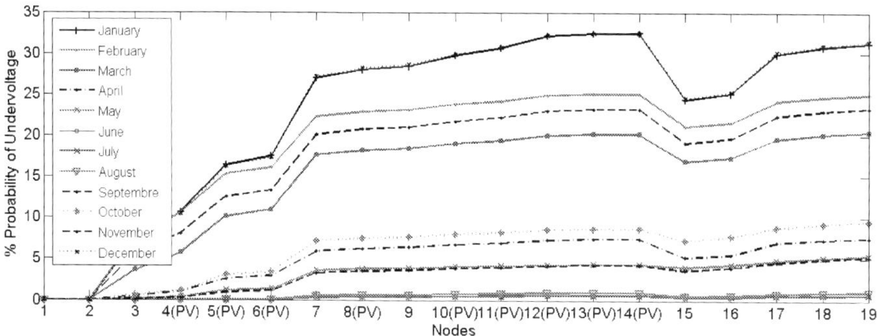

Figure 7. Undervoltage risk at the nodes of the simulated feeder for every month of a typical year.

In Figure 6, it can be seen that nodes adjacent to PV units are also affected by rapid voltage variations and oscillations (caused by the on–off control of the units), which tend to accelerate the degradation of many network components. On the other hand, it has been shown in reference [14] that the number of line congestions and the transformer load are significantly impacted with the emergence of PV generation and electric vehicle (EV)-charging devices. In this way, Figure 8 illustrates the decrease of line congestion when PV generation is included into the grid. Indeed, the power produced by these units is consumed locally by the neighbouring loads, which reduces the line power flows. Conversely, the introduction of EV-charging devices has an adverse impact on the grid. Indeed, these are powerful appliances (typically 230 V and 16 A) that significantly increase the current in different lines. Figure 9 shows the statistical distribution of power flows at the MV/LV substation. One can see that the power at the MV/LV interface decreases when generation is installed into the grid and substantially rises with the integration of EV-charging appliances. Therefore, three different scenarios are investigated in this study. Scenario 1 considers the feeder represented in Figure 4 but with no PV generation. Scenario 2 considers the same feeder with PV units at nodes 4, 5, 6, 13, 14 (rated power 5 kVA) and at node 12 (rated power 2.65 kVA), whereas scenario 3 considers also the EVs connected to the same households than those with PV panels. It should be noted that the EVs are simulated based on typical EV-charging profiles and on the statistical behaviours of current EV owners.

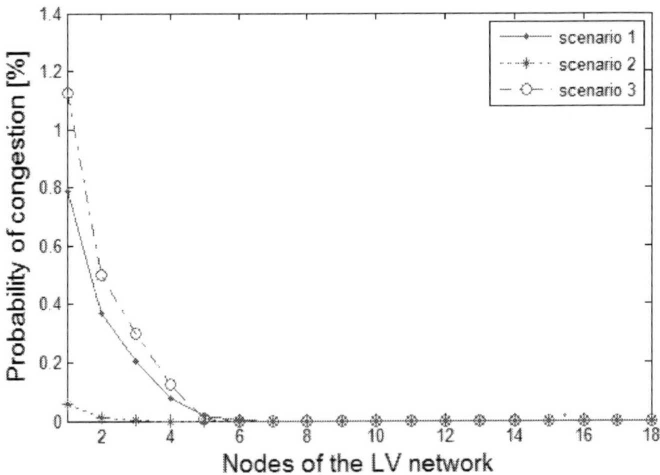

Figure 8. Probability of congestion for each line section of the first feeder with 19 nodes for the month of June.

The previous results highlight the necessity to adopt strategies aiming at limiting these important voltage and current variations. Otherwise, actions can be too restrictive for potential PV hosting capacity of the network or much less optimized in terms of efficiency and cost as roughly represented in Figures 1 and 2. Moreover, the probabilistic framework can also

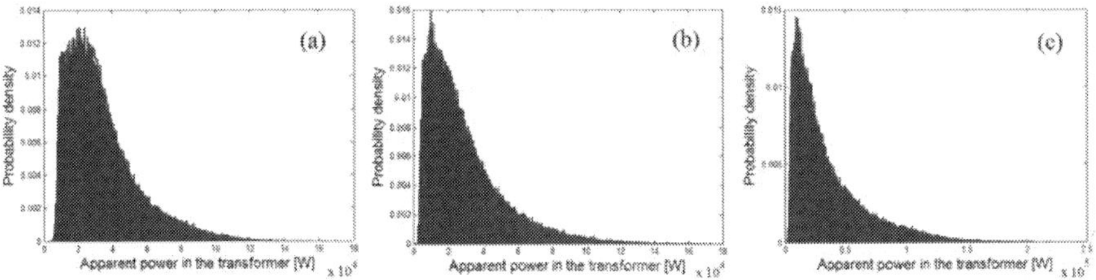

Figure 9. Yearly loading distribution at the transformer for (a) the traditional LV grid, (b) current grids with PV units (b) and (c) future grids with the introduction of EV-charging devices.

simulate future scenarios of PV integration increase, or techno-economically evaluate multiple solutions to network operational problems. It can therefore provide answers for an efficient and secure operation of LV networks, taking into account the optimal level that is represented in Figures 1 and 2.

2.4. Evaluation of the strategies for mitigating the overvoltage risk

Currently, certain DSOs evaluate the potential benefit of load-shifting strategies on mitigating overvoltage risk and increasing the PV energy capture. Incentives that promote self-consumption of the generated PV power by its producer should be designed based on the time variation of the network states. In line with this assumption, the self-consumption rate of 20 residential PV customers has been studied by using the 15-min recordings of the SM2 devices. The self-consumption ratio of each PV customer can be calculated for each quarter of an hour in a day by using both the SM1 and SM2 data and applying the following relation:

$$\text{Ratio}_t = \frac{\left(E_{\text{PV,inj},t} - E_{\text{inj},t}\right)}{E_{\text{PV,inj},t}} \text{ with } t = 1:96 \tag{2}$$

The average self-consumption ratio of three residential PV customers is shown in Figure 10 for each quarter of an hour, based on SM1 and SM2 data for the month of July. According to this diagram, the self-consumption ratio decreases during hours of peak PV injection, differently for each one of the clients. One could therefore assume a quite high potential for strategies that could shift consumption loads towards high PV injection hours. Such strategies could achieve an important mitigation of the overvoltage risk that usually increases during these hours as shown in Figure 5. The design of incentives could be therefore based on a thorough study of the self-consumption behaviour of customers connected to similar LV feeders, by using SM recordings and suitable clustering methodologies.

Apart from load management strategies, certain power or voltage control strategies are currently discussed by the DSOs for facilitating the integration of renewables in the LV network. Among them, reactive power control is a strategy that has been recently implemented

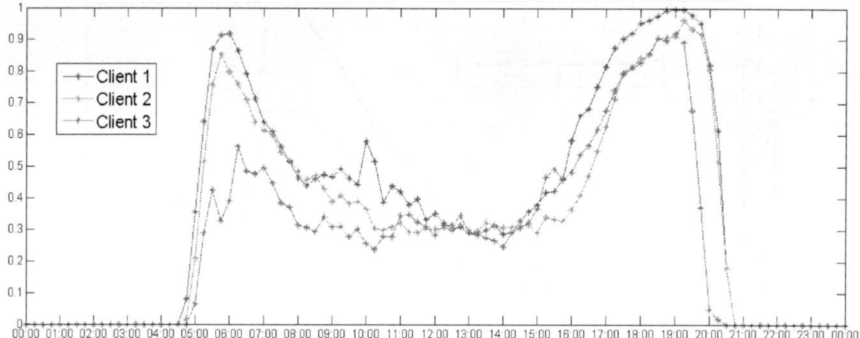

Figure 10. The quarter-hourly average self-consumption ratio of three PV residential PV customers.

in the LV networks of some European countries. In these cases, the distributed PV units have to be equipped with Q/V droop functions in order to provide voltage support as the conventional large power plants. The reactive power of the PV inverter is then adjusted, based on the local voltage, according to a $\cos(\varphi) = f(P)$ function that is defined by the respective national regulations [15].

In this study, a simpler application of this control scheme, that does not consider the actual local voltage, has been simulated in order to evaluate its benefit on the mitigation of the overvoltage risk. The purpose of this example is to illustrate the potential of this probabilistic tool in order to constitute a benchmark for evaluating and comparing voltage control solutions. In the previously presented results, the reactive power generated by PV inverters has been set equal to zero in order to minimize power losses. Turytsin et al. demonstrate that the optimal adjustment of PV inverter reactive power could reduce voltage variations. The maximum available reactive power at each PV node i is calculated as follows [16]:

$$Q_{\text{inv,max},i} = \sqrt{\left(S_{\text{rated},i}^2 - P_{\text{inj},i}^2 \right)}$$

(3)

where S_{rated} is the nominal active power (equal to the apparent power) of the PV unit at node i, whereas $P_{\text{inj},i}$ is the PV active power injection at node i for the system considered state. In order to evaluate the benefit of such a control, the same LV feeder has been simulated with the probabilistic framework, but this time considering that the reactive power of each PV inverter i is equal to 0.1 $Q_{\text{inv,max},i}$. As shown in Figure 11, the overvoltage risk for the scenario with reactive power control has decreased for the month of July.

In European countries where such a control scheme has been implemented in the LV network, the value of the power factor ($\cos(\varphi)$) cannot be lower than 0.95 for an installed power smaller than 6 kVA. Hence, such a voltage support through reactive power can often be inefficient as

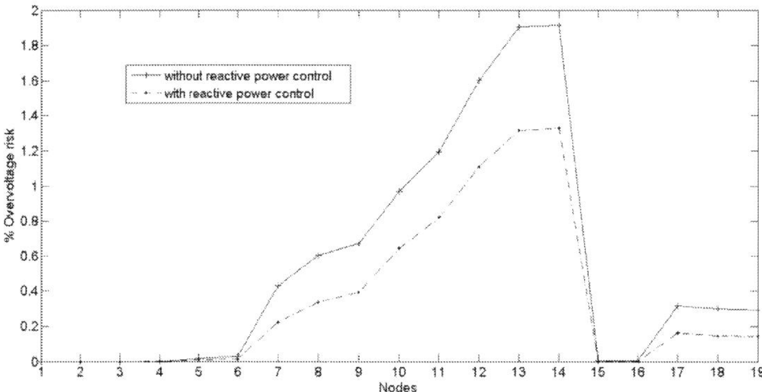

Figure 11. Reactive power control effect on the overvoltage risk for the month of July.

the network voltage is principally influenced by the active power due to the low X/R ratio of the lines [17], [18]. Large amounts of reactive power would be therefore required to influence the voltage. In this context, P/V droop controllers are more efficient and straightforward to provide voltage support in an LV network [19, 21].

Presently, the PV power injection towards the network is solely subject to the maximum power point (MPP) of the PV units. In case of voltage violation, the PV inverters passively undergo the on–off control that is required by the European or national standards, without considering the current network state. In order to address this operational issue, T. L. Vandoorn and colleagues developed a fast-acting primary control scheme based on voltage droops [22]. The droops are applied by P/V controllers requiring neither inter-unit communication nor voltage tracking for synchronization. In this study, P/V droop control is evaluated with the proposed probabilistic framework in the same LV feeder in order to analyse its benefit on the voltage level, the captured renewable energy and the on–off oscillations of the PV inverters. The evaluation considers the time variation of nodal injection and consumption and is therefore based on the analysis of multiple network states and not only the worst-case ones.

The P/V control scheme modifies the injected active power of the PV unit in function of the local voltage in order to prevent and eliminate overvoltage situations. In this way, the total cut-offs of PV units, applied in the conventional on–off control, and their subsequent effects (voltage and current transients as well as significant PV power loss) are avoided. The P/V droop controller adapts P according to the scheme in Figure 1 [22].

For the purpose of evaluating the benefit of the P/V droop control, all PV inverters were considered equipped with a P/V controller. The application of this control in the studied LV feeder, considering SM data for the month of July, demonstrated its efficiency for eliminating the overvoltage risk along the feeder [23]. In Figure 13, the voltage profile at node 14 for the droop control and for the on–off control scenario is indicatively presented for a period of two typical days of July. Generally, the voltage profile of the P/V control scenario coincides with one of the on–off scenarios apart from the three 15-min intervals of the first day and two 15-

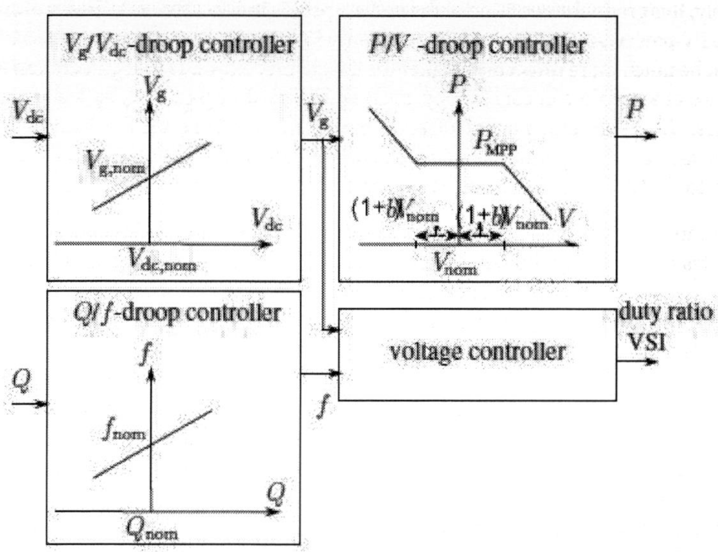

Figure 12. VBD control, active and reactive power controls. Combined operation of the droop controllers to determine the set value of the grid voltage [22].

min intervals of the second day during which the P/V control was enabled. Indeed, during these periods, the rms voltage at node 14 exceeded the reference value of the droop control V_{up} and due to this, the control was enabled in order to modify the injected P of the unit. As a result, the nodal voltage during these intervals was gradually smoothed in order to avoid an eventual cut-off of the unit.

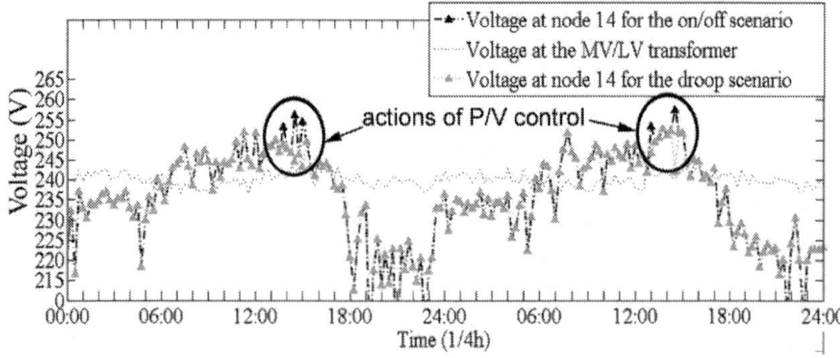

Figure 13. The 15-min voltage at node 14 for the on–off control and the droop scenario during two typical July days.

Obviously, this gradual smoothing of the voltage profile induces the curtailment of an amount of active PV power. As demonstrated in reference [23], the instantaneous curtailed PV energy can often be much more direct and higher for the on–off control. This argument can foster the acceptance of a certain control strategy, such as the P/V droop control, by the concerned PV customers. The long-term impact of such mitigation solutions on the income of the PV producer and on the operation and maintenance cost of the network, which is currently assigned to the DSO, should be evaluated keeping in mind the principle of Figures 1 and 2.

The presented evaluation that takes into account, in a fast optimized manner, multiple network states and not only the worst-case ones can lead to technical solutions that should be tailored to the special needs and necessary security level of each individual network. Moreover, the analysis and design of the LV distribution network with the presented probabilistic framework contribute to a more rigorous analysis of the MV distribution network. Indeed, when studying the MV network, downstream LV feeders are reduced to P–Q buses defined by the power flow at their root nodes. Most often, the time variation of this power flow, due to PV injection and random loads, is not taken into account. Section 2 introduces a probabilistic methodology that aims at analysing the MV distribution network and uses aggregated load and generation profiles at the MV/LV interface obtained thanks to the presented probabilistic framework. The impact of time variation of load and generation profiles in the LV network in this way is considered in the analysis of the MV network as well.

3. Investment planning at the MV level

Originally, distribution networks were oversized in order to easily ensure the connection of new clients without requiring the upgrade or the replacement of the equipment. However, the constant increase of load and generation, although the network infrastructure has not increased accordingly, induces a global system operation closer to its limits. In particular, this may lead to excessive line power flows that can be destructive not only for the network equipment but also for the end-user material. Indeed, important current flows may infer unintended line congestions as well as an extensive increase of the transformer load that can potentially reduce its lifetime [24]. Moreover, as the voltage drop depends not only on the impedance of the line but also on the line power flows, the voltage profile can locally fluctuate outside its acceptable limits. Although the overvoltages are more dangerous for electrical devices, both low- and high-voltage issues have to be taken into account in the regulation of distribution networks. However, this objective of ensuring a good voltage profile can be in conflict with the minimization of line losses that are very costly for the DSO, in the range of 50 euros/MWh, concerning the Belgian situation. Indeed, for a same demand of power, the reduction of line losses is obtained by a voltage rise. Consequently, the DSO will look for a trade-off between the quality of the delivered power and the economic losses.

Currently, several strategies can be used to improve the real-time operation of the grid. The most popular method is the use of on-load tap changer (OLTC) of transformer. The principle is to mechanically adjust the turn ratio of the transformer winding, usually with regard to a local measurement of the current. It should be noted that the dynamic of an OLTC is relatively fast with a total operation time between 3 and 10 s, depending on its design [25].

Then, the reactive power compensation devices are also very useful to get a better voltage profile in a power system with an X/R ratio higher than one. Devices such as the static VAR compensator or static synchronous compensator (STATCOM) have the interesting ability to work in both inductive and capacitive modes. Furthermore, the technology is well developed and financially affordable compared to other techniques [26].

In recent years, the possibility of investing in storage is gaining credibility [27]. Indeed, although the most mature technology (i.e. the pumped hydro) is not appropriate for the needs of distribution systems, other technologies are expected to become increasingly important in the near future. In this way, stationary batteries seem really appealing due to their good flexibility regarding the input and output power. Nonetheless, the cost is still very high and their success will largely depend on the cost reduction. Then, the compressed-air energy storage (CAES) could be a solution in the coming years, but its integration is hampered by its low efficiency and limited operational flexibility. Finally, the hydrogen storage technology is showing promise but is not expected to be sufficiently developed to be competitive before 2020.

Then, the possibility to use the reactive power produced by distributed generation (DG) units is also an investigated alternative [28]. However, this method has a limited margin due to the current legislation restricting the reactive power flows at the connection node of each client (generators as well as industrial customers). Moreover, as there is currently no market for ancillary services at the distribution level, the customers have no incentive to help the DSO in its task of maintaining optimal operating conditions.

Finally, if the power flows far exceed the network capacity, the system reinforcement may constitute the most viable solution.

Note that the DG units connected to the distribution grid are automatically disconnected when the network frequency or the local voltage exceeds the limits defined by the legislation. Therefore, another major asset of a local control of the network operating conditions is the avoidance of such disconnections, which may have to be financially compensated, depending on the contract between the producer and the DSO.

In this study, a new procedure aiming at ensuring the respect of the operational constraints of the distribution grid at any time is implemented. This amounts to avoiding any congestion or voltage violation. The crucial role of maintaining the frequency balance is indeed assigned to the transmission system operator (TSO). The general structure of the proposed method is represented in Figure 14 and is divided into three parts. First, a two-step planning procedure is conducted for identifying the optimal strategies that the DSO has to implement in order to optimally improve the long-term operation of the MV network. As most of the Belgian HV/MV transformers are already equipped with OLTC devices, and given the current context not fostering the reactive control of DG units, the study is mainly focusing on the installation of VAR compensators and storage utilities. In this way, the first part of the planning process evaluates the best positioning of these network regulation devices, whereas the second part consists in sizing the devices by taking cost constraints into account. Obviously, the final choice of the positioning will depend on the geographic and administrative considerations, which will not be considered here, as they have no repercussions on the philosophy and the principle of the method. The second part consists in simulating the real-time centralized control of the system, which is performed by optimizing the command of the network regulation devices in function of the network state. This operation is thus carried out for a large number of simulated states. Finally, the last part is post-processing whose purpose is to compare the probability of constraint violations as well as the total line losses before and after the installation of the network regulation devices.

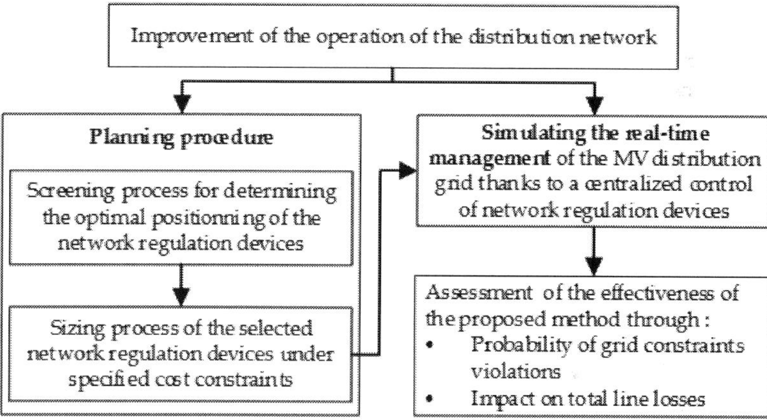

Figure 14. Structure of the proposed procedure for the improvement of operational efficiency in MV distribution grids.

The developed methodologies for the planning procedure and the real-time simulation of the network states are both developed in a non-sequential Monte Carlo procedure due to the lack of real-time measurements, especially at the MV/LV interface. This issue concerns less the industrial area, as quarter-hourly energy flow recordings are mandatory for the Belgian consumers with an installed power higher than 100 kVA. Indeed, these clients are not only invoiced for their actual energy consumption but also for their peak power, regardless of its duration.

When working in a Monte Carlo environment, it is essential that the underlying simulations are fast and efficient in order to avoid an excessive simulation time. Therefore, because of its excellent trade-off between simulation time and quality of the results, the experimental design method was privileged. Indeed, as explained hereafter, this method, which is also referred to as design of experiments, is perfectly suited for conducting the desired studies (i.e. planning procedure and simulation of the centralized control of network regulation devices). Then, as both proposed studies require performing load-flow studies in order to determine the voltage profile and the power losses within the system, a fast, simplified computation, based on the analytical approach presented in reference [29], was implemented. However, in its initial form, this algorithm was designed for a single radial feeder without ramifications; it had to be thus adapted in order to be used on the traditional tree-shape structure of MV distribution networks.

Since the first issues originating from the progressive introduction of renewable energy generation over the past years, the improvement of MV network operation has become a real matter of concern. In this way, in recent studies [30], [31], new voltage-control methods combining the advantages of OLTC action and D-STATCOM response were proposed. However, in practice, this procedure can hardly be implemented as the transformers are not necessarily directly managed by the distribution system operator (i.e. most of the Belgian transformers, even those installed at the distribution level, are the property of the TSO).

This study is intended to be more complete and takes into account the reality of the economic and technical constraints by evaluating the impact of a local centralized control downstream each of the HV/MV transformers. The control is thus at the interface between a decentralized control performed at each connection node of the network regulation devices and a centralized control applied on a national scale.

Practically, this work is structured as follows. First, the principle and the basic concepts of the DOE method are introduced. Then, the proposed methodology is fully presented and applied to an existing MV-Belgian network. Finally, the collected results are exposed and discussed.

3.1. General principle of the DOE method

Design of experiments constitutes a powerful tool to establish and study the effects of multiple inputs on a desired output. The method thus involves two types of variables, the response of interest (y) and the k predictor variables (ξ_1, ξ_2, ..., ξ_k), which are referred to as factors. These are chosen by the experimenter as variables that are supposed to influence the response. The aim of the method is to establish a mathematical description of the system of the form [32]:

$$y = f(\xi_1, \xi_2, ..., \xi_k) + \varepsilon \qquad (4)$$

where ε is the error between the real observed response y and the result given by the model. In order to simplify future calculations, the natural variables (ξ_1, ξ_2, ..., ξ_k) expressed in physical units are transformed in coded dimensionless variables (x_1, x_2, ..., x_k). Moreover, bound constraints need to be defined for each factor. These lower and upper bounds are, respectively, called 'low' and 'high' levels and are characterized in coded variables by '-1' and '$+1$'.

There are different experimental designs, which are commonly divided into two main categories, given the desired objective. The first one constitutes a screening study, which is used to determine the factors that really influence the outcome. This procedure allows therefore keeping only the most influential factors in the context of a selection process. The second one, which is called response surface methodology (RSM), is performed when the objective is to find the optimal settings of the factors in order to obtain the desired system optimization. Both methods can therefore be used in a complementary way.

Each design is characterized by its own mathematical approximation of the response y. Typically, the experimental designs engineered for a screening process are first-order models including the interactions between factors. On the other hand, the designs used in the context of an optimization procedure require flexibility for modelling curvatures in the response surface. Mostly, these are thus second- or even third-order models containing a small number of interactions. Overall, the general form of the model is the following:

$$y = a_0 + \sum_{j=1}^{k} a_j x_j + \sum_{j=1}^{k}\sum_{l=1}^{k} a_{jl} x_j x_l + ... + a_{jl...k} x_j x_l ... x_k + \varepsilon \tag{5}$$

The parameters a_0, a_j, a_{jl}, ... are called the regression coefficients of the model and constitute the unknowns of the problem. In order to solve it, the model (5) is usually written in matrix notation. Considering that n experiments were performed and that the mathematical model encompasses p coefficients, the model can be expressed as:

$$\underset{(n\times1)}{y} = \underset{(n\times p)}{X} \cdot \underset{(p\times1)}{a} + \underset{(n\times1)}{\varepsilon} \tag{6}$$

where, y is called the response vector, X the full design matrix, a the regression coefficients vector and ε the residuals vector. This system consists of n equations for $n + p$ unknowns (p regression coefficients and n residuals). The desired regression coefficients are determined thanks to the least squares method, whose principle is to minimize the sum of squares of the residuals [32]. These estimated coefficients, noted $\bar{a}$, are then given as follows [32]:

$$\bar{a} = \left(X'X\right)^{-1}X'y \tag{7}$$

Finally, the fitted regression model is:

$$\bar{y} = X\bar{a} = X\left(X'X\right)^{-1}X'y \tag{8}$$

where y represents the real value of a response measured following an experiment and $\bar{y}$ the value of the response calculated based on the mathematical model of y (where the least-square estimators $\bar{a}$ were previously determined). The residual ε of the model, that is, the differences between the actual observations y and the corresponding fitted values $\bar{y}$, can be computed as follows:

$$\varepsilon = y - \bar{y} \tag{9}$$

3.2. The screening of significant parameters

The screening study is used to quantify the significance of each factor on the response and is therefore very useful in the context of selection procedure or reduction of the problem size. Here, this method is applied in order to find the best area for installing the network regulation devices. Note that the goal is not to precisely determine an exact location, since the probability that this node can really be used in practice is highly insignificant. Furthermore, this approach allows to substantially decrease the number of factors and to generate important time savings. The underlying aim of the study is thus to avoid the upgrading of the feeders, as such a solution is not economically viable for DSOs, especially considering the current trend to install underground power cables. Practically, two different screening studies are carried out, one for each of the two envisaged control means (i.e. VAR compensators and storage devices). Recall that, due to the relatively low extra costs compared with normal transformers as well as their high efficiency, OLTC devices are already installed in HV/MV substations. Moreover, the control of the reactive power of DG units does not seem feasible at the present time. At the end of the screening study, we will therefore obtain the most favourable areas for installing VAR compensators and storage facilities.

The response y must quantify the effect of the factors on the violations of the operating conditions. The response is thus defined as a weighted sum of the violated constraints within the network. More precisely, the computation consists in the summation of three contributions y_{1a}, y_{1b} and y_2. The first two terms relate to the voltage violations, which are defined according to the European EN 50160 standard and are computed as follows:

$$
\begin{aligned}
y_{1a} &= \sum_{i=1}^{N} \left(V_i - 1.1 V_{nom} \right)^2 \qquad \forall\, V_i > 1.1 V_{nom} \\
y_{1b} &= \sum_{i=1}^{N} \left(V_i - 0.9 V_{nom} \right)^2 \qquad \forall\, V_i < 0.9 V_{nom}
\end{aligned}
\tag{10}
$$

where N is the number of nodes of the studied network. The third term concerns the line congestions, which depend on the thermal limits of the cables.

$$
y_2 = \sum_{l=1}^{L} \left(S_l - S_{max} \right)^2 \qquad \forall\, S_l > S_{max}
\tag{11}
$$

where L is the number of lines. Finally, the response can be expressed as:

$$
y = w_1 \left(y_{1a} + y_{1b} \right) + w_2 y_2
\tag{12}
$$

where w_1 and w_2 represent the respective weights associated to the voltage and power flow violations with $w_1 + w_2 = 1$. When the system is not at risk, the response is therefore equal to zero.

Beyond its simplicity of implementation, the DOE approach presents another highly interesting asset. The coefficients of the mathematical model have indeed a physical meaning, and their values can be interpreted as a quantitative reflection of the effect of the factors on the response. This interesting and useful property is illustrated in Figure 15 for $k = 2$ (factors x_1 and x_2).

$$y(x_1, x_2) = a_0 + \underbrace{a_1}x_1 + a_2 x_2 + \underbrace{a_{12}}x_1 x_2$$

Effect of the factor x_1 Effect of the interaction
on the response y between x_1 and x_2 on y

Figure 15. Representation of the physical meaning of the coefficients of the model as the response of interest.

The number k of factors of the analytical model is thus equal to the number of sub-areas defined within the studied distribution grid. The first step is to arbitrarily divide the network into a small number of representative geographic areas based on its topology. The sensibility of each area for both network regulation devices is evaluated through the impact of a device installed at the closest node of the centre on the considered section. For consistency, the same pre-sizing is considered for all devices of the same type. The general principle of the screening study is shown in Figure 16.

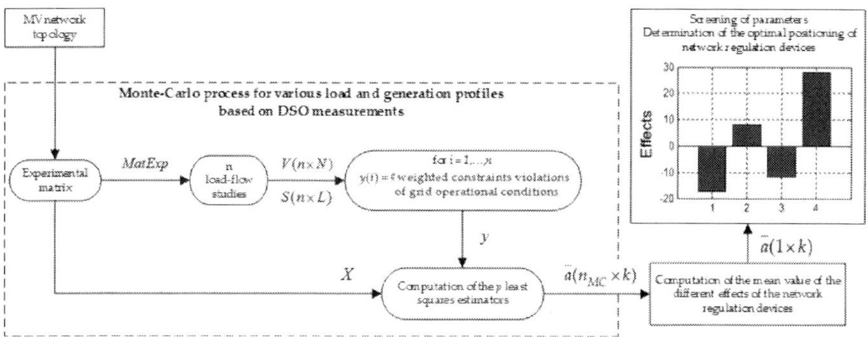

Figure 16. Flowchart of the principle of the implemented screening study.

First, the network topology is used to define all the possible positioning configurations of the network regulation devices. In this study, these are either reactive power compensators or storage devices. This leads to the definition of the experimental matrix $MatExp(n \times k)$, which contains the real values of the k factors and thus defines the n experiments to be performed. An experiment consists in a load-flow calculation, so as to obtain the voltage profile and the power flows in each line. Each experiment is carried out for specific values of the k factors (i.e. network regulation devices), which are defined by the considered experimental design. The full design matrix X is easily inferred from the experimental matrix [32], and knowing the response vector y, the regression variables can be determined by (7).

In order to take the non-deterministic effects of the load and the generation into account, the screening process is implemented in a non-chronological Monte Carlo environment. Such a study allows simulating a large number of operating states of the network. For each of them, the influence of the network regulation devices is evaluated. The process thus simulates the random behaviour of the different MV clients as well as the MV/LV substation by the means of a random sampling of the consumed and/or injected power at each bus. More details about the procedure are given in Section 3.5.1.

The importance of the location of each area for installing network regulation devices is then computed by averaging the effects provided by the screening procedure conducted for each of the n_{MC} Monte Carlo iteration. The results can then be plotted and analysed.

3.3. Sizing procedure of the selected devices

After the determination of the most favourable areas for respectively installing a reactive power compensator and a storage utility, the sizing of these devices is carried out. As one of the objectives of the method is to illustrate the assets of implementing a centralized control at the MV level, the sizing procedure is applied for more than one network regulation device. For simplicity as well as economic reasons, only two devices are considered in this work. Moreover, as the storage technology is still too expensive, the possibility to install two storage utilities is not considered. As a result, two different scenarios are investigated. The first one focuses on the combined installation of a VAR compensator and a storage device, whereas the second one evaluates the possibility to invest in two reactive power compensators.

The sizing of the storage utility concerns only its output power. Indeed, the energy sizing is less critical for the DSO than it would be for an operator eager to maximize its return on investment, since the DSO is not to take the market considerations into account. Its only focus is to use the storage device when it is necessary for alleviating stressed conditions within its grid and to take advantage of the non-critical situations to adjust the amount of stored energy. However, such a control strategy is outside the framework of this study.

As in the screening process, the sizing procedure makes use of Monte Carlo simulations for modelling the different network states. For each simulation, an optimal control of the selected network regulation devices is carried out. Thus, this requires the use of a fast optimization tool, which led to the choice of the response surface methodology. The response is the same as the one used in the screening process (i.e. a weighted sum of the constraints violations); and the response surface thus models the variation of this response with regard to the output power of the two selected regulation devices. The mathematical model used for the study is of the form:

$$y = a_0 + \sum_{j=1}^{k} a_j x_j + \sum_{j=1}^{k}\sum_{l=j+1}^{k} a_{jl} x_j x_l + \sum_{j=1}^{k} a_{jj} x_j^2 + e \tag{13}$$

where y is determined by (7) and the x_j and x_l values are the output powers of the network regulation devices.

The purpose of optimization is to find the configuration of the control devices that eliminates all constraint violations while minimizing their interaction (i.e. power exchanges) with the grid. Mathematically, it amounts to find the intersection between the response surface and the plane $y=0$. The result is thus a line segment and the choice of the optimal point then depends on the considered scenario. In the first scenario, as the storage is currently a much more expensive solution, the optimal configuration is the segment point with the lowest value of active power provided by the storage device. In the second scenario, the aim is to minimize the sum of both reactive power contributions. Indeed, according to the Siemens database [33], the installation price increases approximately linearly with the maximal installed reactive power.

The optimal values given at the end of each Monte Carlo simulation are logged. In this way, at the end of the simulations, the DSO disposes of useful information. Indeed, the output power of devices can be directly translated in installation costs and, given the different values of the regulation devices, the DSO is then able to adapt its investment decision as regards the technical and economic considerations.

3.4. Centralized real-time management of the distribution grid

The method proposes to use a centralized control of each of the network regulation devices of the considered distribution grid. Indeed, the current SVC/STATCOM devices compare the voltage at their connection node to a fixed reference voltage in order to bring back the voltage at the defined reference level. But sometimes, it could be interesting to change that reference depending on the location of the device in the grid. For example, to lower it when there is an increased power generation coming from DG units combined with a low demand. In this context, the output power of the storage devices would be adapted by a remote control of power electronics.

The strategy of the control depends directly on the presence of operational constraints violations. On the one hand, if there are such violations, the aim is to overcome the issue with an optimal use of the network regulation devices. However, with several devices, several configurations can potentially solve the problem. In such a case, the DSO is interested to opt for the configuration that minimizes the line losses. On the other hand, if there is no problem, different strategies can be envisaged. The most straightforward option is to do nothing, which will minimize the number of cycles of the devices and, as a result, increase their lifetime. For the sake of simplicity, this strategy was implemented in this study. The second one is to use the control devices for reducing the line losses. Such an action has to be carried out only if it generates worthwhile financial profit. Finally, if a storage facility is available, its energy level can be adapted in anticipation of the future needs.

The optimization is carried out by using the same response surface methodology than the one used in the sizing procedure (i.e. same response and same variables).

3.5. Application of the DOE method to an existing radial MV grid

3.5.1. Modelling of the MV network

The simulation of the statistical behaviour of the different industrial loads comes from energy flows directly collected at the connection node by the DSO. The power injections at MV/LV substation is provided by the aggregated power profiles determined in the LV planning tool in Section 1. As the PV installations are the property of industrial companies, these are also equipped with SM. This is not the case with most of the wind farms that often belong to independent operators that directly participate in the electricity market and do not depend on the DSO. Consequently, the power generated by wind farms is inferred from openly accessible wind speeds. As data for a large number of years are available, those are perfectly adapted to a Monte Carlo procedure. The wind speed values can then be converted into a value of active power according to the following power curve [34]:

$$
\begin{cases}
P = 0, \, W_t < v_{ci} \\
P = a + b \cdot W_t^2, \, v_{ci} < W_t < v_r \\
P = P_r, \, v_r < W_t < v_{co} \\
P = 0, \, W_t > v_{co}
\end{cases}
\tag{14}
$$

where W_t is the wind speed and v_{ci}, v_r, v_{co} are, respectively, the cut-in, rated and cut-out wind speeds of the power curve. P_r is the nominal power of the wind generator and the parameters a and b are defined as in reference [34].

Therefore, considering the theoretical number of the data available, the same pseudo-chronological Monte Carlo procedure as the aforementioned LV planning tool could be performed. However, several impediments such as the difficulty to gather the data of all clients combined with the noticeable loss of data and measurement errors led to the implementation of a non-sequential Monte Carlo.

Concerning the reactive power of DG installations, it must comply with the operating point of DG as well as the national legislation. This work is based on Belgian rules [3], but could easily be adapted in order to suit other legislation.

The reactive power compensators are modelled by their reactive power consumed (inductive mode) or injected (capacitive mode) into the grid. Then, given the price and the maturity of other storage technologies, the battery is currently the most appropriate solution for the desired application of providing ancillary services to the distribution grid. The pricing of these devices is determined by manufacturers, which will not be revealed here for confidentiality purposes.

As previously mentioned, the permitted range of voltage fluctuations is defined by the EN 50160 standard that allows a variation of ±10% around the nominal value. The thermal limits of power cables are defined as the maximum current that can flow in the conductors without inducing excessive heating. Indeed, the current flowing in the conductors generates heat that needs to be eliminated in the surrounding environment. This diffusion is made difficult by the important thermal resistances of the insulating materials in cables and by the surrounding environment, especially in the case of underground cables. In order to preserve the network security, the DSO must ensure that such situations do not occur.

3.5.2. Definition of the network and of the factors

The DOE method is used on a 28-bus radial network, which is highly critical regarding the risks of congestion and voltage violation, because of the high penetration rate of DG units such as wind farms or large PV plants. These are installed at nodes 6, 10, 15, 23 and 28. This situation is indeed stimulated by the current financial attractiveness of investing in renewable energies. Consequently, in order to avoid an expensive reinforcement of the grid, the introduction of network regulation devices constitutes an interesting solution to investigate. The single-phase diagram of this 20-kV system is shown in Figure 17. The representation of the grid is purposefully divided into eight different sub-areas, which are arbitrarily chosen and correspond to the areas whose influence is evaluated in the screening process. The resistance and reactance per kilometre of each line section are, respectively, 0.098 and 0.106 Ω/km, whereas the distance between nodes varies from 2 to 10 km. The definition of the rated power of each load and DG unit is shown in Table 1.

bus	P_L (kW)	P_G (kW)	bus	P_L (kW)	P_G (kW)
01	0	0	15	0	3900
02	240	0	16	190	0
03	190	0	17	160	0
04	220	0	18	150	0
05	100	0	19	230	0
06	0	2700	20	100	0
07	160	0	21	160	0
08	100	0	22	140	0
09	180	0	23	170	2700
10	260	3900	24	110	0
11	190	0	25	260	0
12	130	0	26	170	0
13	90	0	27	160	0
14	180	0	28	140	3900

Table 1. Data of the 28-bus grid.

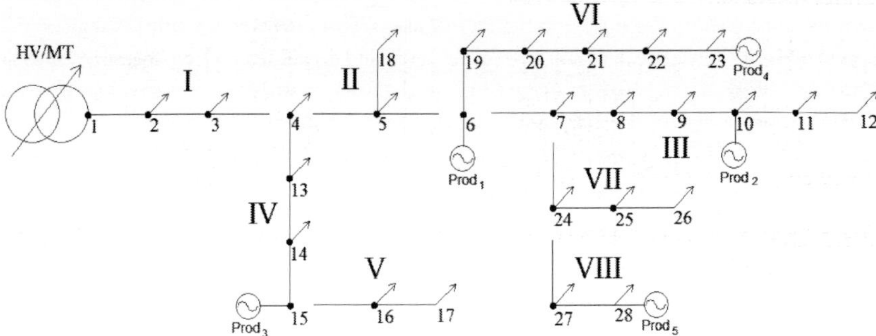

Figure 17. The single-phase diagram of the investigated 28-bus grid.

3.5.3. The screening step

A preliminary analysis of the network behaviour showed that the congestion issues are exclusively observed in areas I and II of the grid, whereas the voltage issues tend to occur at the most distant nodes of the HV/MV substation. This gives useful information even before starting the screening procedure. However, the intuitive solution that consists in positioning one control device at the beginning of the grid and the other one at the extremity is likely to be inefficient. Indeed, the power flows originating from the transformer are representative of the power demand of the downstream loads. Therefore, the benefit of a control device installed in the neighbourhood of the substation is highly localized (i.e. area between the transformer and the control device) and this solution has a good probability to be suboptimal. The determination of the best solution is thus not trivial and requires a more thorough study.

Two screening studies are therefore carried out, both aiming at finding the optimal location to install respectively a reactive power compensator and a storage device. The influence of the eight areas represented in Figure 17 is investigated. For consistency, the same sizing is considered for all devices of the same type. In this way, the maximal sizing of the VAR compensators is adapted to the maximum reactive load of the case study and the maximal output power of the battery is 1 MW, which corresponds to the current upper limit proposed by the manufacturers.

The results of both screening studies are represented in Figure 18. The bar chart representation of the effects is privileged for its simplicity of interpretation. The height of a bar associated to a factor represents indeed the effect of this factor on the response. Therefore, the highest bars correspond to the most influential (and thus efficient) factors.

As expected, the results are quite similar for both scenarios. Indeed, the areas located near the end of the grid are more appropriate than those near the HV/MV transformers in the context of investing in network regulation devices for improving the MV grid operation. This can be explained by the importance of the line impedance. However, while the best place for the

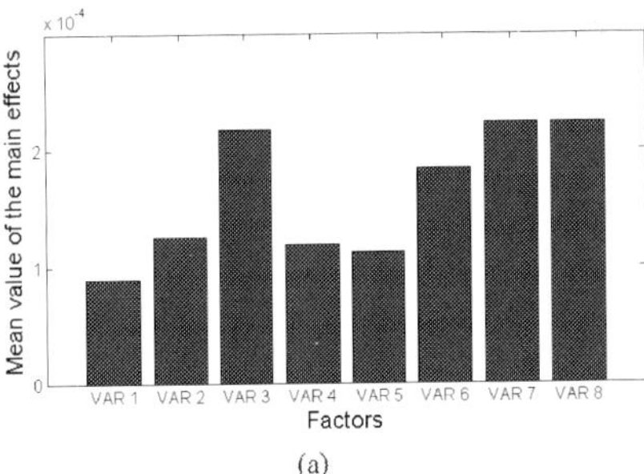
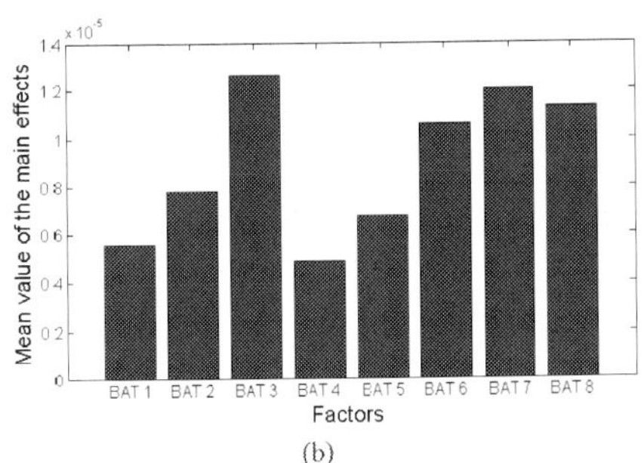

Figure 18. Influence of the eight selected areas for the installation of a VAR compensator (a) and a storage device (b).

storage device is the end of the main feeder in area III, the VAR compensator will have more effect if it is installed near node 28 in area VIII.

3.5.4. The sizing step

The purpose of the sizing step is to use the network regulation devices in order to improve the operation of the grid for a large number of simulated network states. All the optimal values of these devices are then collected as they will constitute the basis for the sizing decision. Indeed, thanks to manufacturer offers, the output power can be associated to a price. The objective is to allow the DSO to make the best decision with regard to the investment costs and the resulting effects on the network operation.

For each network state, the centralized control strategy is carried out by using the aforementioned response surface methodology. The goodness of fit of the model (13) is quantified through the coefficient of determination R^2 [32]. This R^2 coefficient is included in the [0, 1] interval and is equal to one when all the experimental points correspond exactly to those determined by the fitted model. Conversely, if the quality of the model is decreasing, the value of R^2 will drop accordingly.

As previously mentioned, two scenarios are considered here. In the first one, the combined control of a VAR compensator with a storage device is studied while the second one focuses on the optimal regulation of two reactive power compensators. It should be noted that the range of variation of the reactive power compensator is symmetrical $(-Q_{\text{max}}, +Q_{\text{max}})$.

The general model of the polynomial approximation of the response for the first scenario becomes:

$$y = a_0 + a_1.P_1 + a_2.Q_{C2} + a_{12}.P_1.Q_{C2} + a_{11}.P_1^2 + a_{22}.Q_{C2}^2 \tag{15}$$

The optimal values of network storage devices for both defined scenarios are represented in Figure 19.

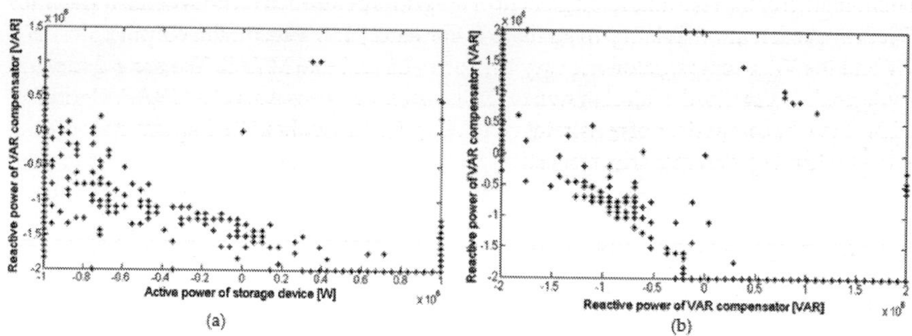

Figure 19. Mapping of the optimal values of the control devices for all simulated network states.

Before interpreting the results, it is important to focus on their accuracy by evaluating the quality of the model used to determine the optimal values of the control devices for each of the Monte Carlo simulation. To that end, the parameter R^2 is assessed for each iteration and the averaged value for both scenarios is then computed. These values are, respectively, equal to 0.88 and 0.89, which indicates a good adjustment of the fitted model with the experimental data.

In the first scenario, as the current maturity of the battery technology does not allow having an output power higher than 1 MW, a significant part of the optimal value is close or equal to the maximum value. However, if the sizing of the VAR compensator is limited to 1.6 MVAR, more than 85% of the constraint violations are eliminated and this solution can therefore constitute a good compromise for the DSO. In the second scenario, it can be seen that a large part of the optimal point corresponds to opposite operating modes of the two VAR compensators (i.e. one is injecting reactive power while the other one is in inductive mode). This can be explained by the conflicting nature of the concurrent elimination of voltage and congestion issues. Indeed, as the line currents are fixed by the power demand, the removal of line congestions requires increasing the voltage. However, if overvoltages are already observed in other parts of the grid, the issue can hardly be solved and necessitates opposite actions of control devices. The nature of their actions then depends on their positioning with regard to the operational constraints violations.

3.5.5. The simulation of the real-time management of the grid

In this method, network regulation devices are used in order to ensure an optimal operation of the network. This objective is translated in the removal of all constraint violations of network operational conditions. The process consists therefore in a state-by-state optimization based on the response surface methodology carried out for a large number of typical years. The effectiveness of the proposed methodology is then tested by taking as evaluation criteria the probabilities of voltage violation and the presence of congestions. The total line losses with and without the centralized control are also compared.

The simulation of the real-time management of the grid is performed for both defined scenarios with their predetermined sizing. In scenario 1, the battery has a maximum output power of 1 MW and the VAR compensator can vary between +1.6 and −1.6 MVAR. The second scenario simulates the centralized control of two reactive power compensators of 2 MVAR. Figures 20 and 21 show the comparison of constraint violations with and without the implemented control for the first and second scenario, respectively.

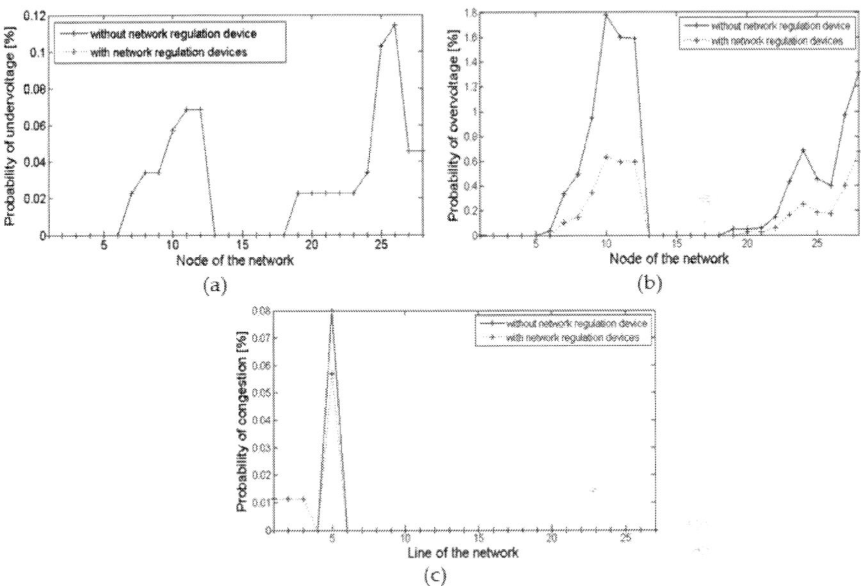

Figure 20. Probability of undervoltage (a), overvoltage (b) and congestion (c) for scenario 1.

In both investigated variants, the undervoltage issues have been completely removed. However, the number of overvoltages has significantly decreased, but they could not be totally eradicated. The probability of congestion has even increased for some nodes, which highlights the difficulty to combine the objectives of simultaneously alleviating overvoltage and congestion issues. In such a case, the DSO would have no other choice than to have recourse to a curtailment of DG units, which may be very expensive. In this context, the research concerning the implementation of load-shifting strategies (e.g. with financial incentives in order to drive customers consuming when there is a lot of generation within the system) is of major importance.

The impact of control strategy on the line losses is represented in Table 2. The results are averaged to represent the total losses of a representative year. Recall that their minimization was not included in the objective function of the study since the reduction of the number of operating cycles of the control devices was privileged.

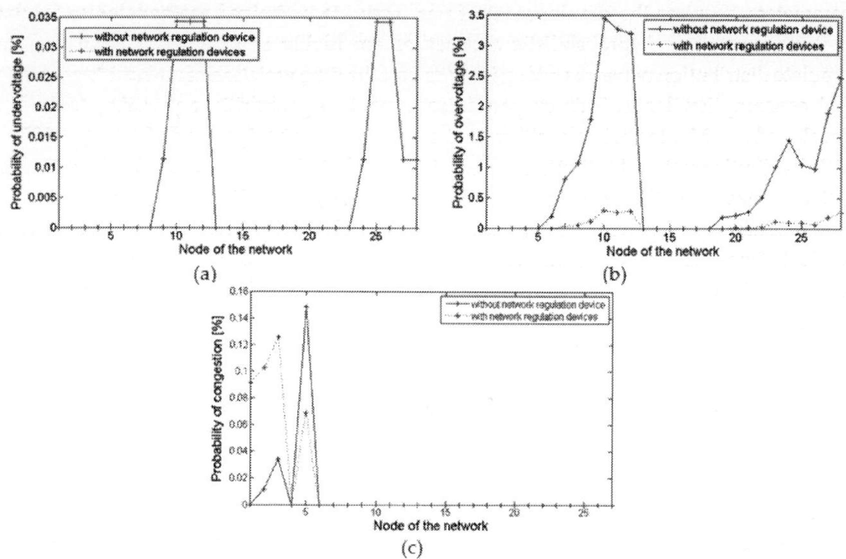

Figure 21. Probability of undervoltage (a), overvoltage (b) and congestion (c) for scenario 2.

	Scenario 1		Scenario 2	
	Without control	With control	Without control	With control
Total line losses [MW]	632.89	822.73	665.30	843.07

Table 2. Comparison of the power losses with and without network regulation devices for both scenarios for a typical year.

It can be seen that the amount of power losses increases with the network regulation devices. This can be explained by the presence of several DG units near the end of the grid. Indeed, in order to avoid the overvoltages inferred by the generation of these installations, the storage and the VAR compensators consumed locally active or reactive power. This energy was used by some loads that consequently need to be supplied by the HV/MV transformer. The power flows at the MV output of the substation are thus increased and, as the line losses are proportional to the square of the electrical current, these losses, together with the probability of congestion, rise significantly.

4. Conclusions and perspectives

The natural evolution of distribution networks with high integration of distributed renewable energy sources and random nature of consumptions loads, both uncertain and time variable parameters, requires the development of new accurate optimized methodologies for their analysis. To this end, probabilistic approaches are highly recommended since they can simulate distribution networks taking into account the time variation of renewable generation and consumption loads. This chapter introduced two probabilistic planning tools, both developed in a Monte Carlo algorithm environment, for the analysis and planning of LV and MV distribution networks. Section 1 presented a probabilistic framework that analyses in a fast, optimized way the voltage profile along LV feeders, taking into account the uncertainty of their loading parameters node by node, based on real SM data. It can therefore be deployed for the techno-economic evaluation and refinement of solutions to operational problems in LV feeders with high PV penetration. The only prerequisite for its deployment is the availability of SM measurements in the studied LV feeder. This prerequisite goes along with the perspective of a wide rolling out of SM devices in the European LV networks. Section 2 introduced a new method intended to constitute an efficient and realistic planning tool for the DSO. In this way, the study takes three of the main objectives of the DSO into account. Indeed, the screening study aims at finding the best area for installing network regulation devices for minimizing the probability of violations of the operational constraints. Then, the sizing procedure is more complete by accounting for the investment costs. Finally, the centralized control of the number of operating cycles could also easily be implemented for reducing the line losses if such an action turns out to be financially worthwhile.

Author details

Jean-François Toubeau*, Vasiliki Klonari, Jacques Lobry, Zacharie De Grève and François Vallée

*Address all correspondence to: Jean-Francois.TOUBEAU@umons.ac.be

Electrical Engineering Department, University of Mons, Mons, Belgium

References

[1] P. Lemaire, "Electricity market changes, impact on thermal generation", SRBE/KBVE seminar, 24 Sept. 2014.

[2] EN50160, Voltage characteristics of electricity supplied by public electricity networks, 2012.

[3] SYNERGRID, Prescriptions techniques spécifiques de raccordement d'installations de production décentralisée fonctionnant en parallèle sur le réseau de distribution, C10/11, 2012.

[4]   M.H.J. Bollen, F. Hassan, Integration of Distributed Generation Sources in the Power System, 1st ed. IEEE Press Series on Power Engineering, Hoboken, New Jersey: Wiley, 2011.

[5]   EPIA, Global Market Outlook for Photovoltaics 2014-2018, 2014.

[6]   S. Conti, S. Raiti, "Probabilistic load flow using Monte Carlo techniques for distribution networks with photovoltaic generators", Solar Energy, vol. 81, no. 12, 1473–1481, 2007.

[7]   F.J. Ruiz--Rodriguez, J.C. Hernandez, F. Jurado, "Probabilistic load flow for radial distribution network with photovoltaic generators", IET Renewable Power Generation, vol. 6, no. 2, 110–121, 2011.

[8]   R. Billinton, R. Karki, "Reliability/cost implications of utilizing photovoltaics in small isolated power systems", Reliability Engineering & System Safety, vol. 79, no. 1, 11–16, 2003.

[9]   R.M. Moharil, P.S. Kulkarni, "Reliability analysis of solar photovoltaic system using hourly mean solar radiation data", Solar Energy, vol. 84, no. 4, 691–702, 2010.

[10]   EPIA, Connecting the Sun: Solar Photovoltaic on the Road to Large-Scale Grid Integration, 2012.

[11]   F. Vallee, V. Klonari, T. Lisiecki, O. Durieux, F. Moiny, J. Lobry, "Development of a probabilistic tool using Monte Carlo simulation and smart meters measurements for the long term analysis of low voltage distribution grids with photovoltaic generation", International Journal of Electrical Power & Energy Systems, vol. 53, 468–477, 2013.

[12]   V. Klonari, F. Vallee, O. Durieux, Z. De Greve, J. Lobry, "Probabilistic modeling of short term fluctuations of photovoltaic power injection for the evaluation of overvoltage risk in low voltage grids", EnergyCon, Croatia, 2014.

[13]   V. Klonari, J.-F. Toubeau, O. Durieux, Z. De Gréve, J. Lobry, F. Vallée, "Probabilistic analysis tool of the voltage profile in low voltage grids", In Cired, France, 2015.

[14]   J.-F. Toubeau, V. Klonari, Z. De Gréve, J. Lobry, F. Vallée, "Probabilistic study of the impact on the network equipment of changing load profiles in modern low voltage grids", accepted for presentation in ICREPQ'15, Spain, 2015.

[15]   CEI 0-21, Regola tecnica di riferimento per la connessione di Utenti attivi e passivi alle reti BT delle imprese distributrici di energia elettrica, 2012.

[16]   K. Turytsin, P. Sul, S. Backhaus, M. Chertkov, "Options for control of reactive power by distributed photovoltaic generator", Proceedings of IEEE, vol. 99, no. 6, 1063–1073, 2011.

[17] B. Blazic, B. Uljanic, B. Bletterie, K. De Brabandere, C. Dierckxsens, T. Fawzy, W. Deprez, "Active and autonomous operation of networks with high PV penetration", MetaPV FP7 project, 2011.

[18] C. Dierckxsens, K. De Brabandere, W. Deprez, "Economic evaluation of grid support with photovoltaics", MetaPV FP7 project, 2012.

[19] J. Au-Yeung, G.M.A. Vanalme, J.M.A. Myrzik, P. Karaliolios, M. Bongaerts, J. Bozelie, W.L. Kling, "Development of a voltage and frequency control strategy for an autonomous LV network with distributed generators", UPEC 2009, Scotland, 2009.

[20] H. Laaksonen, P. Saari, R. Komulainen, "Voltage and frequency control of inverter based weak LV network microgrid", in FPS, The Netherlands, 2005.

[21] A. Engler, O. Osika, M. Barnes, N. Hatziargyriou, DB2 Evaluation of the local controller strategies, www.microgrids.eu/micro2000, 2005, accessed in June 2014.

[22] T.L. Vandoorn, J. De Kooning, B. Meersman, L. Vandevelde, "Voltage-based droop control of renewables to avoid on-off oscillations caused by overvoltages", IEEE Transactions on Power Delivery, vol. 28, no. 2, 845–854, 2013.

[23] V. Klonari, J.-F. Toubeau, T. L. Vandoorn, B. Meersman, De Grève, Zacharie, J. Lobry, and F. Vallée, "Probabilistic Framework for Evaluating Droop Control of Photovoltaic Inverters," *Electr. Power Syst. Res.*, vol. 129, pp. 1–9.

[24] P.T. Staats, W.M. Grady, A. Arapostathis, R.S. Thallam, "A procedure for derating a substation transformer in the presence of widespread electric vehicle battery charging", IEEE Transactions on Power Delivery, vol. 12, no. 4, Oct. 1997.

[25] D. Dohnal, "On-load tap-changers for power transformers", Maschinenfabrik Reinhausen GmbH, 2013.

[26] J. Dixon, L. Moran, E. Rodriguez, R. Domke, "Reactive power compensation technologies: state-of-the-art review", IEEE Transactions on Industry Applications, vol. 93, no. 12, 2144–2164, Dec. 2005.

[27] C. Pieper, H. Rubel, "Revisiting energy storage. There is a business case", Report of the Boston Consulting Group, Feb. 2011.

[28] V. Calderaro, G. Conio, V. Galdi, G. Massa, A. Piccolo, "Optimal decentralized voltage control for distribution systems with inverter-based distributed generators", IEEE Transactions on Power Systems, vol. 29, no. 1, 230, 241, Jan. 2014.

[29] E. Tzimas, I. Papaioannou, A. Purvins, "Demand shifting analysis at high penetration of distributed generation in low voltage grids", Electrical Power and Energy Systems, 2012.

[30] B. Bakhshideh Zad, J. Lobry, F. Vallée, "Coordinated control of on-load tap changer and D-STATCOM for voltage regulation of radial distribution systems with DG

units," 3rd International Conference on Electric Power and Energy Conversion Systems, Istanbul, Turkey, October 2013.

[31] J.-F. Toubeau, F. Vallée, Z. De Grève, J. Lobry, "Optimal allocation process of voltage control devices and operational management of the voltage in distribution systems using the experimental design method", Proceedings of IEEE International Energy Conference (ENERGYCON), 1083–1090, 13–16 May 2014.

[32] D.C. Montgomery, Design and Analysis of Experiments, 5th ed. Hoboken, New York: Wiley, 2001.

[33] A.J. Wood, B.F. Wollenberg, Power Generation, Operation, and Control, New York: John Wiley & Sons, Inc., pp. 430–432, 1996.

[34] L-B. Fang, J-D. Cai, "Reliability assessment of microgrid using sequential Monte Carlo simulation", Journal of Electronic Science & Technology, vol. 9, no. 1, 31–34, March 2011.

Chapter 11

Simulation Methods for the Transient Analysis of Synchronous Alternators

Jérôme Cros, Stéphanie Rakotovololona, Maxim Bergeron, Jessy Mathault, Bouali Rouached, Mathieu Kirouac and Philippe Viarouge

Additional information is available at the end of the chapter

http://dx.doi.org/10.5772/61604

Abstract

The integration of unconventional renewable energy sources on the electrical grid poses challenges to the electrical engineer. This chapter focuses on the transient modeling of electrical machines. These models can be used for the design of generator control, the definition of the protection strategies, stability studies, and the evaluation of the electrical; mechanical; and thermal constraints on the machine. This chapter presents three modeling techniques: the standard d-q equivalent model, the coupled-circuit model, and the finite element model (FEM). The consideration of magnetic saturation for the different models is presented. The responses of the different models during three-phase, two-phase, and one-phase sudden short circuit are compared.

Keywords: Power generator, Synchronous alternator, electrical machine modeling, dq model, coupled circuits, finite element analysis

1. Introduction

There are mainly two types of large synchronous alternators having a wound rotor: the round rotor and the salient pole rotor [1]. Generally, these synchronous alternators are directly connected to the grid in parallel, with many other alternators. In this case, the stator voltage and the frequency (f) are imposed by the grid and the generators work at a fixed speed (n) depending on the number of magnetic pole pairs in the machine (p); ($n = 60 \cdot f / p$). For example, a round rotor synchronous generator with $p = 1$ would have a speed of $n = 3600$ rpm on an $f = 60$ Hz electrical grid. Synchronous generators with $p = 1, 2$ are known

as turbo generators and present a round-shaped rotor. The prime mover of large turbo generators are generally steam turbines, which operate at high speeds. In other cases, the prime mover runs at low speeds, so that salient pole synchronous generators are used. The largest salient pole generators are used in dams for the production of electricity with hydraulic turbine. The optimal speed of a hydraulic turbine is relatively slow; thus, the generator has a high number of poles.

The growing demand for energy and care for environment protection is spurring the development and integration of unconventional renewable energy sources such as solar, wind, and tidal power. The unconventional renewable energy brings challenges on the generator design as the energy production cannot always be done at fixed prime mover speed. In this case, different kinds of generators can be used, and generally, they are associated with power electronic converters to connect to the grid. For example, the main type of generators used for speed-varying applications is the doubly fed induction generators (DFIG) [2-3]. The stator winding of these machines is directly connected to the grid, but there is a three-phase winding on the rotor that is fed through power electronics. Such machine designs can create a rotating field, going faster, slower, or at the constant speed than the rotor. Thus, the stator frequency can be imposed regardless of the rotor speed. For a given generator, the larger the speed deviation is allowed, the larger the rating of the power electronics must be in order to deliver the rated power. Conventional synchronous generators are also used for variable speed applications by using a power electronic converter connected to the stator winding.

The integration of unconventional renewable energy must not be made at the cost of a less reliable power distribution. A reliable power distribution network is characterized by a robust design of its equipment, control, and protection strategies [4-5]. The voltage and power distribution on the grid is maintained mostly by the proper control of torque and excitation of the generators. Accurate steady-state and transient models of the electrical machine are required [4-6].

Modeling a generator is a real challenge, as an accurate representation of the machine must deal with many internal considerations, such as the saturation effects, nonlinearity of the material's properties, rotational effects, and the large number of complex current paths inside the generator. Such a detailed analysis provides information on the stator voltage harmonics, copper/iron losses, and torque ripples. This detailed analysis is of high importance for the reliable design at all operating points. However, the performance of a model in terms of accuracy and computational effort depend on its simplifying assumptions.

Transient machine models can be divided into three types: phase-domain model, d-q model, and finite element model. The phase-domain model is an analytical approach based on the voltage and flux-current equations of the synchronous machine expressed in the fixed a-b-c reference frame [7-9]. It requires precise knowledge of the time-dependent inductance matrix, which can be a difficult task. Hence, many researchers use time-stepping finite element analysis to determine the unknown matrix for a certain discretization [9]. This approach is known as "coupled finite element state space approach."

The d-q model is also based on the voltage and flux-current equations but expressed in a virtual reference frame rotating at synchronous speed [7-8]. Based on certain assumptions, this model represents synchronous machines by two magnetically decoupled equivalent circuits comprising lumped *R-L* parameters. Many identification methods are found in the literature for the d-q model parameters, such as the *standstill time response* (SSTR) tests, the *rotating time-domain response* (RTDR) tests, the *standstill frequency response* (SSFR) tests [10], the *open-circuit frequency response* (OCFR) tests, and the *on-line frequency response* (OLFR) tests. More recent approaches include *online measurements* techniques. These methods are well reviewed in the literature [11].

Finally, the finite element model is used in solving numerically the general equations of electromagnetics [12]. It is the most accurate of the three types, but also the most expensive in terms of computational effort. Nonlinear phenomena such as saturation and skin effects can be taken into account by this model.

In this chapter, the three types of transient electrical machine models are presented and compared. Understanding these differences, identification methods, and scope of these models provide useful information to the electrical engineer to determine the problems that can be treated with each model. The chapter is divided into following sections:

- Presentation of the generator used for experiments and model validations

- The standard d-q equivalent model and the SSFR-identification method

- The detailed coupled-circuit model

- The finite element model

- Improvements to the models in order to consider the saturation

- Comparison of the models in order to simulate different kind of short circuits

2. Synchronous generator used for this study

The machine used for the experiments is a three-phase 5.4-kVA round rotor synchronous generator. The stator has 54 slots. The rotor has three concentric coils per pole and a total of 24 copper damper bars connected by short-circuit rings (Figure 1).

The stator lamination stack is shown in Figure 2. The winding of this machine is of a special construction; it is made of rectangular conductors, and there are 14 conductors per slot distributed in four layers. They are soldered in the extremity in order to put the individual conductors in series to form coils. It can be seen that the stator slots are almost closed near the air gap (Figure 2). Data of the generator nameplate are given in Table 1. Figures 3 and 4 present, respectively, half of the rotor and stator laminations. The main dimensions are given in millimeter. Figure 5 presents the stator-winding repetitive section.

Figure 1. Rotor of the synchronous generator

Figure 2. Stator on the side of end-winding connections

Power	5.4 kVA
Electrical frequency	60 Hz
Line-to-line voltage	280 V
Stator phase current (star connected)	11.1 A
Rotor field current for the nominal no-load voltage	0.5 A
Speed	1800 rpm

Table 1. Nominal parameters

Phase winding; number of turns: 156	Phase resistance: 0.156 Ω
Field winding; number of turns: 2208	Field resistance: 21.5 Ω
Damper bars resistivity: 22 nΩ.m	DC bar resistance: 21 μΩ
Contact ring resistance: 170 μΩ	End-field winding: 0.8 Hz

Table 2. Main parameters

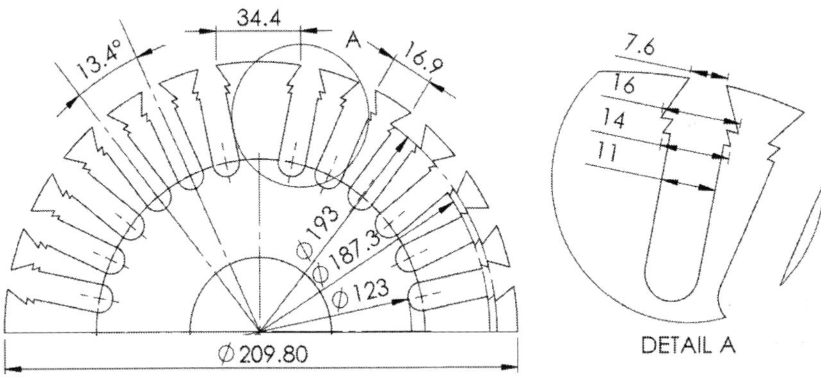

Figure 3. Rotor steel laminations

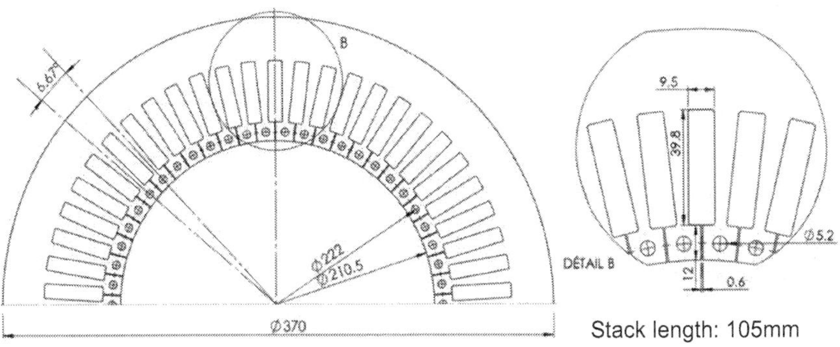

Figure 4. Stator steel laminations

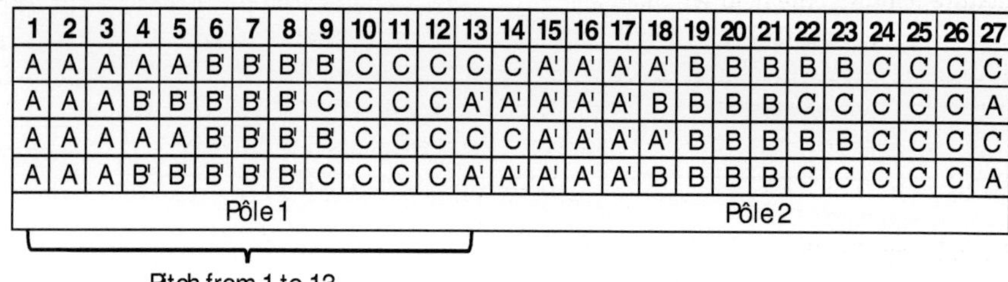

Figure 5. Stator-winding repetitive section

3. d-q equivalent model

3.1. Modeling method

The d-q model uses a transformation, known as the d-q transformation or Park transformation, in order to obtain a reference frame rotating with the rotor instead of the fixed stator frame [8]. The assumptions associated with the use of d-q transformation are presented below [7]:

- Magnetic saturation effects are neglected. By choosing two orthogonal axes, it is assumed that currents flowing in one axis do not produce flux in the other axis.

- The armature windings are sinusoidally distributed along the air gap so that each phase winding produces a sinusoidal magnetomotive force (mmf) wave. Hence, space harmonics are negligible and the induced voltage is purely sinusoidal.

- The stator slots do not cause appreciable variation of the rotor inductances with rotor position.

- Magnetic hysteresis is negligible.

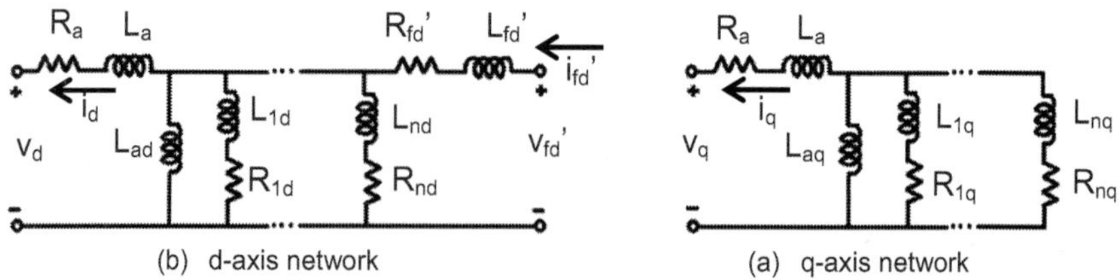

Figure 6. General d-q model equivalent circuit

Under these assumptions, synchronous machines are modeled by two independent, magnetically uncoupled circuits in *d-* and *q*-axes (Figure 6). Direct-axis model includes the *d*-axis armature winding, the field winding, and any additional equivalent damper windings. The damper bar elements (L_{nd}, R_{nd}) represent equivalent current paths in the d-axis damper bars or in the rotor steel body. The number of assumed rotor circuits determines the order of the model. The same approach applies to the q-axis network, except that the field winding is not present. The d-q model parameters are the R-L lumped circuit elements (Figure 6) as viewed from the stator. A detailed development of the d-q model is presented in [7].

3.2. Model application

A second-order d-q model was chosen for the synchronous generator at hand [13]. Its parameters are experimentally identified with two methods: the well-known standstill frequency response (SSFR) tests and the reverse identification method. We then compare some simulation and experimental results.

3.2.1. Parameter values identification from SSFR tests

Test procedures and recommendations when performing SSFR tests are well documented in the IEEE Std 115-2009 [14]. Tests in d- and q-axes are carried out separately in specific rotor positions. The simple network theory establishes relationships between the d-q model parameters $\{R_a, L_a, L_{ad}, L_{1d}, R_{1d}, L_{fd}, R_{fd}, L_{aq}, L_{1q}, R_{1q}, L_{2q}, R_{2q}\}$ and six transfer functions known as $Z_d(s)$, $L_d(s)$, $sG(s)$, $Z_{afo}(s)$, $Z_q(s)$, and $L_q(s)$. These transfer functions are evaluated experimentally or by finite element simulation when the exact geometry and material properties of the machine under study are available. Figure 7 shows a comparison between transfer functions obtained from experimental SSFR and from 2D-FE-simulated SSFR using *Flux 2D* by *Cedrat* [15] for the laboratory machine under study. We see that the curves are very similar. It took approximately 18 minute to perform the simulation, while experimental tests needed 3 hour to complete.

For $Z_d(s)$, $L_d(s)$, and $sG(s)$, the field winding is shorted and the rotor is aligned along the direct axis. For $Z_{afo}(s)$, the rotor is in d-axis, but the field winding is left open. Finally, for $Z_q(s)$ and $L_q(s)$, the rotor is short-circuited and aligned along the quadrature axis. Once all the transfer functions are obtained, the model parameters are determined by curve-fitting technique.

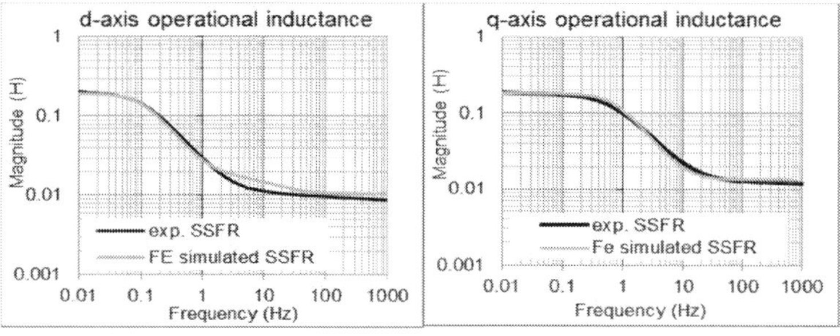

Figure 7. Comparison of experimental and simulated transfer functions

3.2.2. Parameter values identification from reverse identification

The method is based on the analysis of the model's response after a sudden three-phase short circuit and a sudden line-to-line short circuit. These two tests are usually used to fully describe the machine behavior in both axes. From an initial guess of the equivalent circuit parameters $\{R_a, L_a, L_{ad}, L_{1d}, R_{1d}, L_{fd}, R_{fd}, L_{aq}, L_{1q}, R_{1q}, L_{2q}, R_{2q}\}$, the tests are simulated using the Matlab/SPS block "SI Fundamental Synchronous Machine." Then the armature and field current waveforms are compared to experimental data in order to calculate the sum of errors between simulation and experience. The circuit parameters are then iteratively modified using optimization process until the error is minimized.

The machine is modeled by a "SI Fundamental Synchronous Machine" block. It is a d-q model having one equivalent damper winding along d-axis and two equivalent damper windings along q-axis. The nominal field current (I_{fn}) has to be defined as the field current corresponding to nominal voltage on the air gap line (I_{fq}) in order to be consistent with (L_{ad}) and the linear approximation used in this study.

The speed and field voltage supply of the experimental tests must be imposed in the simulation. A clock and lookup tables are used to assign test data values to each simulation time step. The initial rotor position and fault time are chosen such that the experimental and simulated short circuits are synchronized. When the synchronous machine block is used in discrete simulation, a small parasitic resistive load is required at the machine terminals to avoid numerical oscillations. Figure 8 shows the Simulink block diagram used to simulate a short circuit.

The optimization process is implemented with the "fmincon" function of the optimization tool of Matlab. It allows finding the minimum value of constrained nonlinear multivariable problems. There are no particular constraints imposed in the optimization problem, except that all variables must be nonnegative. The field and armature currents of both three-phase and line-to-line short circuits are optimized simultaneously. Results are presented in Table 3.

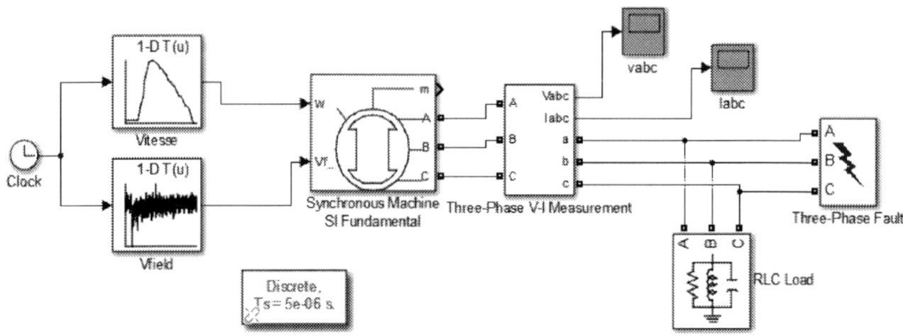

Figure 8. Simulation block diagram

3.3. Validations

The parameters identified by the SSFR and reverse identification methods are presented in Table 3. To evaluate the transient responses of the d-q model, we used the Matlab/Simulink block "SI Fundamental Synchronous Machine" with the parameter values identified using experimental SSFR and reverse identification methods. We compared the field and armature responses to a three-phase short circuit and to a line-to-line short circuit (both with neutral not connected). The field current was approximately equal to 0.22 A (0.44pu). As seen in Figures 9a and 10a, the model is able to reproduce the armature current's waveform. In Figure 10b, the field current's general behavior is well represented. However, the oscillations are less damped in the case of simulated data compared to test data.

Parameter	Experimental SSFR	Reverse identification	SI unit
L_{ad}	171.0	170.6	mH
L_{aq}	176.9	179.0	mH
R_a	156	156	mΩ
L_a	9.1	9.5	mH
R_{fd}	187	186	mΩ
L_{fd}	2.95	2.26	mH
R_{1d}	712	390	mΩ
L_{1d}	0.692	0.113	mH
R_{1q}	1.616	0.871	Ω
L_{1q}	132	0.98	mH
R_{2q}	1.17	2.379	Ω
L_{2q}	3.7	81	mH

Table 3. d-q model parameters of the laboratory machine

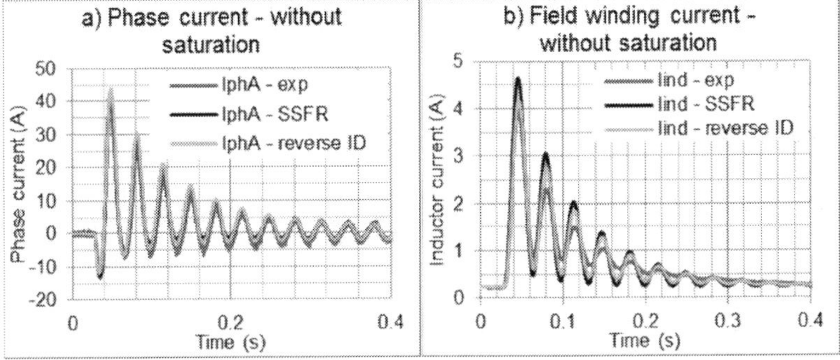

Figure 9. Three-phase short circuit with a field current equal to 0.22 A

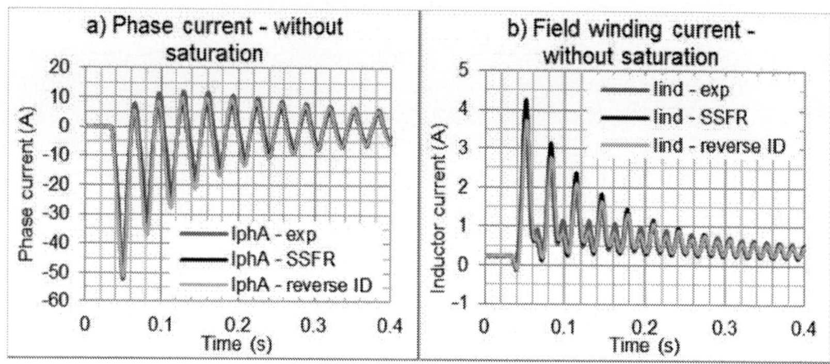

Figure 10. Line-to-line short circuit with a field current equal to 0.22 A

4. Coupled-circuit model

This section presents the coupled-circuit modeling method and illustrates its application for the simulation of the generator under study. This method allows the modeling of the magnetic couplings between every electrical circuit in a machine. Compared to the d-q standard model presented earlier, it is a higher-order model that can take account of the spatial harmonics [8].

4.1. Modeling method

The behavior of a polyphase machine can be represented by several electrical circuits that are magnetically coupled. Depending on the discretization level, each individual circuit can be equivalent to a winding or a coil or a conductor or a part of a massive conductor having a constant current density. Generally, the selected circuits correspond to the stator phases, the rotor field, and the rotor damper winding [16]. The magnetic saturation is not taken into account in this section, but an improvement of the model using a simple method is presented in section 6.

The method we present does not introduce a reference frame transformation and the stator has a fixed angle. Each electrical circuit is linear and can be modeled by a resistance and several inductances that depend on the rotor angular position θ. The mutual inductances between the stator and the rotor alternate relative to the rotor position. Other inductances may also vary around a mean value relative to the space harmonics caused by rotor or stator slotting or saliency.

The set of circuit equations can be written in a vector matrix form as shown in (1). The elements of matrix $L\,(\theta)$ provide the magnetic couplings in the stator, in the rotor, and between them (2):

$$[V] = [R][I] + \frac{d\{[L(\theta)][I]\}}{dt} \tag{1}$$

$$L(\theta) = \begin{bmatrix} [L_S]_{n \times m} & [M_{RS}]_{n \times m} \\ [M_{RS}]_{m \times n} & [L_r]_{m \times n} \end{bmatrix} = \begin{bmatrix} \begin{bmatrix} L_{s1} & M_{s12} & \cdots & M_{s1n} \\ M_{s21} & \ddots & \ddots & \vdots \\ \vdots & \ddots & \ddots & M_{sn\text{-}1n} \\ M_{sn1} & \cdots & M_{snn\text{-}1} & L_{sn} \end{bmatrix} & \begin{bmatrix} M_{s1r1} & M_{s1r2} & \cdots & M_{s1rm} \\ M_{s2r1} & \ddots & \ddots & \vdots \\ \vdots & \ddots & \ddots & M_{sn\text{-}1rm} \\ M_{snr1} & \cdots & M_{snrm\text{-}1} & M_{snrm} \end{bmatrix} \\ \begin{bmatrix} M_{r1s1} & M_{r1s2} & \cdots & M_{r1sn} \\ M_{r2s1} & \ddots & \ddots & \vdots \\ \vdots & \ddots & \ddots & M_{rm\text{-}1sn} \\ M_{rms1} & \cdots & M_{rmsn\text{-}1} & M_{rmsn} \end{bmatrix} & \begin{bmatrix} L_{r1} & M_{r12} & \cdots & M_{r1m} \\ M_{r21} & \ddots & \ddots & \vdots \\ \vdots & \ddots & \ddots & M_{rm\text{-}1m} \\ M_{rm1} & \cdots & M_{rmm\text{-}1} & L_{rm} \end{bmatrix} \end{bmatrix} \tag{2}$$

To represent the movement and the variation of current, we may express (1) in the following manner, where Ω is the rotor angular speed:

$$[V] = [R][I] + \Omega \left\{ \frac{d[L(\theta)]}{d\theta_m} \right\}[I] + [L(\theta)] \left\{ \frac{d[I]}{dt} \right\}. \tag{3}$$

The voltage on each circuit must be imposed to solve (3) and find the derivative of all currents (4):

$$\frac{d[I]}{dt} = [L(\theta)]^{-1} \cdot \left[[V] - \left([R] + \Omega \frac{d[L(\theta)]}{d\theta_m} \right)[I] \right] \tag{4}$$

The circuit equations to account the stator star or delta connection and the damper bars connected in a grid or cage design. Figure 11 presents an example of circuit for the rotor damper bar connections connected by shorting-ring impedances (R_{ri}, L_{ri}) in a cage design. The unknowns of the circuit problem are the mesh currents (J_i), and the matrix $[L \ (\theta)]$ and $[R]$ must be modified as detailed in Ref. [16].

Now that the circuit model is defined, the values of the inductance $[L \ (\theta)]$ and resistance $[R]$ matrix have to be identified. The identification of the resistance matrix is straightforward and depends on the section of the conductor and its length. The values in the inductance matrix must be identified for a given number of angular positions (θ). Some authors propose an analytical approach to estimate the inductance curves using winding functions [8, 17]. It assumes that the permeance of iron is infinite. Another way to avoid additional simplifications is to use a finite element method [18], as proposed in this study.

For analysis purposes, the time required for this identification is quite well compensated by its increased precision. The identification is done with 2D finite element simulation with static

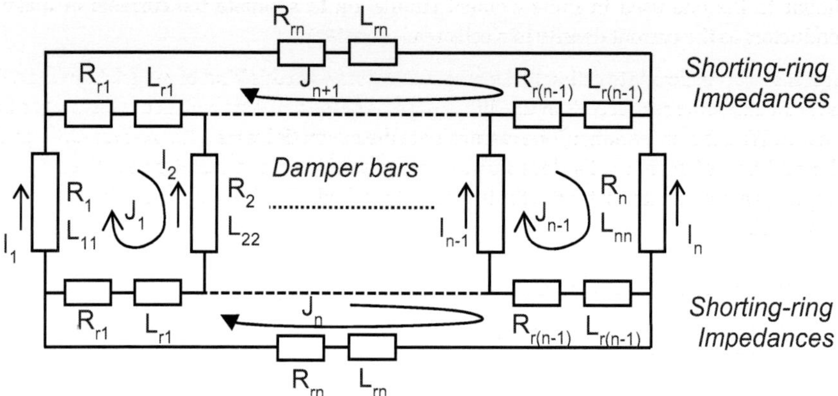

Figure 11. Equivalent circuit for a rotor cage for the damper winding

magnetic field resolution. The permeability of the magnetic material must be a constant (no magnetic saturation) and the extremity effects (magnetic flux leakage in the third direction) are neglected. They could be considered in a 3D simulation. This method allows a fast, general, and accurate identification of the inductance curves.

From the inductance curves, we can evaluate their first and second derivatives curves [18]. Linear interpolation in these curves is an efficient method to estimate the inductance value and its derivative value for any rotor position. Knowing the inductance values of two discrete rotor positions (θ_0 and θ_1), one can compute the inductance value for any rotor position θ between θ_0 and θ_1 using (5):

$$\text{for } \theta_0 < \theta \leq \theta_1 \qquad L(\theta) = L(\theta_0) + \left(\theta - \theta_0\right) \cdot \frac{L(\theta_1) - L(\theta_0)}{\theta_1 - \theta_0} \tag{5}$$

The coupled-circuit model can easily be implemented in Matlab-Simulink that provides many numerical methods to solve the set of differential equations. A variable-step solver like a fifth-order ode15s is efficient in terms of simulation time, accuracy, and stability. The simulation time is also improved using the rapid accelerator mode of Simulink. The linear interpolation in a table of discrete values is easy to implement with the Simulink tools. Using an identification of inductance curves with 2D FEM, the coupled-circuit model has the advantage of combining the precision of the finite element spatial representation for the magnetic couplings, with high efficient variable-step solvers.

One can notice that the assumption related to the proximity and skin effects is not a constraint in this model. Indeed, it is always possible to divide a massive conductor into several elementary circuits carrying different currents [18]. The order of the model is increased, but the simulation of transients is still faster than with the finite element method. This approach is similar to the one used in finite element simulation to simulate the currents in massive conductors as the current density in a cell area is constant.

It seems not practicable to estimate the magnetic losses with this kind of model. However, the user can add some search coils in the different parts of the machine, which can be used as flux sensors. With that information, one can evaluate the magnetic losses with the Bertotti formula. The addition of search coils does not make the study more difficult as all the magnetic couplings are calculated with the finite element method. This increases slightly the order of the model.

4.2. Coupled-circuit model application

To evaluate the inductance curves of the generator under study, we used a 2D finite elements software (Flux 2D of Cedrat [15]). To reduce the size of the study domain, and reduce the simulation time required for inductance identification, the magnetic periodicity is considered. In this case, users should control the multiplying factors and calculate the inductances for a machine part only. This management of multiplying factors is simple to integrate in the coupled-circuit model. This allows more options for the winding connections between different parts of the machine.

The generator under study has a magnetic periodicity on half of its domain. Consequently, we have to identify 16 coupled circuits: 3 circuits for the stator phases, 1 circuit for the field, and 12 circuits for the damper bars. Thus, the skin effect in the damper bars is neglected. The number of rotor positions should be chosen high enough to allow the representation of the no-load voltage harmonic content. The influence of the discretization of the inductance curves with different numbers of rotor positions (20, 10, and 5 rotor positions per stator slot pitch) is analyzed in [18]. The use of 20 rotor positions within one stator slot pitch allows a very precise prediction of the harmonics due to the slot pulsation field. However, the shapes are quite well preserved even with five rotor positions per stator slot pitch. The higher harmonic content can be readily seen on the curve of the field self-inductance.

4.3. Validations

This section compares the waveforms obtained with the coupled-circuit model to the experimental ones.

The no-load voltage of the generator has an important harmonic content (third harmonic in the line-neutral voltage waveform and slot harmonic). The model is able to reproduce the shape of the curves accurately, but there is a small difference on the signal amplitude caused by the linear permeability used in the FE model (Figures 12 and 13).

When all phase ends are short-circuited with themselves, the phase current is trapezoidal (Figure 14) because of the third harmonic of the no-load voltage. Figure 15 presents the current waveform of a sustained three-phase short circuit, with a star connection and without the neutral–ground connection. In that case, the harmonic current multiples of three cannot circulate and the shape of the short-circuit current becomes more sinusoidal. The simulated waveforms are similar to the experimental data for both cases.

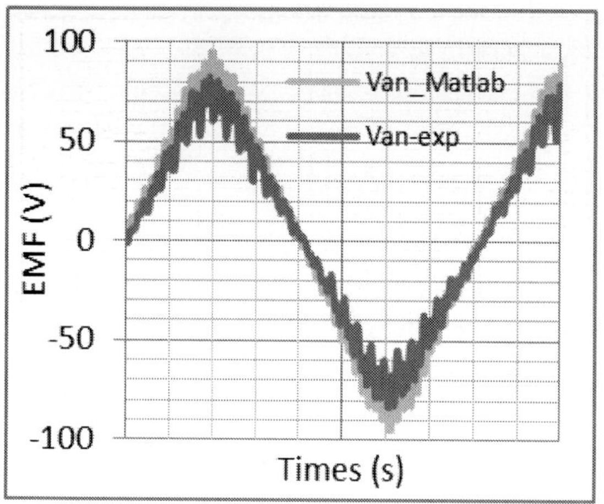

Figure 12. No-load voltage (V_{LN}) (field-winding current: 0.22 A; speed: 914 rpm)

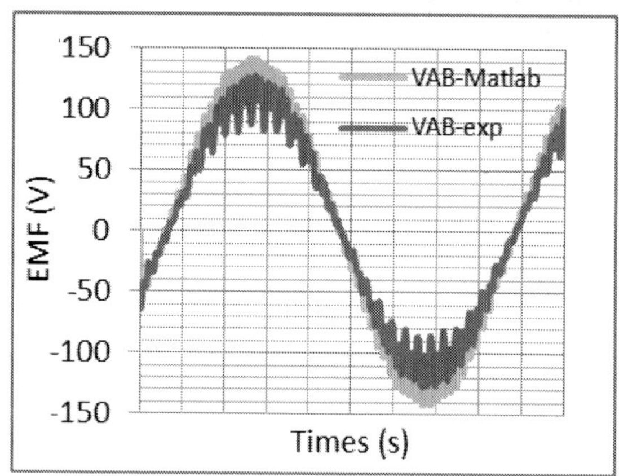

Figure 13. No-load voltage (U_{LL}) (field-winding current: 0.22 A; speed: 914 rpm)

Here, we compare the sudden short-circuit response of every phase with itself. To compare the transient responses, we have synchronized the signals and imposed the experimental speed and field voltage. Prior to the sudden short circuit, the machine is operated at 914 rpm without load and the field current is 0.22 A (0.44 pu). At this point of operation, the magnetic saturation is negligible. Figure 16 compares the response of the phase and field current during the transient. The waveforms are well reproduced by the simulation. The current overshoot is around 2 pu for the stator winding and 8 pu for the field winding.

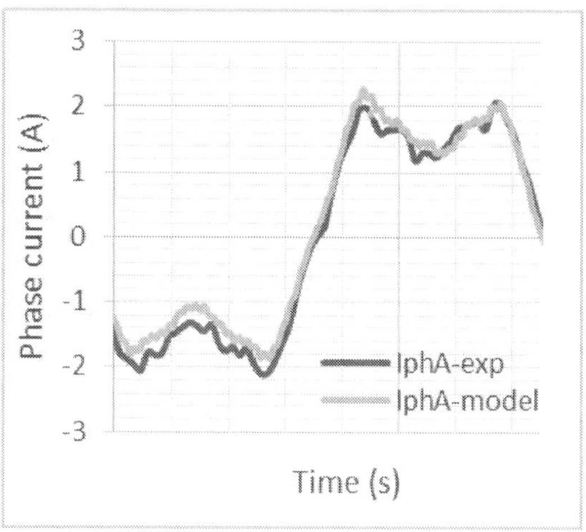

Figure 14. Steady-state current with each stator phase shorted to itself

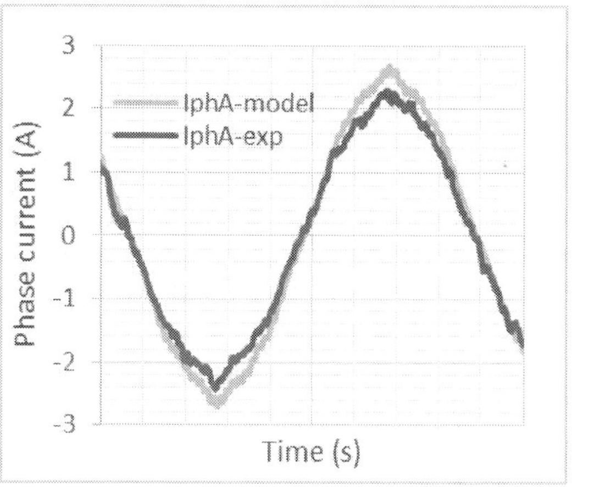

Figure 15. Steady-state current with shorted star connection

Figure 17 compares the results of a three-phase short circuit without the neutral connection. The currents and the time constant are nearly the same as in Figure 16, but the shape of the phase current is more sinusoidal.

A phase-to-phase short circuit is presented in Figure 18. This short circuit generates a negative sequence current in the stator that induces an alternative voltage at twice the stator electrical frequency in the rotor winding. This explains the field current oscillations at the end of the transient period.

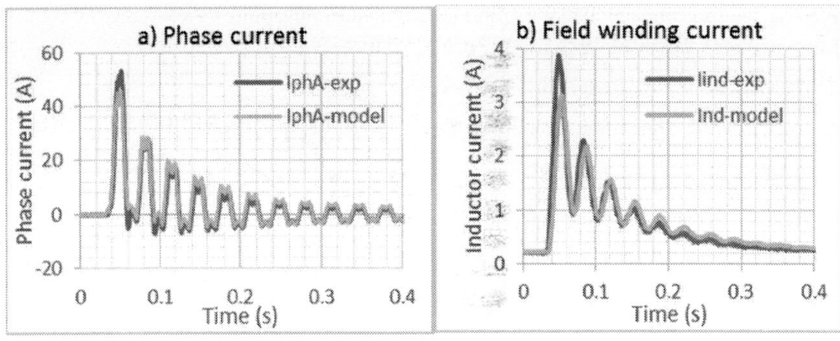

Figure 16. Transient currents when each stator phase is short-circuited to itself

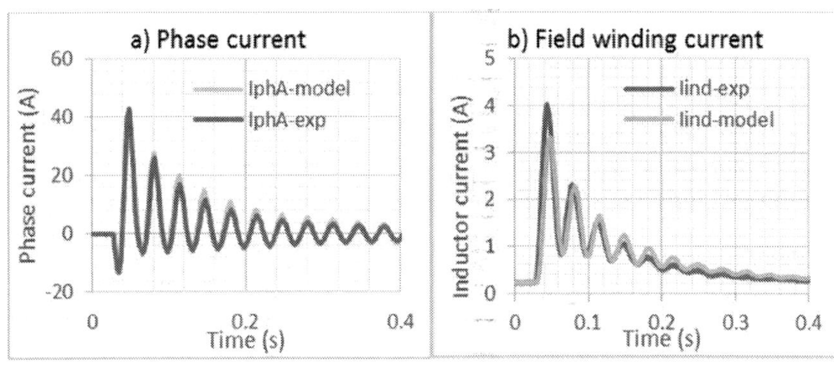

Figure 17. Transient currents for a three-phase short circuit without neutral connection

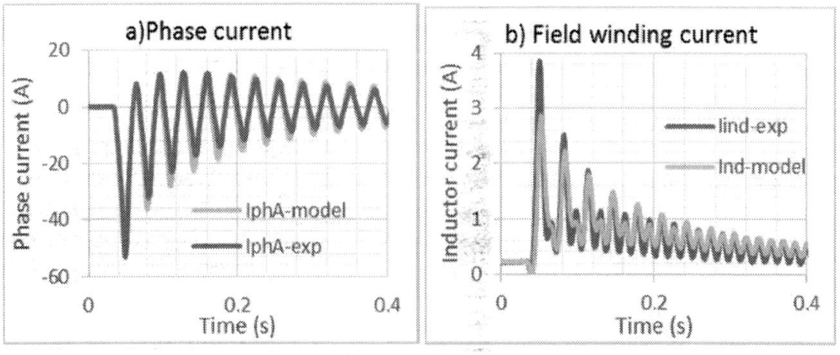

Figure 18. Transient currents for line-to-line short circuit without neutral connection

If we add a neutral connection to the ground, current harmonic multiples of three can circulate and the shape of the phase current is trapezoidal (Figure 19). This is also true for a single-phase short circuit (Figure 20).

These results show that the coupled-circuit model is very performant to simulate various transient responses by taking into account the time and space harmonics of the machine. The assumptions that neglect the magnetic saturation and the skin effect in the damper circuit appear to be valid with a low field current (0.44 pu). Comparisons with a higher field current (1 pu) are presented in section 7 for all the proposed models.

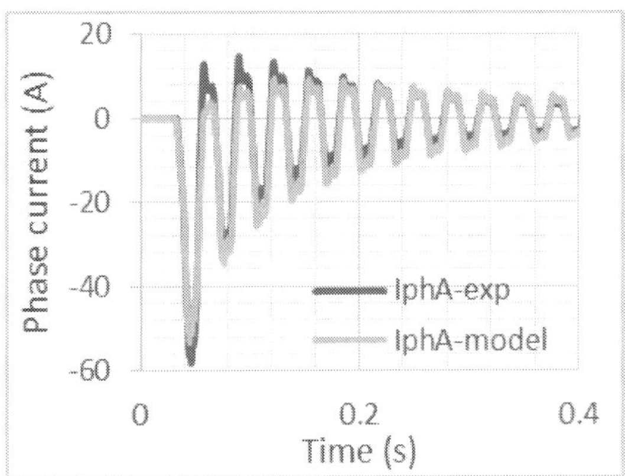

Figure 19. Line-to-line short circuit with neutral connection

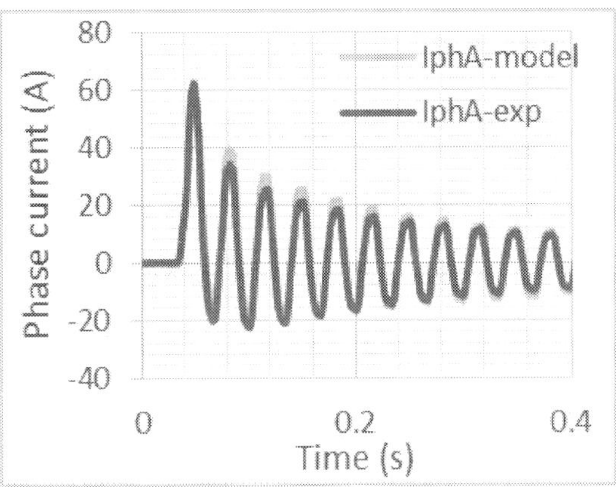

Figure 20. Phase current with one phase shorted to the neutral

5. Finite element model

The electromagnetic field computation based on Maxwell's equations using the finite element method is the state of the art in high-precision modeling of electrical machine. While being precise, it is the highest time-consuming simulation method. For the simulation of electrical machines, it is generally assumed that the displacement currents can be neglected and the electromagnetic problem can be studied with magnetostatic or magnetodynamic simulations [12]. This section discusses the modeling method and compares the simulation results of generator under study to the experimental measurements.

5.1. Modeling method

Most electrical machines can be studied with a 2D approximation [12]. That approximation is valid when the geometry of a machine section does not change along the axial direction. The problem is solved using the magnetic vector potential that has only one component in the z direction. Some boundary conditions (Dirichlet, Neumann, periodicity, etc.) also minimize the size of the study domain. Finite element simulation can consider the movement of the rotor, $B(H)$ curves of nonlinear magnetic materials, and massive conductor (proximity and skin effects). The conductors can be connected to each other's with an external electrical circuit. The currents in all conductors and the copper losses can be calculated with greater precision. For a given rotor position, it is also possible to calculate the torque by the Maxwell stress tensor or the virtual works and to estimate the magnetic losses in the machine. Time-stepping computations are based on Euler method using a fixed time step. More details on the finite element method applied to electrical machines are provided in [12].

5.2. Model application

We used a commercial software, Flux 2D of Cedrat [15], for the finite element modeling of the generator. Figure 21 shows the periodic 2D sketch of the resolution domain. Massive conductors of the damper winding are modeled using the copper conductivity. The external circuit links the damper bars with constant resistances that represent the contact resistances and the shorting ring.

5.3. Validations

To consider the magnetic saturation, we used an analytic saturation curve [15]. This is a combination of a straight line and an arc tangent curve made with two parameters ($\mu_r = 4000$, $B_s = 1.5$ T). The stator and rotor are made of the same magnetic material. The simulations are made in magnetodynamic with a fixed time step of 0.25 ms (135 pts per electrical period at 30 Hz). The experimental speed variations and the field voltage are imposed as inputs. Experimental stator voltage waveforms are also imposed after the sudden short circuit.

The steady-state waveforms of no-load electromotive force (emf) and short-circuit phase current with or without neutral connection are very close to the experimental results, and that is true for all points of operation; the saturation is well represented. We compare the same

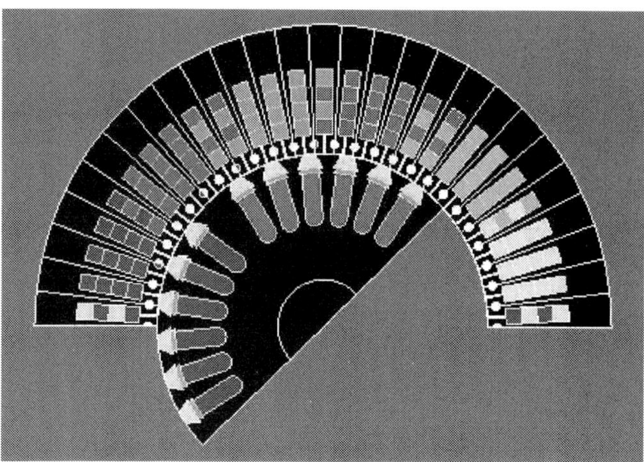

Figure 21. 2D transverse sketch

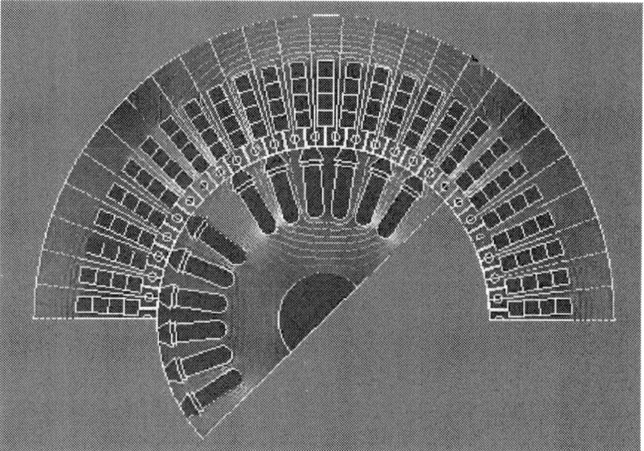

Figure 22. Flux densities (if = 0.22 A)

results as for the previous models using the experimental waveforms with a field current of 0.22 A (0.44 pu). The phase current transient responses during a three-phase short circuit (Figure 23) and a two-phase short -circuit (Figure 24) are compared to the experimental measurements. By comparing these results with the ones obtained with the previous models, it can be seen that the shape of the signals (peak value, time constant, oscillations) is always well reproduced. The most significant difference is observed on the oscillation amplitude of the field current during a line-to-line short circuit (Figure 24b).

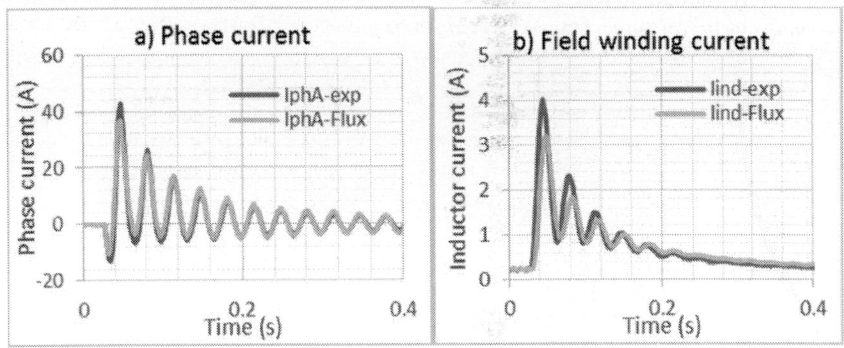

Figure 23. Phase current of a sudden three-phase short circuit without neutral

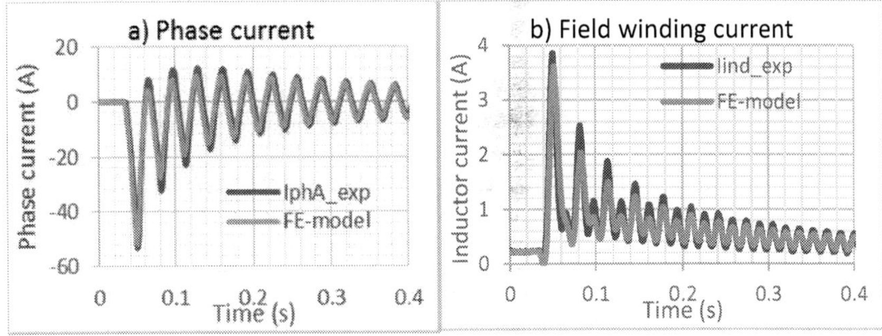

Figure 24. Transient currents for line-to-line short circuit without neutral

6. Improved circuit models considering magnetic saturation

In synchronous machines, the magnetic saturation is clearly visible on the no-load characteristic. It is possible to approximate the no-load characteristic with the air gap line and neglect the saturation phenomenon For machine operating under a low field current. We have shown that this approximation is valid for the transient simulation of the generator with a field current equal to 44% of its nominal value.

When the saturation takes effect, that approximation is no longer true [20]. The no-load voltage amplitudes are greatly overestimated with a linear model, and this leads to incorrect steady-state and transient simulation results. Figure 25 compares the field current response of both circuit models (d-q and coupled circuits) during a three-phase short circuit at the nominal field current (0.5 A). The current waveforms obtained are less damped than the experimental result. This overestimation of the currents leads to incorrect prediction of losses and transient torques.

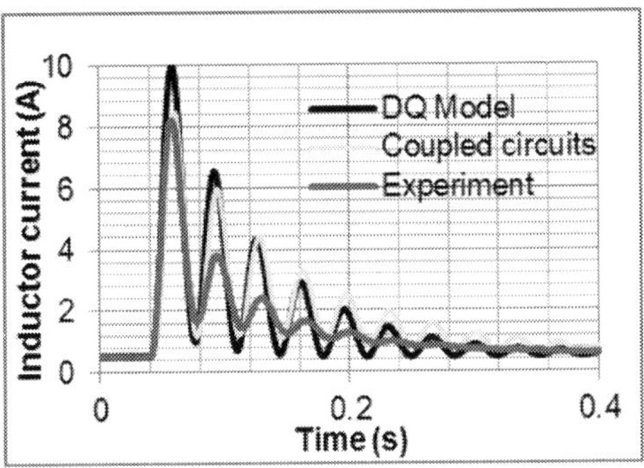

Figure 25. Overestimation of field current without considering the saturation

Many methods are proposed to take into account the magnetic saturation in the d-q model [7, 20-22]. It is generally assumed that the leakage inductance does not vary regarding the level of saturation. Only the magnetizing inductances ($B(H)$, L_{ad}) are affected. It is also assumed that the stator armature reaction has no influence on the magnetic saturation. Consequently, the magnetic saturation can be deduced from the machine no-load operation.

The simplest method uses two saturation factors (L_{aq}, k_{sd}) to modify the value of the magnetizing inductances (k_{sq}, L_{ad}) (6). The factor (L_{aq}) is function of the excitation current (k_{sd}). It is computed from the no-load characteristic and the air gap line. For a given field current, we compute the ratio between the no-load voltage (i_f) and the equivalent voltage on the air gap line: E_v(7).

$$\begin{cases} L_{ad\,sat} = k_{sd}(i_f) \cdot L_{ad} \\ L_{aq\,sat} = k_{sq}(i_f) \cdot L_{aq} \end{cases}$$

(6)

$$k_{sd}(i_f) = \frac{E_v(i_f)}{E_{linear}(i_f)}.$$

(7)

For the salient pole machine, the coefficient ($k_{sd}(i_f) = \frac{E_v(i_f)}{E_{linear}(i_f)}$.) is usually set equal to 1 for all operating conditions as the path of q-axis flux is mostly in the air [7]. In the case of round rotor alternator, the coefficient (k_{sq}) is often assumed equal to (k_{sq}) as the q-axis saturation data are usually not available. In reality, magnetic saturation creates a cross-coupling between d- and q-axes [22]. However, the differences in stability performance when considering cross-magnetization have not been demonstrated for large generators [7]. In the case of the coupled-circuit model, we have applied the same simplified method using only one saturation coefficient equal to (k_{sd}). All parameters of the inductance matrix have been multiplied by this coefficient.

For the generator under study, we computed the variations of (k_{sd}) using the no-load characteristic up to twice the nominal field current. When the field current is greater than this value, the saturation coefficient is kept constant at its smallest value.

Figure 26 presents the simulation results when the magnetic saturation is taken into account. The d-q model current response is improved, but it is the coupled-circuit model that provides the best results. This improvement is impressive as the magnetic saturation is roughly taken into account.

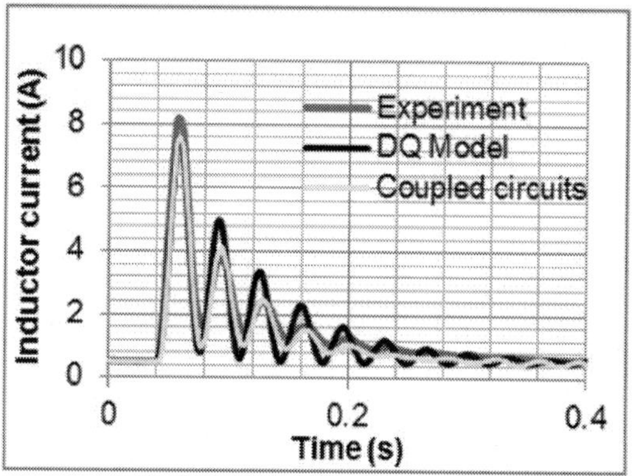

Figure 26. Field current when considering the saturation

Note that the field current peak reaches 16 times its nominal value. At that specific moment, the finite element analysis shows that the magnetic saturation is predominantly located in the stator and rotor tooth tips because of the magnetic stator reaction (Figure 27).

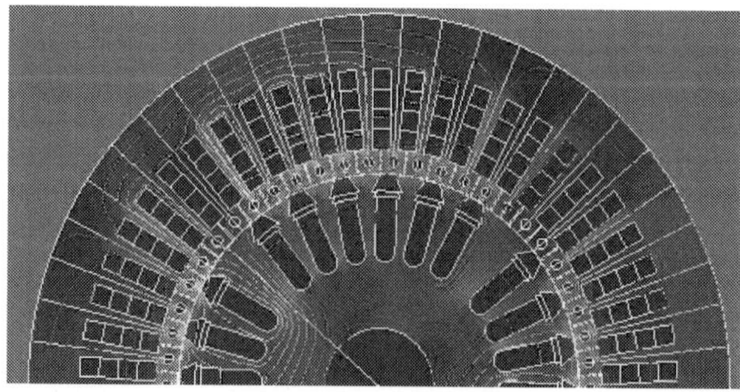

Figure 27. Magnetic flux densities at 0.055 s (field current peak)

7. Sudden short-circuit comparisons

This section compares the ability of the d-q, coupled-circuit, and finite element models to reproduce the current responses during sudden short circuits. We consider four kinds of short circuits: the sudden two-phase and three-phase short circuits without neutral connection, a two-phase short circuit with neutral–ground connection, and a single-phase short circuit with neutral–ground connection. We also compare several waveforms that have not been measured like the damper bar currents, the electromagnetic torque, and the damper circuit losses.

7.1. Three-phase short circuit without neutral connection at full voltage

Right before both sudden short circuits, the machine is at 900 rpm without load and at rated field current. Solely for the purpose of comparing the current waveforms, the experimental rotor field voltage waveform and the experimental speed are used as inputs in the simulations. In addition, at the moment of the short circuit and until the end of the simulation, the measured voltage at the short circuit (at the breaker terminals) is applied in the simulation. The initial values for the rotor position and field current are imposed to the experimental ones. The time required for the simulation is of a few seconds for the d-q model, 15 second for the coupled circuit, and 2 hour 45 minute for the FE simulation. Figures 28 to 30a compare the phase current during the three-phase short circuit.

The experimental phase current peaks at 7 pu during this short circuit. The response of the d-q and the coupled-circuit models are similar, and the differences with the experimental waveforms appear to be related to the speed variations and the precision on the initial conditions. Other factors could cause this difference such as magnetic saturation and skin effect. The finite element simulation is more accurate to consider these phenomena, but there is no significant improvement on the phase current response. The fixed step, the simple Euler method, and the precision of material properties could explain the discrepancies. The moment of the short circuit is between two time steps of 0.25 ms. Figure 30b shows the field current waveform that can be compared to Figure 26.

When comparing the waveforms of damper bar currents (Figure 31), it can be seen that the result of the coupled-circuit model matches closely with those of the finite elements. The d-q model does not provide that information. The experimental damper bar currents were not measured on the generator.

A comparison of simulated electromagnetic torque is shown in Figure 32. Figure 33 is a comparison of total copper losses in the damper bar circuit.

As the d-q model neglects all voltage and slotting harmonics, the high-frequency torque oscillations are filtered (Figure 32). The torque calculation methods of the two other models (FE and coupled circuit) lead to similar waveforms. It suggests that the electromagnetic torque peaks at 180 Nm (6.4 pu) compared to 110 Nm for the d-q model. One can also conclude that the assumptions used in the coupled-circuit model are valid. The finite element method takes account of the skin effect in the bars. This explains that the damper losses are near twice larger than the other models' estimations (Figure 33).

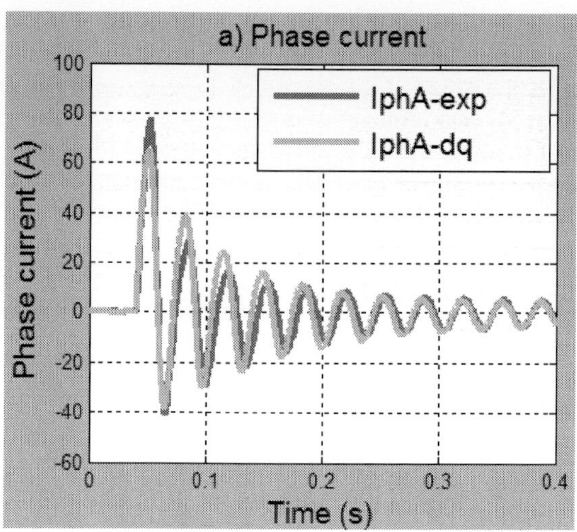

Figure 28. d-q model/experiment

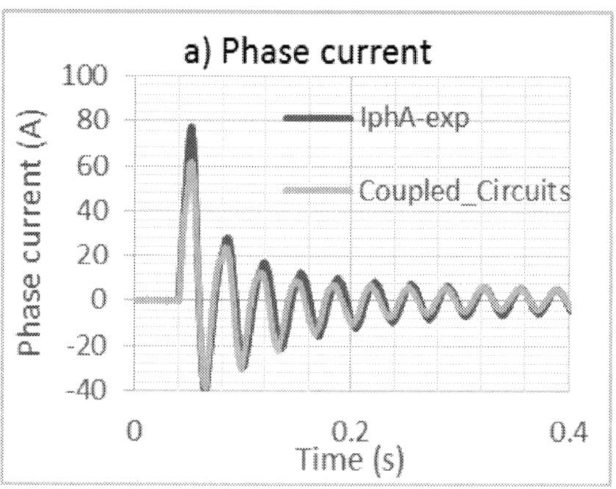

Figure 29. Coupled circuit/experiment

7.2. Two-phase short circuit without neutral connection at full voltage

The same conditions are used for the simulation of the sudden two-phase short circuit as for the previous section on the sudden three-phase short circuit.

From Figures 34 to 36, the phase and field currents are compared. A similar error can be seen on the evaluation of the first current peak. The possible advantage of a finite element simulation in obtaining these current waveforms is not entirely clear.

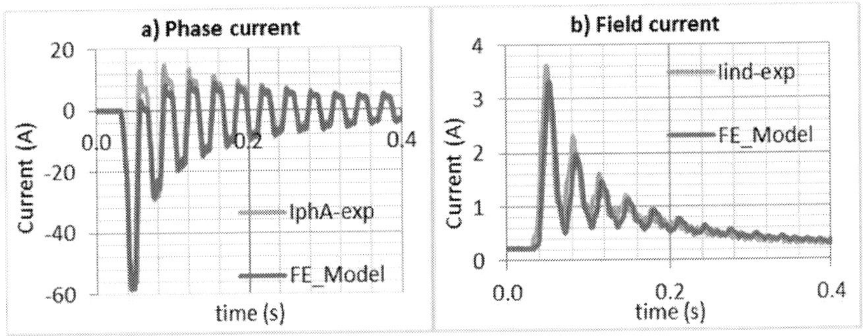

Figure 30. Three-phase short circuit; finite element model vs experiment

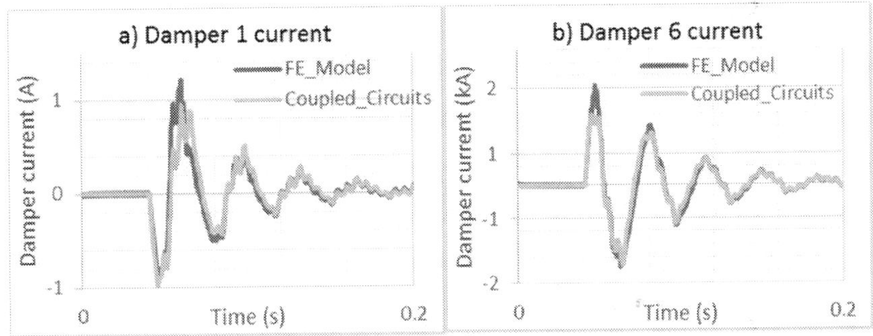

Figure 31. Three-phase short circuit; damper bar currents

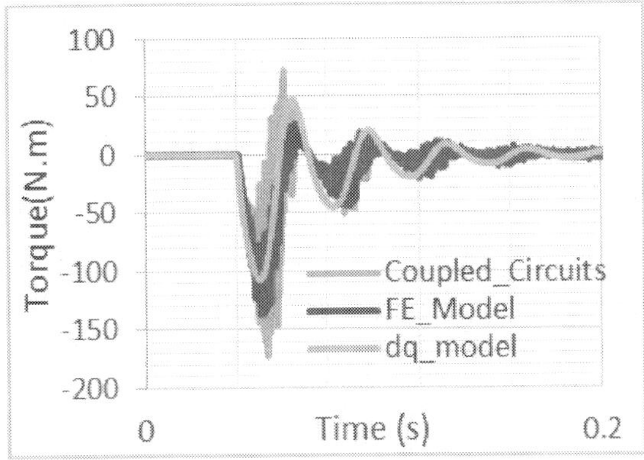

Figure 32. Electromagnetic torque

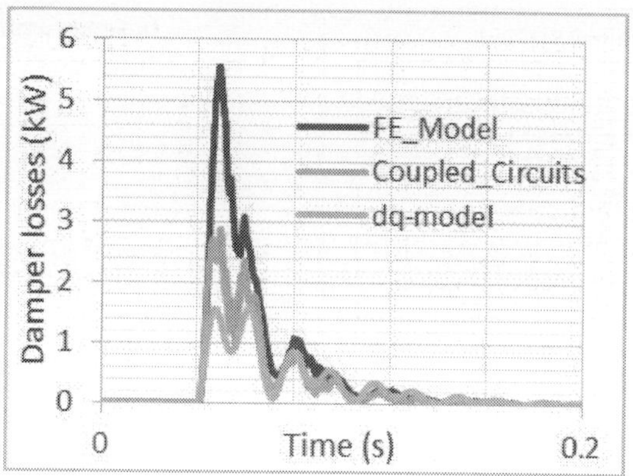

Figure 33. Total copper losses of damper circuit

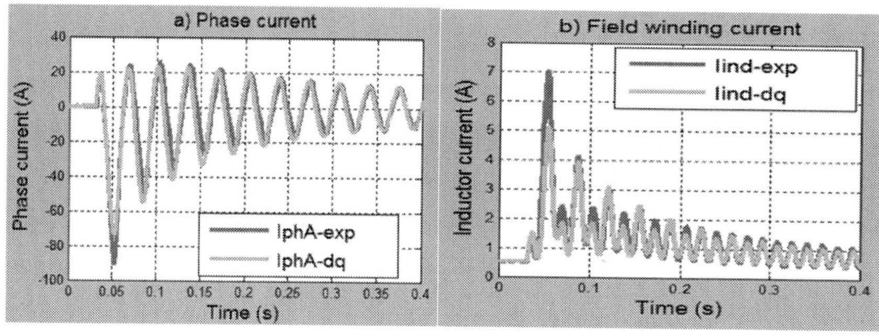

Figure 34. Two-phase short circuit; d-q model with saturation vs experiment

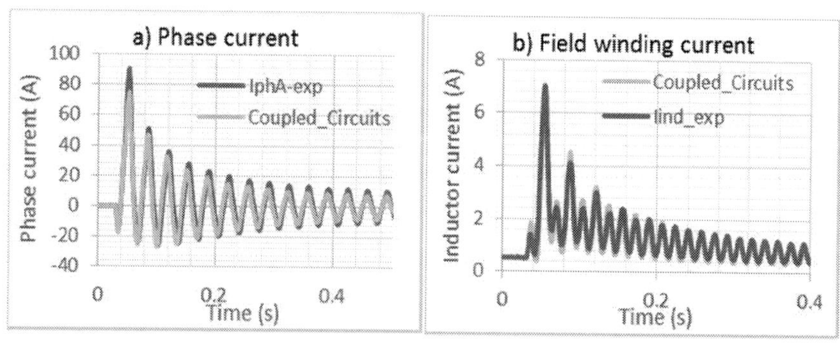

Figure 35. Two-phase short circuit; coupled circuit model vs experiment

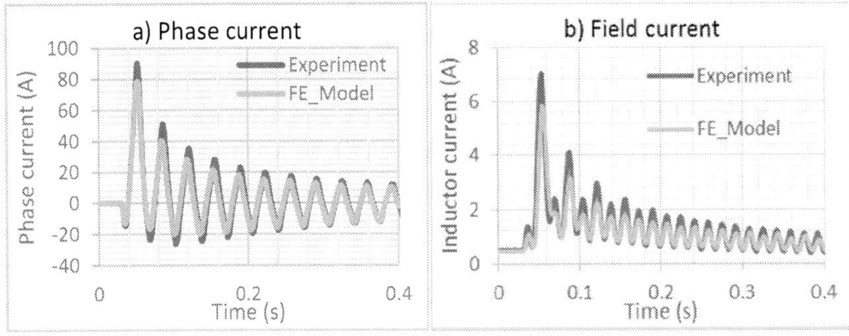

Figure 36. Two-phase short circuit; finite element model vs experiment

Comparing the waveforms of the damper bar currents shows that the coupled-circuit model provides results very close to the finite element model (Figure 37).

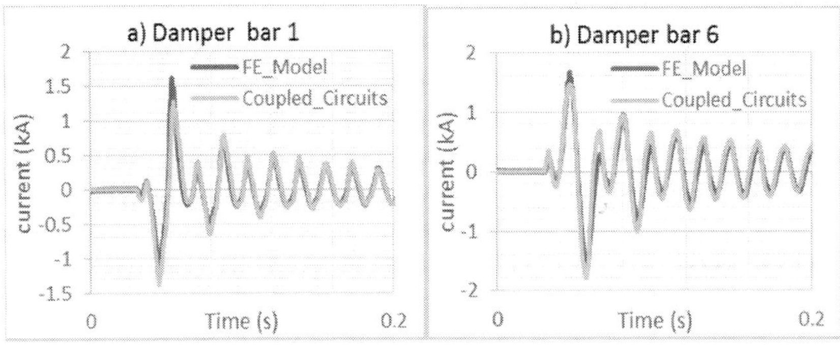

Figure 37. Two-phase short circuit; damper bar currents

Figure 38 shows a comparison of the electromagnetic torque, and the general pattern is the same for all simulation methods. The two first periods are detailed on the second figure. The torque oscillations of high slotting frequency are overestimated by the coupled-circuit model. However, the waveforms of the finite element and coupled-circuit models are very similar after several periods when the current is lower (Figure 39). Some local magnetic saturation could explain this attenuation of torque oscillations, but the numerical methods used may be also involved.

Figure 40 presents the damper circuit losses. The highest peak value (7 kW) is obtained with the finite element method, during the first period, according to the skin effect. The d-q model underestimates these losses, but the differences are reduced afterward.

We conclude that a simple magnetic saturation modeling method provides valid current waveforms, and the skin effect has little influence on the current waveforms. The largest differences between the models are related to the electromagnetic torque and the damper bar losses.

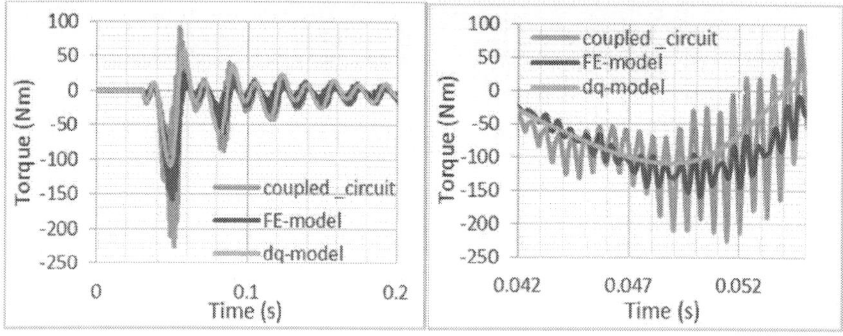

Figure 38. Electromagnetic torque (general pattern and detail 1)

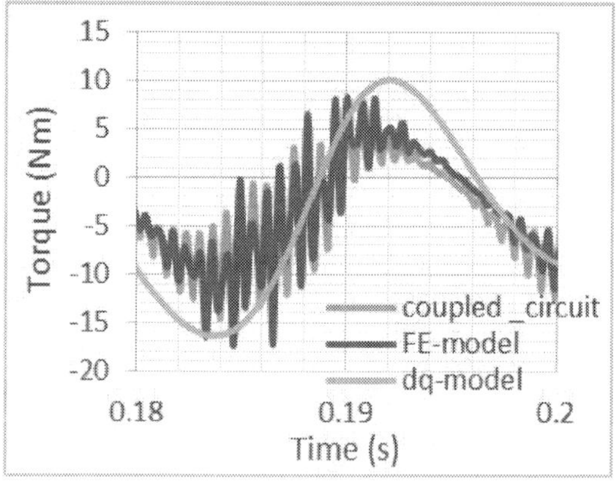

Figure 39. Torque (detail 2)

7.3. Two-phase short circuit with neutral–ground connection

The neutral connection to the ground changes the current waveform because the harmonics multiple of three have high amplitudes in the phase emf voltage. It should be noted that the Matlab d-q model cannot simulate short circuits with neutral connection because the emf harmonics and homopolar current are neglected.

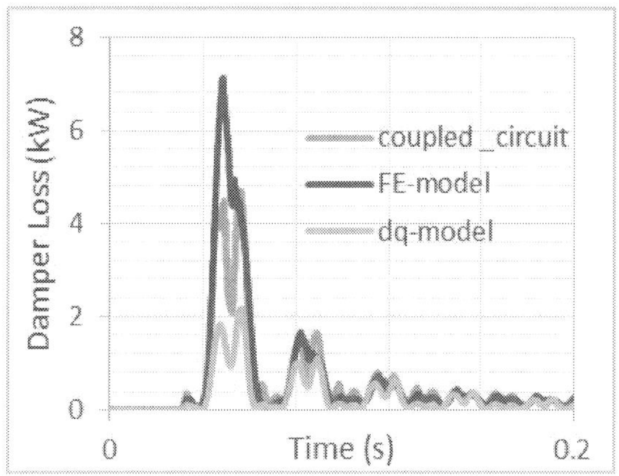

Figure 40. Copper losses of damper circuit

Figures 41 and 42 compare the two-phase short-circuit responses at low field current (0.22 A–0.44 pu). The results are better with the finite element model (Figure 42) than the coupled-circuit model (Figure 41). Explanations are provided in section 7.1.

Figure 43 shows that the damper current waveforms calculated by the finite element model are close to the ones provided by the coupled-circuit model. Again, the coupled-circuit model underestimates these currents.

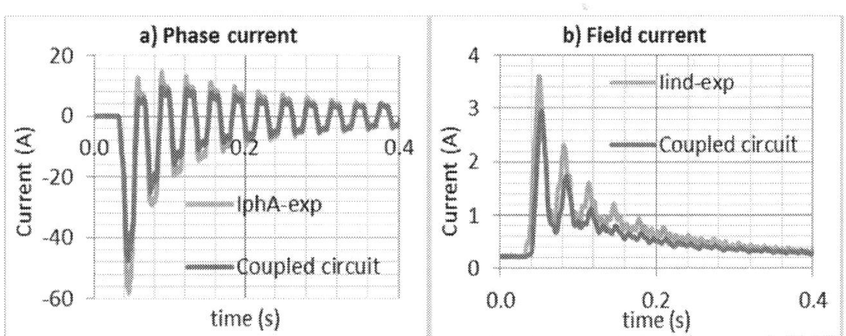

Figure 41. Two-phase short circuit; coupled circuit model vs experiment

7.4. Single-phase to ground short circuit

The same conditions are used for the simulation of the sudden single-phase short circuit as for the previous section. In this case, there is a difference of 20 A on the first cycle of the phase

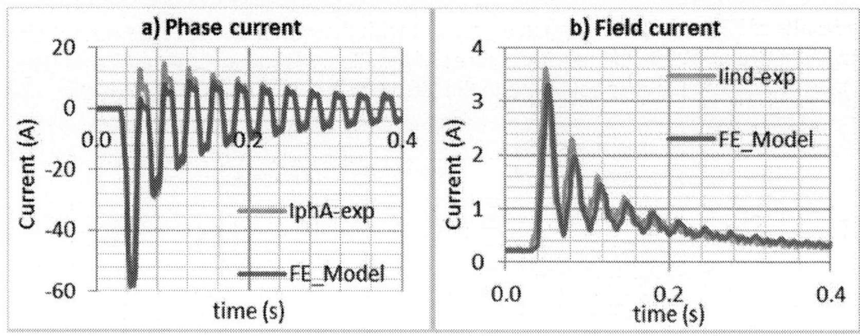

Figure 42. Two-phase short circuit; finite element model vs experiment

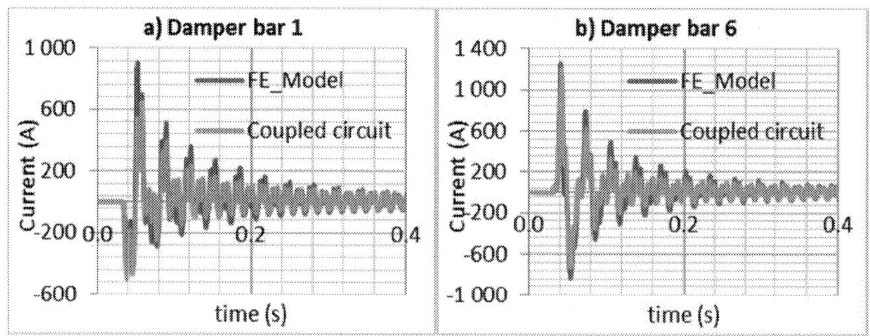

Figure 43. Two-phase short circuit; damper bar currents

current between the coupled-circuit model and the experimental measurements (Figure 44a). However, the field current waveforms are close (Figure 44b).

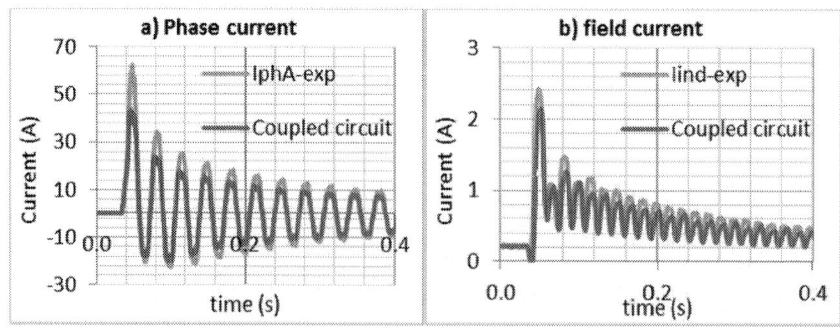

Figure 44. Single-phase short circuit; coupled circuit model vs experiment

The results of FE model show a good agreement with the experimental phase current wave-form (Figure 45a). The field current response of the FE model appears to be filtered (Figure 45b). This phenomenon may be related to the discretization chosen during simulation.

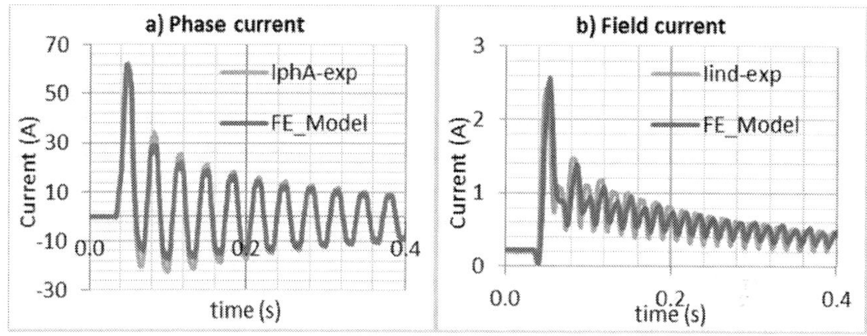

Figure 45. Single-phase short circuit; finite element model vs experiment

8. Conclusion

We have presented three electrical machine models:

- The d-q model is widely used for protection, control, and stability studies. This model is available from Matlab-Simulink library. One way to identify the d-q model parameters without short-circuit tests is to carry out a test of standstill frequency response (SSFR). This test is performed using an experimental setup. However, the parameters may also be estimated by a finite element method in complex when the exact geometry and material properties of the machine are available. This model is fast and represents accurately the waveforms of the currents in the stator and the field winding.

- The coupled-circuit model is a high-order circuit model that considers all space harmonics in the machine and all winding configurations. We presented a method that uses a 2D finite element method to compute the magnetic couplings and evaluate discrete inductance variations according to the rotor position. This model is easy to implement in Matlab-Simulink. It has the advantage of combining the precision of the finite element spatial representation for the magnetic couplings, with efficient variable time-step solvers. It is a bit slower than the d-q model, but gives better accuracy on the phase and field signal waveforms. In addition, this model can estimate the currents in other circuits, such as the damper bars. This improves the estimation of transient torque and provides results closer to the finite element method.

- The finite element model using a magnetodynamic solver with the rotor movement is the more general approach to obtain local magnetic data in various parts of the machine. This analysis requires a perfect knowledge of the machine's geometry and material properties. The main drawback of this method is the simulation time. This model is the most accurate with regard to the local magnetic saturations and the damper bar losses with the skin effect. It also gives the best estimation of magnetic losses.

We showed that all three models are efficient to estimate the transient response of an electrical generator. Comparisons were made with a laboratory generator by performing various types of short circuit at various excitation levels. It was shown that the magnetic saturation must be taken into account for a generator in normal operation.

Acknowledgements

This research was supported by the National Sciences and Engineering Research Council (NSERC) of Canada and Alstom Énergies Renouvelables Canada Inc.

Author details

Jérôme Cros*, Stéphanie Rakotovololona, Maxim Bergeron, Jessy Mathault, Bouali Rouached, Mathieu Kirouac and Philippe Viarouge

*Address all correspondence to: cros@gel.ulaval.ca

LEEPCI, Laval University, Quebec, Qc, Canada

References

[1] I. Boldea, Synchronous Generators, 1st ed. CRC Press, USA, 2005.

[2] D. Aguglia, P. Viarouge, R. Wamkeue, J. Cros, "Optimal Selection of Drive Components for Doubly-Fed Induction Generator Based Wind Turbines," InTech, ISBN 978-953-307-221-0, 4 April 2011. http://www.intechopen.com/books/wind-turbines

[3] D. Aguglia, P. Viarouge, R. Wamkeue, J. Cros, Determination of fault operation dynamical constraints for the design of wind turbine DFIG drives. Mathematics and Computers in Simulation, vol. 81, no. 2, p. 252–262, 2010.

[4] "IEEE Guide for Synchronous Generator Modeling Practices and Applications in Power System Stability Analyses," IEEE Std 1110-2002 Revis. IEEE Std 1110–1991, pp. 0_1–72, 2003.

[5] "Approved IEEE Recommended Practice for Excitation Systems for Power Stability Studies (Superseded by 421.5-2005)," IEEE Std P4215D15, 2005.

[6] "IEEE Guide for Test Procedures for Synchronous Machines Part I Acceptance and Performance Testing Part II Test Procedures and Parameter Determination for Dynamic Analysis." 2011.

[7] P. Kundur, Power system stability and control. New York: McGraw-hill, 1994.

[8] T.A. Lipo, Analysis of Synchronous Machines, 2nd ed. CRC Press Taylor & Francis Group, USA, 2012

[9] E. Deng, N.A.O. Demerdash, A coupled finite-element state-space approach for synchronous generators. I. Model development. IEEE Transactions on Aerospace and Electronic Systems, vol. 32, no. 2. pp. 775–784, 1996.

[10] I. Kamwa, P. Viarouge, H. Le-Huy, E.J. Dickinson, A frequency-domain maximum likelihood estimation of synchronous machine high-order models using SSFR test data. IEEE-Energy Conversion, vol. 7, no. 3, pp. 525–536, 1992.

[11] M. Ghomi and Y. N. Sarem, Review of synchronous generator parameters estimation and model identification. 42nd International Universities Power Engineering Conference, 2007. UPEC 2007, pp. 228–235, 2007.

[12] JPA. Bastos, N. Sadowski, Electromagnetic modeling by finite element methods, CRC press, USA, 1, avril 2003

[13] I. Kamwa, P. Viarouge, H. Le-Huy, E.J. Dickinson, Three-transfer-function approach for building phenomenological models of synchronous machines. IEE Proceedings-Generation, Transmission and Distribution, vol.141, no. 2, pp. 89–98, 1994.

[14] "IEEE Standard Procedures for Obtaining Synchronous Machine Parameters by Standstill Frequency Response Testing (Supplement to ANSI/IEEE Std 115-1983, IEEE Guide: Test Procedures for Synchronous Machines)," IEEE Std 115A-1987, p. 0_1, 1987.

[15] Flux2D Cedrat, user guides and technical paper on synchronous motor, 2015, http://www.cedrat.com/

[16] X. Luo, Y. Liao, H. Toliyat, A. El-Antably, T.A. Lipo, Multiple coupled circuit modeling of induction machines. Transaction on IEEE Industry Applications Society, vol 31, no2, pp. 311–318, 1995.

[17] A. Tessarolo, Accurate computation of multiphase synchronous machine inductances based on winding function theory. IEEE Transactions on Energy Conversion, vol. 27, no. 4, pp. 895–904, 2012

[18] J. Mathault, M. Bergeron, S. Rakotovololona, J. Cros, P. Viarouge, "Influence of discrete inductance curves on the simulation of a round rotor generator using coupled circuit method", Electrimacs 2014, 2014 May 19–22, Valencia (ESP).

[19] M. Benecke, G. Griepentrog, A. Lindemann, Skin effect in squirrel cage rotor bars and its consideration in non-steady state operation of induction machines, PIERS online, vol. 7, p. 5, 2011.

[20] T.W. Nehl, F.A. Fouad, N.A. Demerdash, Determination of saturated values of rotating machinery incremental and apparent inductances by an energy perturbation method. IEEE Transactions on Power Apparatus and Systems, vol. PAS-101, no. 12, pp. 4441–4451, 1982.

[21] J.E. Brown, K.P. Kovacs, P. Vas, A method of including the effects of main flux path saturation in the generalized equations of AC machines. IEEE Transactions on Power Apparatus and Systems, vol. 1, pp. 96–103, 1983.

[22] S.A. Tahan, I. Kamwa, A two-factor saturation model for synchronous machines with multiple rotor circuits. IEEE Transactions on Energy Conversion, vol. 10, no. 4, pp. 609–616, 1995.

Chapter 12

Photovoltaic Power Plant Grid Integration in the Romanian System–Technical Approaches

Dorin Bică, Mircea Dulău, Marius Muji and Lucian Ioan Dulău

Additional information is available at the end of the chapter

http://dx.doi.org/10.5772/62739

Abstract

Technical issues of the integration of distributed generation (GD) in distribution networks are related to effects on power service quality (supply reliability, network voltage changes), power flows and network losses. The aim of this chapter is to perform a technical and economic evaluation of the impact that the connection of a Photovoltaic (PV) Power Plant has upon the electric energy distribution network in the distribution operator (DO) area. Additionally, the study also examines the effects on final consumers. The approach starts from general considerations and it continues with their practical application in a case study.

Keywords: impact on the electric grid, power flow, power quality, PV Power Plant, reliability

1. Introduction

Over the past decades, worldwide interest in renewable energy sources has risen significantly. Limitation of fossil fuels (oil and gas), such as the increasing cost of these primary energy sources and their impact on the climate change have stimulated interest in the area of alternative electrical energy supplies. Currently, the share of decentralized power systems in the electricity infrastructure has increased considerably [1-8].

The alignment of Romania's legislation to the European energy policy is achieved by investments mainly focused on building photovoltaic plants with megawatts generation capacities. The need to maintain the performance standards in electricity supply to consumers by distribution operators (DO), to which most of photovoltaic (PV) power plants are connected, requires an analysis of the impact that they have upon power distribution network.

The chapter aims to perform a technical and economical impact that the connection of a PV power plant has upon the electric energy distribution network in the DO area.

Other researches in this field are focused on integration, design or impact [1-7] of PV power plants [1-7].

2. Photovoltaic (PV) power plants situation at the Romania national level

Nationally, until 21 January 2014, according to Transmission and System Operator (TSO) - TRANSELECTRICA records, the PV power plants in use in our country are connected to the electrical distribution network with a maximum injection power of 835,076 MW, while those connected to the electrical transport network have an injection power of 24,562 MW. The PV power plants have a geographical distribution that covers most of the country, except for the mountain areas or the north and northeast parts of the country (Figure 1).

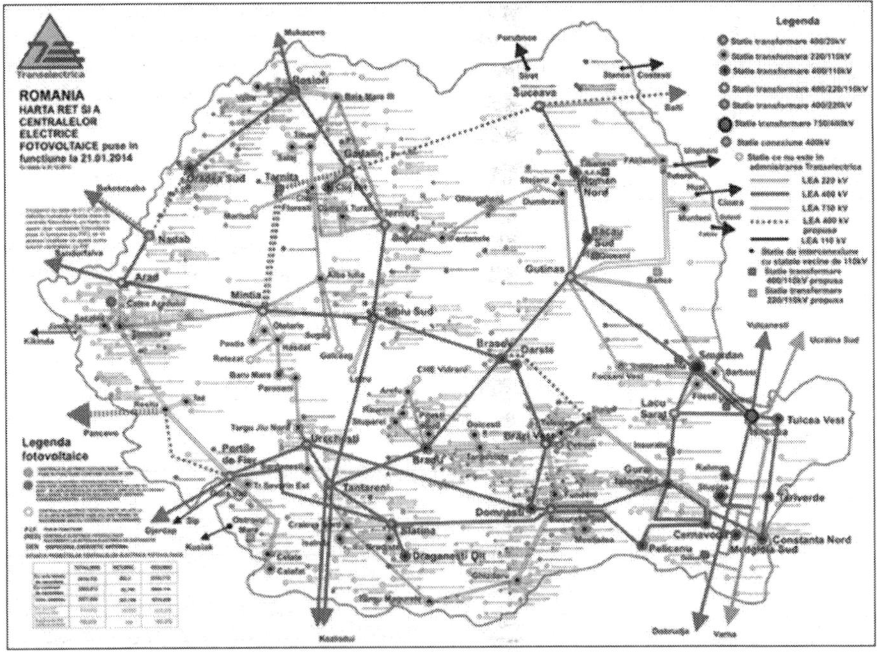

Figure 1. Transmission grid and PV power plants map.

During winter, for the National Power System (NPS), the consumption is considered as peak load, so according to data from TSO on 21 January 2014, the consumption at national level was in the range of 8212 MW to 8762 MW, a value of cumulative electricity generation of all producers, resulting in a surplus of 549 MW of power.

These readings are instantaneous values recorded at 1:04 p.m., the time when the production of electric power from PV power plants has the highest relevance, being this particular time of a winter day when the PV power plants produce the maximum power reported of installed capacity.

Figure 2 represents the production of electricity for the timeframe 9:00 a.m. to 7:00 p.m., according to generating technology types.

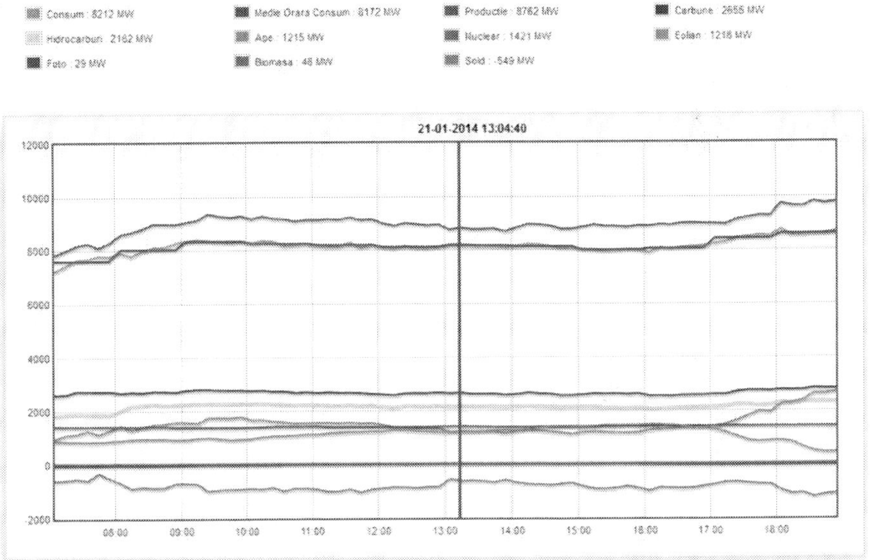

Figure 2. Generation-consumption instantaneous values for 21 January 2014.

Considering the entire winter period, when annual consumption is at maximum level and the PV power plants electricity production is 30% lower than the installed power, the PV power plants contribution to the total power compared to other types of producers is small: PV power plants – 0.33%, coal – 30.95%, hydrocarbons – 24%, hydro power plants – 14%, nuclear power plants – 16.2%, wind power plants – 14% and biomass – 0.52%.

The situation during the summer, when consumption is considered as base load, the differences are not significant, PV power plants operating at full capacity, noting that national consumption drops to an average of about 6000 MW.

Because a significant number of PV power plants, with an installed capacity of around 5500 MW, are not in use, their impact on the overall National Power System (NPS) cannot be analyzed in conjunction with the impact of other producers in the system.

Currently, in the Mureş County there are 15 PV power plants in operation, with a total installed power (P_i) of 25,626 MW, of which 13 are connected to the (DO) - medium voltage network (20 kV) - and 2 PV power plants are connected to the low voltage network of the same DO (Table 1).

P_i (MW)	<1 (MW)	1-2 (MW)	2-3 (MW)	3-4 (MW)
Number of PV	7	1	5	2

Table 1. Number of PV power plants in Mureş County and installed capacity

3. Minimal technical requirements of PV power plants for safe operation

The PV Power Plants, in addition to the Technical Norms governing the conditions for connecting the users to public electricity distribution ("Technical Code of Transport Network" [9] and "Technical Code of Electrical Distribution Networks" [10]), should also comply with the Technical Standard: "Requirements for the Connection to Public Electricity Networks for Photovoltaic Power Plants" [11] issued by Regulatory Authority for Energy (ANRE).

To ensure the safety operation of both the NPS and the PV power plants, the PV power plants fall into two categories, depending on the installed power: non-dispatchable power plants (NDPV) (that complies with the installed power $P_i \leq 5$ MW) and dispatchable power plants (DPV) ($P_i > 5$ MW). Their connection to public electricity networks requires mandatory compliance with certain technical requirements before putting them into service and afterwards.

The following necessary requirements must to be respected by DPV power plants:

- DPV power plants must be capable, at the common connection point (CCP), for an indefinite time, to produce simultaneously active and reactive maximum power according to the weather conditions, with respect the equivalent PQ performance diagram, the frequency range 49.5-50.5 Hz and the admissible range of voltage;

- All the inverter components of a DPV power plant must have the following capabilities:

  - To remain connected to the network and run continuously, in the frequency range 47.5-52 Hz;

  - To remain connected to the main electrical network when a variation of up to 1 Hz/second of frequency occurs;

  - Continuous operating voltage in the range of 0.9-1.1 of nominal voltage;

- DPV power plants and the inverters components must remain functional during voltage drops and voltage deviations (Figure 3);

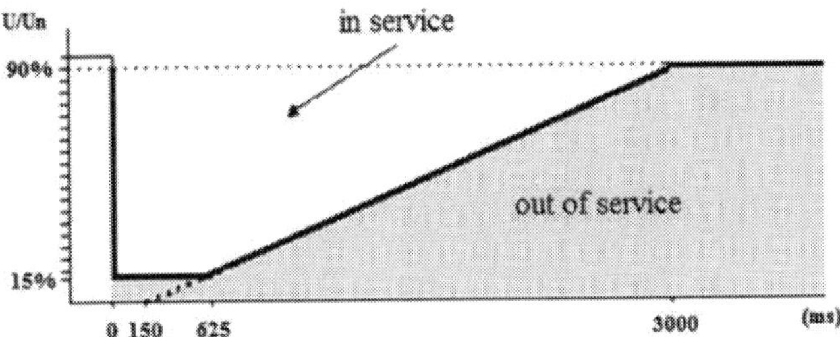

Figure 3. Magnitude of voltage drops for safety operation of DPV power plants and inverters.

- DPV power plants will be provided with automatic active power-frequency regulator. This will act as a frequency-active power response curve, exemplified in Figure 4, where P_m represents the instantaneous value of power. The coordinates of points A, B, C, D and E depend on the frequency value, on the active power that the plant can generate, and on the limit of active power preset value, in the ranges: A (50-47 Hz), B (50-47 Hz), C (50-52 Hz) and D and E (50-52 Hz). The position of the points must be set as requested by the network operator with a maximum error of ± 10 mHz;

- During voltage drops, all the inverters must inject maximum reactive power for a minimum of 3 seconds without exceeding the operating limits of DPV power plants;

- Modification of the active power generated because of frequency deviations will be made, as much as possible, in solar radiance conditions by adjusting the generated active power. If the frequency reaches a higher value (corresponding D-E on the characteristic curve), the DPV power plant can be disconnected;

- The active power generated by a DPV power plant must be limited to a preset value: DPV power plants must ensure that the active power at the CCP operate inside of ± 5% of the installed power of the power plant and must have the ability to set the deviation rate of active power at a value required by the TSO (MW/minute);

- DPV power plants must be equipped with reliable and secure protection systems, to protect against faults in electrical network, as well as against faults in the NPS;

- DPV power plants holder is required to protect the photovoltaic panels, inverters components and auxiliary installations against damage that may be caused by faults in their own installations or by the impact of the grid, to the correct operation of triggering the PV power plant protections or the network failures;

- In steady-state operation of the network, DPV power plants must ensure a voltage deviation inside its limits (between ± 5% of the rated voltage).

- The solution for the connection of PV power plants should not allow the operation in islanded mode. This involves disconnection of the PV power plants from the network as a result of the disconnection of the power supply stations to which the plant is connected.

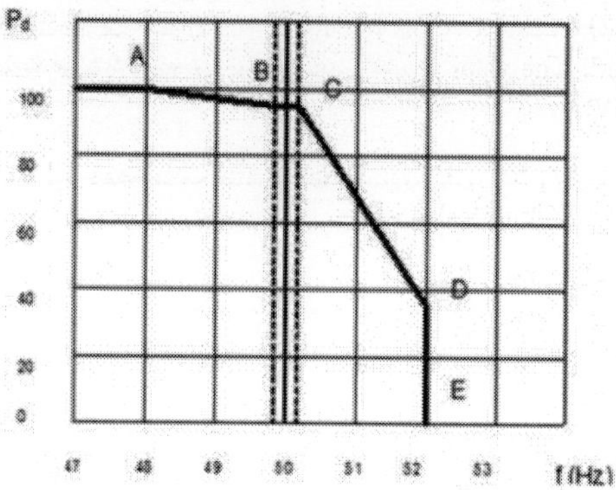

Figure 4. Active power and frequency deviations.

4. Case study: Chirileu PV power plant 3.2 MW

As a particular case study we analysed the impact of a photovoltaic power plant on the grid to which it is connected. For this study we selected the Chirileu PV Power Plant with an installed capacity of 3.2 MW connected to the 20 kV overhead electric line (OEL), Ungheni-Cipău and supplied from 220/110/20 kV Ungheni station system [12].

4.1. General data of Chirileu PV power plant

The PV power plant is located on an area of approximately 56,000 m² in the Chirileu village, Mureş County. The plant is located 9.6 km from 220/110/20 kV Ungheni station and 12.7 km from Cipău supply point.

The general technical data of the photovoltaic power plant are as follows:

- Installed power, P_i = 3.2 MW ;

- Apparent power, S = 3.33 MVA ;

- Nominal voltage (voltage in connection point), V = 20 kV;

- Net power injection in distribution energy network (DEN), 3.2 MW;

- Power factor, $cos\ \Phi$: 0.96 capacitive / 0.96 inductive;

- Estimated annual energy supplied to the system, W = 4.262 MWh;

- Installed power usage time, T = 3.65 hours / day (1332 hours/year);

- Daily operating mode in summer, T_a = maximum 12 hours/day; daily operating mode in winter: T_w = maximum 2 hour/day.

To emphasize the production capacity of the PV Plant, a power and energy production report for 11 July 2013 (Table 2) cumulative for the entire plant and of all inverters operating at full capacity, was generated. The total energy production was 22.27 MWh with operation at nominal power (3.2 MW).

The daily power and energy production graphics of the Chirileu PV Power Plant are presented in Figures 5 (summer) and 6 (winter).

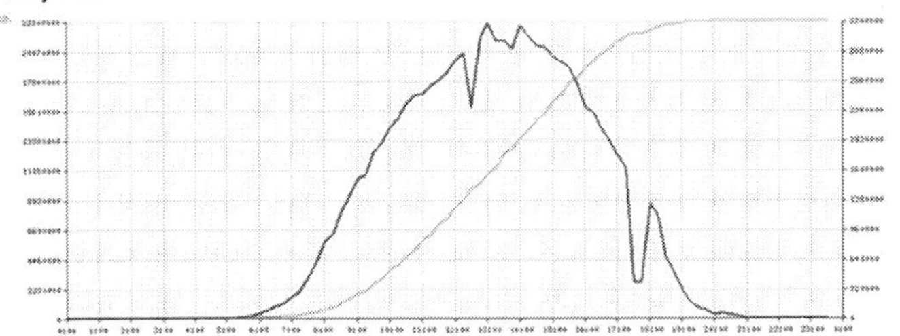

Figure 5. PV power plant energy production chart – Summer.

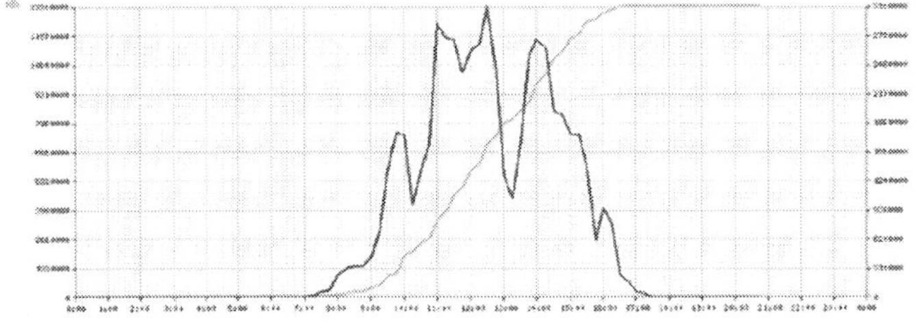

Figure 6. PV power plant energy production chart – Winter.

Energy/Power recorded data		
Hour	Energy (Wh)	Power (W)
2013-07-11 05:40:00	5	100
2013-07-11 06:00:00	4174	29,800
2013-07-11 07:00:00	104,825	190,600
2013-07-11 08:00:00	482,515	661,200
2013-07-11 09:00:00	1,488,494	1,360,000
2013-07-11 10:00:00	3,135,034	1,951,200
2013-07-11 11:00:00	5,451,184	2,478,400
2013-07-11 12:00:00	7,868,025	2,733,500
2013-07-11 13:00:00	10,346,612	1,450,500
2013-07-11 13:10:00	10,745,461	3,174,300
2013-07-11 14:00:00	13,160,949	2,937,500
2013-07-11 15:00:00	15,774,228	2,875,300
2013-07-11 16:00:00	18,364,122	2,401,000
2013-07-11 17:00:00	20,422,601	1,847,600
2013-07-11 18:00:00	21,387,377	477,400
2013-07-11 19:00:00	22,091,317	413,900
2013-07-11 20:00:00	22,234,959	57,500
2013-07-11 21:00:00	22,265,127	1900

Table 2. Energy/power recorded data of PV power plant (11 July 2013).

4.2. Network topology and characteristics

4.2.1. Network topology and characteristics of 20 kV Ungheni-Cipău OEL

The 20 kV OEL Ungheni-Cipău, 20.7 km long, comprises a conductor AlOl 70/12 mm^2 (between towers 1 and 127) and AlOl 50/5 mm^2 (between towers 128 and 239) (Figure 7).

The power transport capacity of the 70 mm^2 OEL cross-section conductor is 7.8 MVA, and the economic power transport capacity is 7.2 MVA. The power transport capacity of the 50 mm^2 and OEL cross-section conductor is 6.1 MVA and the economic power transport capacity is 5.8 MVA.

The 20 kV OEL parameters (Ungheni-Cipău) are as follows:

- Length: 20.7 km;

- Resistance (per phase, per km): 0.43m (Ω/km);

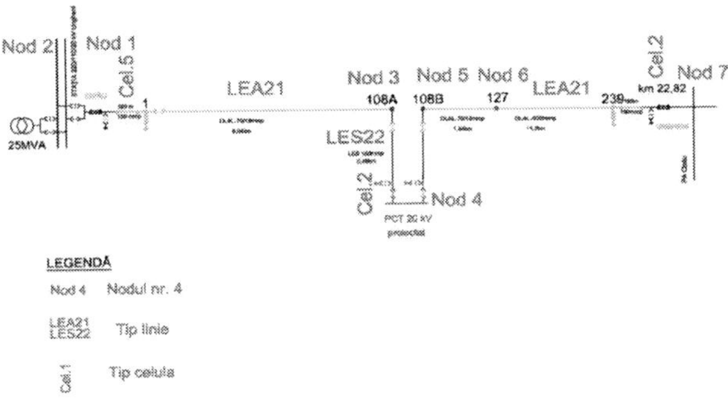

Figure 7. 20 kV OEL Ungheni-Cipău.

- Reactance (per phase, per km): 0.373 (Ω/km);

- Geometric mean distance (GMD): 1764 (mm).

4.2.2. Loads for 220/110/20 kV Ungheni station and 20 kV Ungheni-Cipău OEL

In 220/110/20 kV Ungheni system station from Figure 8 (that belongs to TSO), 110/20 kV, 25 MVA transformer (Transformer 1)- had, in January 2013, a 13.91 MW (winter peak load-WPL) and a 11.43 MW (summer base load-SBL) in July 2013. The 20 kV Ungheni-Cipău OEL had a 2.74 MW (WPL) in January 2013 and 1.94 MW (SBL) in July 2013.

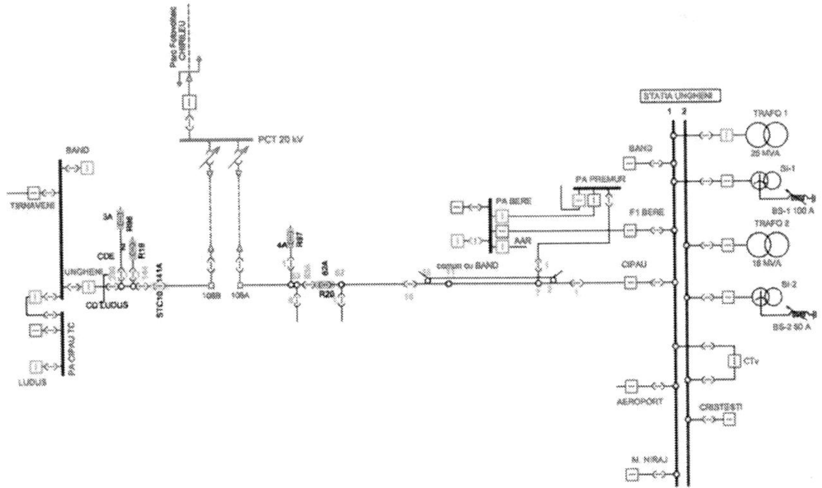

Figure 8. Ungheni system station and OEL 20 kV Ungheni-Cipău.

At 20 kV, Ungheni-Cipău OEL is connected to 39 consumers with a total installed apparent power 9.89 MVA.

The power dispatcher of the DSO provides that the nominal voltage on 20 kV Ungheni-Cipău station should be maintained at around 20.6 kV. This value is imposed so that the voltage drop should not exceed (-5%) the nominal value of the standard voltage.

4.2.3. Reserve loads for 20 kV OEL and Transformer 1 of Ungheni 110/20 kV Station

The load reserves of both line and the Transformer 1 from the station are presented in Table 3, considering the maximum power supplied by Chirileu PV Power Plant and the admissible load. The OEL relates to economic transport capacity (70 mm^2 cross section) and the optimal charging state (appropriate for minimum power losses) of transformer, considered at 75% of nominal apparent power (25 MVA).

Equipment	Existing load (MW)	PV Power Plant contribution (MW)	Allowable load (MW)	Reserve load (MW)
Ungheni–Cipău 20 kV OEL **(WPL)**	2.74	-3.20	6.62	0.68
Ungheni–Cipău 20 kV OEL **(SBL)**	1.74	3.20	6.62	1.68
220/110/20 kV Ungheni Station Tranformer 1 – 25 MVA **(WPL)**	13.91	-3.20	18.86	1.75
220/110/20 kV Ungheni Station Tranformer 1 – 25 MVA **(SBL)**	11.43	-3.20	18.86	10.63

Table 3. Load reserves of both line and the Transformer 1.

For the admissible load of 20 kV OEL, maximum load duration (T_M) of 2000 hours/year was considered, which covers the T_M duration required by the PV Power Plant, of 1332 hours/year.

4.3. Technical impact of PV Power Plant upon distribution power network

4.3.1. Setting the voltage for the connection of the PV Power Plant

As a first step for connecting the plant to a transmission or distribution electric network, the calculation of the voltage step at the connection point is required, and that is done by calculating the moment of load, M_l.

$$M_l(MVAkm) = S_{max}(MVA) \cdot L(km) = 33.4(MVAkm) \tag{1}$$

For a moment of load M_l = 33.4 (MVAkm) and the maximum power supplied inside limits (2.5 -7.5 MVA), according to Table 4, the PV Power Plant is within Class C, so it can be connected to the MV network (20 kV).

Class of power plant	Maximum load S_{max} (MVA)	Moment of load M_l (MVAkm)	Voltage step at connection point (kV)	User connection possibilities	
				Directly to grid network voltage(kV)	Through transformers
A	Over 50	Over 1500	400 220 110	- 220 110	400/110 kV 220/110 kV 220/MV kV 110/MV kV
B	7.5–50	Maximum 1500	110	110	110/MV kV
C	2.5–7.5	30–80	110 20	110 (20)	110/MV kV 20/6 (10) kV 20/0.4 kV
D	0.1–2.5	3-8	20 10 6	6-20	20/0.4 kV 10/0.4 kV 6/0.4 kV
E	0.03-0.1	Maximum 0.05	0,4	0.4	MV/0.4 kV
F	< 0.03	-	0,4	0.4	-

Table 4. Voltage step levels at connection point according to the required power.

The technical standards establish important criteria that are taken into account for the connection of users to public electricity networks, which determine the voltage step level and the connection point:

- The cost of works needed to achieve the connection;

- The technical requirements regarding operation and safety level of the electrical network;

- The requirements to maintain the power quality of the transmission and distribution network appropriate for all the consumers.

4.3.2. Calculation of network safety level in connection point

To calculate the reliability of equipments in the connection point, it is necessary to take into account the bus safety indicators of Ungheni 220/110/20 kV station (Table 5).

The parameter values of the reliability indicators for the equipments (cell line, measure, voltage transformer, current transformer) are presented in Table 6.

Station	Bus	Rate of failure (λ)	Rate of repair (μ)	Maximum duration of an interruption (h)
Ungheni 1	1-20 kV	0.007920	10,583.1	1.44
Ungheni 2	2-20 kV	0.00528	10,583.1	1.44

Table 5. Bus safety indicators of Ungheni 220/110/20 kV station

No.	Equipment/20kV network element	Rate of failure $(\lambda)*10^{-4}$	Rate of repair $(\mu)*10^{-4}$
1	Disconnector	0.01	500
2	Circuit breaker, recloser	0.15	750
3	MR	0.14	750
4	Voltage transformer	0.011	250
5	Current transformer	0.003	500
6	Bus1 (New)	0.035	450
7	Bus 2 (Old)	0.022	600
8	20kV OEL	0.05	1200
9	20kV UEL	0.3	150

Table 6. Reliability indicators for the equipments

The calculation of the duration of interruption was performed according to the "Standard on Calculation Methods for Safety Operation of Power Plants and Devices" NTE 005/06/00 [13], using a specialized software (Figures 9 and 10).

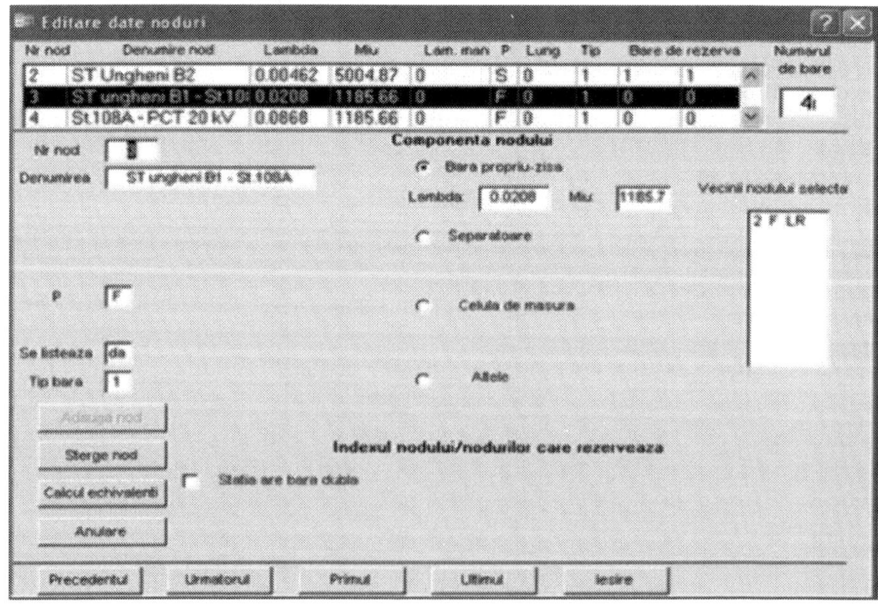

Figure 9. Graphical use interface-Buses data.

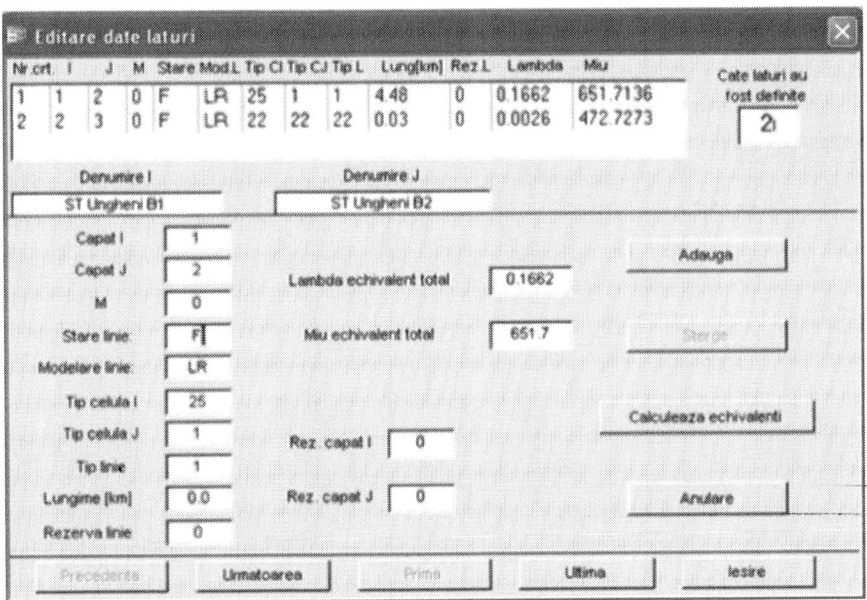

Figure 10. Graphical use interface-Branches data.

The following main results are highlighted:

- In the connection point, the maximum duration of interruption is 20.58 hours;

- The maximum duration of interruption in case of damage, in the Ungheni 220/110/20 kV station increases from 1.44 hours to 2.49 hours ;

- The maximum duration of interruption in the 20 kV connection point increases to 22.38 hours.

It can be concluded that the impact of Chirileu PV Power Plant connection to 20 kV Ungheni-Cipău OEL, in terms of reliability, is significant both in Ungheni station and in 20 kV Cipău connection point.

4.4. Load flow, power losses and voltage level analysis in steady-state conditions

To evaluate the impact of Chirileu PV Power Plant on NPS (in particular upon the 220/110/20 kV Ungheni station, Ungheni-Cipău 20 kV OEL and consumers connected to the electric line), the following input data must be taken into account:

- Network topology and characteristics of the area: Ungheni Station-Ungheni-Cipău 20 kV OEL – consumers;

- Typical daily loads in Ungheni station and typical daily loads on 20 kV Ungheni-Cipău OEL;

• Electrical energy production capacity of Chirileu PV Power Plant.

To perform the voltage drops, power flows and power losses, the maxim value of winter peak load (WPL) and summer base load (SBL), both in Ungheni station and 20 kV Ungheni-Cipău OEL, are considered in steady-state conditions.

Load flow studies for the power system's steady-state operation and development planning are performed to investigate MW and MVAR in the branches of the network and busbars (nodes) voltages [8].

In this case we analysed the evolution of voltage levels on the 20 kV Ungheni station bus, at the common connection point and at the end side of OEL, for two specific steady-state conditions:

• Peak load on 20 kV Ungheni-Cipău OHL, PV Power Plant in service;

• Base load on 20 kV Ungheni-Cipău OHL, PV Power Plant out of service.

Calculation of the power flow for 20 kV Ungheni -Cipău OEL is performed using NEPLAN software package [14] for the system (Figure 11), which allows for the design, simulation and management of electrical network. The solution of the power flow is obtained by means of the extended Newton-Raphson method due to the possibility of leading to an easier convergence process.

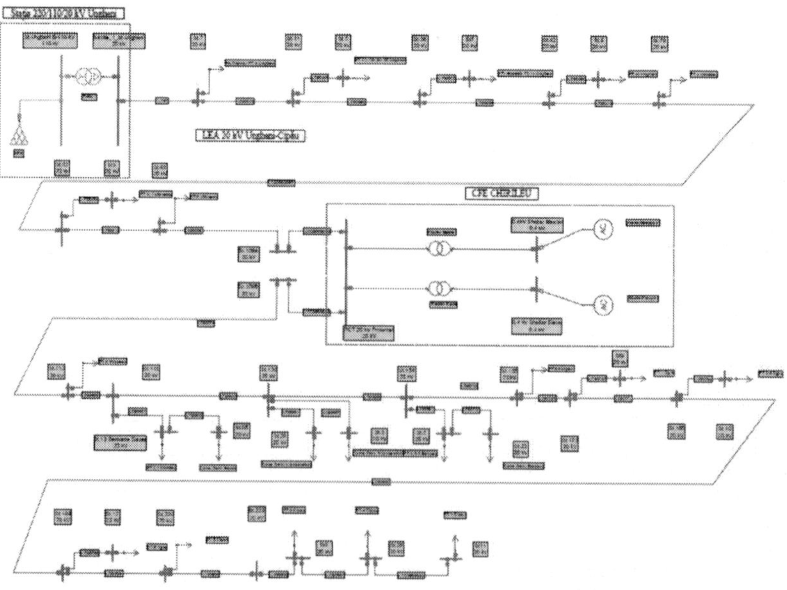

Figure 11. Line diagram of system in Neplan.

4.4.1. Power flow and power losses results

The summary simulation results for the two regimes, winter peak load (WPL) and summer base load (SBL), for two cases, PV Power Plant out of service and PV Power Plant in service, are presented in Tables 7-10.

P Loss (kW)	93.79	
Q Loss (kVAr)	348.79	
P Imp (kW)	2770.04	
Q Imp (kVAr)	1134.29	
P Gen (kW)	2770	
Q Gen (kVAr)	1134.29	
P Load (kW)	2676.2	
Q Load (kVAr)	785.5	
Iron Losses (kW)	31.68	
Un (kV)	**20**	**110**
P Loss Line (kW)	60.65	0
Q Loss Line (kVAr)	51.08	0
P Loss Transformer (kW)	0	33.14
Q Loss Transformer (kVAr)	0	297.1

Table 7. WPL-PV Power Plant out of service

P Loss (kW)	62.43	
Q Loss (kVAr)	413.18	
P Imp (kW)	-461.18	
Q Imp (kVAr)	1211.83	
P Gen (kW)	0	
Q Gen (kVAr)	0	
P Load (kW)	3137.5	
Q Load (kVAr)	-426.33	
Iron Losses (kW)	31.69	
Un (kV)	**20**	**110**
P Loss Line (kW)	30.5	0
Q Loss Line (kVAr)	24.49	0
P Loss Transformer (kW)	0	31.92
Q Loss Transformer (kVAr)	120.75	267.93

Table 8. WPL - PV Power Plant in service

P Loss (kW)	66.59	
Q Loss (kVAr)	310.25	
P Imp (kW)	1834.09	
Q Imp (kVAr)	1365.24	
P Gen (kW)	1834.09	
Q Gen (kVAr)	1365.24	
P Load (kW)	1767.5	
Q Load (kVAr)	1054.99	
Iron Losses (kW)	31.66	
Un (kV)	**20**	**110**
P Loss Line (kW)	34.11	0
Q Loss Line (kVAr)	28.37	0
P Loss Transformer (kW)	0	32.84
Q Loss Transformer (kVAr)	0	281.88

Table 9. SBL-PV Power Plant out of service

P Loss (kW)	82.3	
Q Loss (kVAr)	439.96	
P Imp (kW)	-1350.2	
Q Imp (kVAr)	1508.09	
P Gen (kW)	0	
Q Gen (kVAr)	0	
P Load (kW)	3117.7	
Q Load (kVAr)	-453.1	
Iron Losses (kW)	31.65	
Un (kV)	**20**	**110**
P Loss Line (kW)	50	0
Q Loss Line (kVAr)	42.6	0
P Loss Transformer (kW)	0	32.29
Q Loss Transformer (kVAr)	119.89	277.46

Table 10. SBL-PV Power Plant in service

4.4.2. *Voltage level analysis*

According to the Performance Standard for Electricity Distribution Service, the voltage deviation and admissible limits are presented in Table 11.

Phenomenon	Admissible limits
Limits for contractual voltage U_c magnitude at MV and HV	Contractual voltage U_c inside voltage limits (± 5 %) of nominal voltage
Flicker	$P_{lt} \leq 1$, during 95% of a week
Fast voltage change magnitude in steady-state conditions	± 5 % of nominal voltage U_n at LV ± 4 % of U_c at MV and HV
Unbalance (negative component) - K_n	LV and MV, $K_n \leq 2$ %, for 95% of a week HV, $K_n \leq 1$ %, for 95% of a week
Frequency	50 Hz ± 1 % for interconnected network 50 Hz +(4% /- 6 %) for isolated network

Table 11. Voltage deviation limits

The results related to the voltage levels are presented in Table 12.

Loads Network buses	Winter peak load-PV Power Plant out of service	Winter peak load-PV Power Plant in service	Summer base load-PV Power Plant out of service	Summer base load-PV Power Plant in service
20 kV Ungheni Station bus	20.5 kV	20.51 kV	20.49 kV	20.48 kV
Common connection point 20 kV	19.94 kV	20.61 kV	20.03 kV	20.50 kV
PT 1 Cipău – end side of OEL	19.74 kV	20.40 kV	19.86 kV	20.33 kV

Table 12. Buses voltage magnitude results

The power flow results for the two regimes, winter peak load (WPL) and summer base load (SBL) are presented in Table 13.

Power flow on 20 kV Ungheni Station bus		
Steady-state condition	P (kW)	Q (kVAr)
Winter peak load - PV Power Plant out of service	2736.89	836.58
Winter peak load - PV Power Plant in service	-493.24	943.39
Summer base load - PV Power Plant out of service	1801.01	1083.36
Summer base load - PV Power Plant in service	-288.45	1214.85

Table 13. Power flow results

Due to transit of power produced by the plant, which is supplied on the 20 kV Ungheni Station bus, the power losses in the electric line increase, as presented in Table 14.

Steady-state condition	Active power losses (kW)
Winter peak load - PV Power Plant out of service	60.65
Winter peak load - PV Power Plant in service	30.5
Summer base load - PV Power Plant out of service	50
Summer base load - PV Power Plant in service	34.11

Table 14. Power losses results

4.4.3. Financial analysis - impact of Chirileu PV power plant

The interruptions of electricity supply service result in financial losses, from the producer's point of view. According to the results of calculation of reliability indicators, the maximum duration of an interruption, in case of a fault to the point of connection, is about 20 hours.

According to the DSO performance standards, a specific quality indicator is calculated: EENS - electrical energy not supplied, defined as the total energy not supplied by the producer (PV Power Plant) to the users, due to interruptions:

$$EENS = \sum_{i=1}^{n} P_i \cdot D_i = 1.4 \cdot 16 = 22.4 (MWh) \tag{2}$$

where:

P_i - non-delivered power, due to interruptions (the average of the active power non- generated by PV Power Plant (1.4 MW);

D_i - interruption duration that is considered relevant during the time when PV Power Plant is out of service (e.g., summer, 16 hours/day).

The price of a green certificate is 130 Lei/MWh, and the producers receive two green certificates for each MW and an extra 170 Lei/MWh, the price of energy traded on the free market.

$$Losses(Lei) = 22.4 MWh \cdot (2 \cdot 130) + 22.4 \cdot 170 = 9632(Lei / interruption) \tag{3}$$

5. Conclusions

Electricity produced by conventional, large scale central generation requires HV transmission networks such as HV, MV and LV distribution networks to reach its consumers, while DG, often located at the MV busbars and closer to load, requires transporting facilities, may reduce network losses and increase service quality.

The work presents and explores both of general and particular issues related to:

- State-of –the art of PV power plants at Romanian national level and specifically at the study area: electricity balance, structure, connection and integration in the electrical network.

- Technical conditions required for the connection of PV Power plants such as electric power networks in order to reach the criteria of reliability, power quality and economy.

- Case study regarding the impact of a PV power plant on the electric grid to which it is connected. The study is performed upon the photovoltaic power plant with an installed capacity of 3.2 MW connected to the 20 kV OEL and supplied from a 220/110/20 kV station.

The results provide a complete overview of the situation from a technical (safety and power quality) and economical point of view. The results may be also be used as a basis for development plans of future Distributed Generation Sources.

Author details

Dorin Bică[1*], Mircea Dulău[1], Marius Muji[1] and Lucian Ioan Dulău[2]

*Address all correspondence to: dorin.bica@ing.upm.ro

1 "Petru Maior" University of Tîrgu Mureş, Tîrgu Mureş, Romania

2 Technical University of Cluj Napoca, Cluj Napoca, Romania

References

[1] Begovic MM, Kim I, Novosel D, Aguero, JR, Rohatgi A. „Integration of Photovoltaic Distributed Generation in the Power Distribution Grid", 2012 45th Hawaii International Conference on System Science (HICSS), pp. 1977–1986, ISSN : 1530-1605, E-ISBN: 978-0-7695-4525-7, Print ISBN: 978-1-4577-1925-7, DOI: 10.1109/HICSS. 2012.335.

[2] Eltawila MA, Zhaoa Z. „Grid-Connected Photovoltaic Power Systems: Technical and Potential Problems—A Review", Renewable and Sustainable Energy Reviews, Volume 14, Issue 1, January 2010, pp. 112–129, DOI:10.1016/j.rser.2009.07.015.

[3] Yang B, Li W, Zhao Y, He X. „Design and Analysis of a Grid-Connected Photovoltaic Power System", IEEE Transactions on Power Electronics, Volume 25, Issue: 4, pp. 992–1000, ISSN: 0885-8993, DOI:10.1109/TPEL.2009.2036432.

[4] Fazliana A, Kadir A, Khatib T, Elmenreich W. „Integrating Photovoltaic Systems in Power System: Power Quality Impacts and Optimal Planning Challenges", Interna-

tional Journal of Photoenergy, Volume 2014, Article ID 321826, 7 pages, http://dx.doi.org/10.1155/2014/321826.

[5]   Sioshansi FP. „Smart Grid: Integrating Renewable, Distributed & Efficient Energy", Elsevier, 2012, ISBN: 978-0-12-386452-9.

[6]   Hossain J, Mahmud A. „Renewable Energy Integration: Challenges and Solutions", Springer, 2014, ISBN: 978-981-4585-27-9.

[7]   Dulău LI, Abrudean M, Bică D. „Effects of Distributed Generation on Electric Power Systems", The 7th International Conference Interdisciplinarity in Engineering, (INTER-ENG 2013), Procedia Technology, Vol. 12, 2014, pp. 681-686, ISSN 2212 – 0173.

[8]   Eremia M. (Editor), Song YH, Hatzyargyriou N, Buta A, Cârțină Gh. „Electric Power Systems. Volume I. Electric Networks", Publishing House of the Romanian Academy, Bucharest, 2005.

[9]   "Technical Code of Transport Network" issued by Regulatory Authority for Energy (ANRE). Available from: www.anre.ro/download.php?f=f62EiQ%3D%3D&t=vdeyut7dlcecrLbbvbY%3D [Accessed: 2015-01-10].

[10] "Technical Code of Electrical Distribution Networks" issued by Regulatory Authority for Energy (ANRE). Available from: www.anre.ro/download.php?f=f657gw%3D%3D&t=vdeyut7dlcecrLbbvbY%3D [Accessed: 2015-01-10].

[11] "Requirements for the connection to public electricity networks for photovoltaic Power Plants" issued by Regulatory Authority for Energy (ANRE). Available from: www.anre.ro/download.php?f=f6yEgw%3D%3D&t=vdeyut7dlcecrLbbvbY%3D [Accessed: 2015-01-10].

[12] Georgescu O, Jeflea I, Moldovan C, Marchievici E. „Comparative Review Between Project Solution and Real Operating Conditions Regarding the Voltage Drops and Power Losses For Photovoltaic Park Case from Chirileu", Journal Of Sustainable Energy Vol. 4, No. 3, September, 2013, ISSN 2067-5534.

[13] Standard on calculation methods for safety operation of power plants and devices NTE 005/06/00. Available from: http://www.scribd.com/doc/75749380/Nte-005-Normativ-Privind-Metodele-Si-Elementele-de-Calcul-Al-Sigurantei-in-Function-Are-a-Instalatiilor-Energetice#scribd [Accessed: 2015-01-10].

[14] www.neplan.ch